AF608757

Charged Particle Beam Physics

An Introduction for Physicists and Engineers

Sarvesh Kumar
Inter University Accelerator Center, New Delhi, India

Manish K. Kashyap
Jawaharlal Nehru University, New Delhi, India

Authors

Sarvesh Kumar
Inter University Accelerator Center,
New Delhi
India

Manish K. Kashyap
Jawaharlal Nehru University,
New Delhi
India

Cover Design: Wiley
Cover Image: © TK

Library of Congress Card No.: applied for

British Library Cataloguing-in-Publication Data
A catalogue record for this book is available from the British Library.

Bibliographic information published by the Deutsche Nationalbibliothek
The Deutsche Nationalbibliothek lists this publication in the Deutsche Nationalbibliografie; detailed bibliographic data are available on the Internet at http://dnb.d-nb.de.

Print ISBN 9783527414048
ePDF ISBN 9783527832958
ePub ISBN 9783527832972
oBook ISBN 9783527832965

Typesetting Lumina Datamatics, Noida, India
Printing CPI Group (UK) Ltd, Croydon CR0 4YY

C9783527414048_080825

To Our Father Late Shri Mahender Pal

Contents

Foreword

There exist in the literature many books and treatises on Charged Particle Beam Physics and, in general, on accelerator physics. So when the first author, Sarvesh, approached me to write the foreword to this book, I was very curious to find out what distinguishes the present one from previous books. After going through the manuscript, I found that almost all aspects of the physics of particle beams have been presented, keeping the beginning student in mind. Beginning with a brief history of the development of accelerators, topics like ion sources and accelerating and beam transport devices, along with their working principles, are described in an easy style with just enough detail for a first introduction to the subject. Numerical problems of practical relevance are included at the end of each chapter, which will aid the learner in understanding the concepts. Another nice feature of the book is the inclusion of diagnostic methods for charged particle beams, vacuum devices, and the measurement of vacuum. References to detailed works on each topic are provided for those interested in pursuing the subjects in detail.

This book provides excellent material for an introductory course in accelerator physics and would also be a handy reference for the relevant concepts for the practitioners.

Amit Roy

Preface

The motivation for writing "*Charged Particle Beam Physics: An Introduction for Physicists and Engineers*" is to simplify the process of learning accelerator physics. The book directly addresses the fundamental topics required in this field. Although my background is primarily in laboratories utilizing electrostatic and linear accelerators, this book also caters to circular accelerator laboratories, addressing common elements found across various accelerator systems. The beam optics examples provided focus on single-pass configurations and typically deal with standard beam transport lines.

This book represents the culmination of years of extensive research, reflection, and collaboration in accelerator physics. During this time, I had the privilege of visiting Fermilab, USA, and CERN, Geneva, where the insights gained from these experiences, have been invaluable. The process of writing this book has been both challenging and rewarding, prompting me to explore new perspectives and deepen my understanding of the subject.

Charged Particle Beam Physics: An Introduction for Physicists and Engineers is structured into eight chapters, each focusing on a key aspect of accelerator physics. I have incorporated beam optics code simulations to visually illustrate basic concepts in beam optics. Below is a summary of the chapters:

1. **Chapter 1:** Introduction to different types of accelerators and the historical evolution of accelerator technology.
2. **Chapter 2:** Basics of particle sources and their working principles.
3. **Chapter 3:** Fundamentals of beam optics and an introduction to phase space formalism.
4. **Chapter 4:** Motion of charged particles in various magnetic fields.
5. **Chapter 5:** Motion of charged particles in various electric fields.
6. **Chapter 6:** Motion of charged particles in radio frequency fields.
7. **Chapter 7:** Determining basic beam parameters using standard beam diagnostics utilized worldwide.
8. **Chapter 8:** Understanding vacuum systems in accelerators, covering vacuum physics, pumps, and gauges.

This book is grounded in my 18 years of experience at the Inter University Accelerator Center, New Delhi, where I have primarily focused on beam optics for designing particle accelerators. I have consistently enjoyed working on hands-on calculations and beam optics simulations using various simulation and tracking codes. Achieving accurate transverse and longitudinal parameter matching for particle beams is a key part of designing effective particle accelerators. It requires careful estimation of beam matching parameters alongside the design parameters of beam optical devices. The conversion of effective beam optics layouts into their physical counterparts must be performed with precision to ensure that the accelerator functions as closely as possible to the ideal model.

The book is written in a broad, accessible format, intended to serve as a primary source and is mainly for undergraduate student to get quick access to the field and also extend the outreach toscientists and engineers worldwide. It is particularly useful for beginners in accelerator physics, providing a clear and concise introduction to key topics. Additionally, it can serve as a practical reference for those involved in designing accelerator components, providing quick insights and essential information before proceeding with computational simulations. The student working at school level in beamline project can quickly get familiar to fundamental concepts in accelerator physics.

Apart from basic physics to undergraduate students, the aim of this book is also to explore the design of accelerator machines and their basic components. The academic community working in accelerator physics often needs to perform fundamental calculations before embarking on detailed simulations, whether for beam optics or the design of beam optical devices. Based on my experience, I have compiled many of the essential elements of accelerator physics into a single resource, allowing students to easily grasp the basic principles. This book serves as both a detailed glossary of accelerator terms and a practical guide for performing the basic calculations associated with most of the accelerator elements.

The content covers electrostatic, magnetostatic, and radiofrequency devices used in accelerator physics, with an emphasis on both transverse and longitudinal beam dynamics. The introductory chapters provide a broad overview of the principles behind various particle accelerators and ion sources. Different types of ion sources are characterized by their beam emittance, current, and profiles, all of which present significant challenges for beam extraction and transport to the target. Accordingly, this book focuses heavily on beam physics, particularly on the optical components that are common across most accelerators.

Beam physics is the core of accelerator physics, involving both theoretical calculations and the practical challenges associated with implementing beam optics designs to transport particles effectively to their intended target. The book covers the design of essential beam optical components – both passive and active. Topics include the design and function of dipole magnets, quadrupole magnets, steerers, scanners, electrostatic dipoles and quadrupoles, sextupoles, radio frequency cavities, bunchers, choppers, electrostatic accelerating tubes, and more.

The transfer matrix approach is used extensively to calculate design parameters for beam optical components, followed by discussions of the hardware design and practical difficulties encountered. To further illustrate the concepts, the open-source beam optics code COSY Infinity is employed to demonstrate first-order beam optics through various optical components. The designs are explained in a general manner so that readers can adapt them to their specific requirements. To access the COSY Infinity code, registration is required on the website of the corresponding research group at the Center for Dynamical Systems, Michigan State University, as detailed below:

https://www.bmtdynamics.org/cosy/register.htm

Registration instructions are available at:

https://www.bmtdynamics.org/cosy/

For any queries related to COSY Infinity, please contact:

support@cosyinfinity.org

I would like to express my sincere gratitude to Dr. Martin Berz and Dr. Kyoko Makino, Department of Physics, Michigan State University, for granting permission to use the COSY Infinity code to generate the beam optics diagrams used throughout this book.

As you read through the pages of *Charged Particle Beam Physics: An Introduction for Physicists and Engineers*, my hope is that you will find the knowledge and insights that will help you to achieve a particular goal or understanding as primary source of information on the exact topic of accelerator physics. This book is meant to be a resource, a guide, and a companion in your journey through world of particle accelerators.

Thank you for taking the time to engage with this work. I am excited to share it with you and look forward to the conversations and ideas that it may inspire. I dedicate this book to the application of particle accelerator technology for the betterment of human life and the peaceful advancement of mankind.

New Delhi, India
August 31, 2024

Sarvesh Kumar

About the Authors

Dr. Sarvesh Kumar was an experimental plasma and accelerator physicist with nearly two decades of experience at the Inter University Accelerator Centre (IUAC), New Delhi. He has participated in significant national and international projects. Notably, he contributed in the design, fabrication, and commissioning of three major accelerator-based facilities: Low Energy Ion Beam Facility (LEIBF), High Current Injector (HCI) for heavy ions and Delhi Light Source (DLS), supported by internationally peer-reviewed and national publications. He had worked with Pelletron accelerators for beam tuning operation. His deep perseverance in developing skills of teaching and education in applied physics led him to author this book, aiming to bridge theory with real-world applications in plasma and accelerator physics. His passion lies in connecting with young graduate/postgraduate students, encouraging out-of-the-box thinking, and guiding them in translating innovative ideas into practical systems. Dr. Kumar had served as a Scientific Associate at CERN, Geneva, designing the RFQ upgrade for LINAC4 accelerator, contributing overall to the Large Hadron Collider project. He combines academic rigor with hands-on innovation in the charged particle, cost-effective accelerators design. He had designed many achromatic bends at IUAC Delhi for heavy ions and electron beams which are now in place and serving the user community of IUAC. His strength lies in designing compact plasma chambers and ion beam systems suitable for university level teaching laboratories. A firm believer in teamwork awakefield acceleration, plasma instabilities, particle accelerator designs, beam optics and beam-plasma interaction.

Prof. Manish K. Kashyap, currently at School of Physical Sciences, Jawaharlal Nehru University (JNU), New Delhi, India, has rich teaching experience of 19 years. He has supervised nine PhD and one post-doctoral students till date, and seven PhD students are still working under his supervision. His areas of interest are Condensed Matter Physics, Plasma Physics, and Accelerator Physics. His work integrates theoretical modeling with experimental perspectives to enhance beam quality and stability in ion accelerators. He has explored novel acceleration mechanisms and diagnostics, advancing both fundamental and applied aspects of beam-plasma systems. He has published 110 research papers in international journals/conference proceedings, earning widespread recognition in the scientific community. His research continues to impact the development of next-generation accelerators and plasma-based technologies in India and abroad. He also works on perovskite solar cells and 2D van der heterostructures.

Acknowledgments

I am grateful to the CERN Accelerator School reports, the resources from the United States Particle Accelerator School, and the various authors of beam dynamics codes, accelerator physics resources worldwide, all of which have made it possible to write this book in a form that is accessible to beginners in accelerator physics, including school and undergraduate students in physics and engineering.

I am deeply thankful to the Director of the Inter University Accelerator Center (IUAC), New Delhi, whose encouragement and unwavering support have been instrumental in bringing this book to fruition. I also wish to express my sincere gratitude to Jawaharlal Nehru University, New Delhi and South Asian University, New Delhi, for providing the research collaboration that made this work possible.

I am especially grateful to our research group working in accelerator and plasma physics: Prof. Jyotsna Sharma (South Asian University, New Delhi), Dr. Ankur Taya (Mukand Lal National College, Yamuna Nagar, Haryana), Dr. Renu Singla (Daulat Ram College, University of Delhi, Delhi), Ms. Niketan Jakhar and Mr. Chandan Thakur (Jawaharlal Nehru University, New Delhi), and Mr. Sandeep Kumar (Indian Institute of Technology, New Delhi) who have provided me an excellent environment to develop this book on accelerator physics.

Lastly, I would like to extend my heartfelt thanks to my colleagues at IUAC, who were my first teachers in accelerator physics. Their guidance and instruction have been a blessing, and without their early teachings, this book would not have been possible.

I am deeply grateful to my family for their unwavering support, giving me the freedom and time to fully dedicate myself to the writing of this book.

Sarvesh Kumar

Acronyms

ACH	Achromat
ASM	Analyzing cum Switching Magnet
AT	Accelerating Tube
BPM	Beam Profile Monitor
DTL	Drift Tube LINAC
ECR	Electron Cyclotron Resonance
EL	Einzel Lens
EQD	Electrostatic Quadrupole Doublet
EQT	Electrostatic Quadrupole Triplet
FFC	Fast Faraday Cup
HCI	High Current Injector
HEBT	High Energy Beam Transport
IH	Inter Digital H type
LEBT	Low Energy Beam Transport
LINAC	Linear Accelerator
MEBT	Medium Energy Beam Transport
MHB	Multi-harmonic Buncher
MQD	Magnetic Quadrupole Doublet
MQT	Magnetic Quadrupole Triplet
RFQ	Radio Frequency Quadrupole
ST	Steerer Magnet

Fundamental constants used in particle accelerators.

Constant	Symbol	Value
Speed of light	c	2.99792458×10^{8} m/s
Elementary charge	e	$1.602176634 \times 10^{-19}$ C
Mass of electron	m_e	$9.10938356 \times 10^{-31}$ kg
Mass of proton	m_p	$1.67262192369 \times 10^{-27}$ kg
Planck's constant	h	$6.62607015 \times 10^{-34}$ J · s
Reduced Planck's constant	$\hbar$	$1.054571817 \times 10^{-34}$ J · s
Electric constant (permittivity of free space)	ε_0	$8.854187817 \times 10^{-12}$ F/m
Magnetic constant (permeability of free space)	μ_0	$4\pi \times 10^{-7}$ N/A^2
Fine-structure constant	α	$\approx \frac{1}{137}$
Boltzmann constant	k_B	1.380649×10^{-23} J/K
Avogadro's number	N_A	$6.02214076 \times 10^{23}$ mol^{-1}
Rydberg constant	R_∞	$1.0973731568508 \times 10^{7}$ m^{-1}
Fermi coupling constant	G_F	1.1663787×10^{-5} GeV^{-2}
Gravitational constant	G	6.67430×10^{-11} m^3 · kg^{-1} · s^{-2}
Stefan–Boltzmann constant	σ	$5.670374419 \times 10^{-8}$ W m^{-2} · K^{-4}

About the Companion Website

Charged Particle Beam Physics: An Introduction for Physicists and Engineers
This book is accompanied by a companion site:

https://www.wiley.com/go/Kumar_1e

This website includes:

- Answers
- Videos

1

Introduction to Charged Particle Beams

"If you want to find the secrets of the universe, think in terms of energy, frequency, and vibration."

—*Nikola Tesla*

In accelerator physics, a particle beam is typically defined as a collection of like-charged particles, all moving with momentum predominantly in one direction compared to the other two transverse directions. This characteristic allows the beam to be transported over long distances using electromagnetic fields and further accelerated to energies reaching several teraelectronvolts (TeV) in modern accelerators. Naturally occurring particle beams exist in space, commonly referred to as cosmic rays or solar particles. A specific example is the stream of charged particles, such as solar wind or proton beams, emitted by the Sun. These particles are captured by Earth's magnetic field, resulting in collisions with particles in Earth's upper atmosphere. In the Earth ionosphere, charged particles from the solar wind, guided by Earth's magnetic field, collide with oxygen and nitrogen atoms in the atmosphere, exciting them and releasing energy as colorful light displays known as the aurora, or Northern Lights.

Curiosity: *Have you ever wondered how particle accelerators can speed up tiny particles like protons or electrons to nearly the speed of light? Why do they need a vacuum—can't particles just move through air like anything else? To push particles forward, accelerators use special devices called RF cavities that work like swings, giving the particle a timed "kick" each cycle. But as particles gain speed, why do we need magnets to bend and focus their paths—why don't they just go straight? And did you know that your microwave oven uses a tiny kind of particle accelerator called a magnetron? What if, instead of electric fields, we could use sound waves or even gravity to accelerate particles—could that work someday? If batteries produce voltage, why can't we just use a giant one to reach GeV energy levels? Inside circular accelerators, how do particles manage to stay in sync with the accelerating fields without flying off-track? When too many particles gather close together, do they repel each other and cause the beam to spread out? It's amazing that the first particle accelerator, built in the 1930s, could fit on a tabletop! And why do fast-moving particles give off brilliant light—called synchrotron radiation—when they are forced to turn? Finally,*

Charged Particle Beam Physics: An Introduction for Physicists and Engineers, First Edition.
Sarvesh Kumar and Manish K. Kashyap.

Companion Website: https://www.wiley.com/go/Kumar_1e

imagine this: could we one day shrink a whole accelerator down to fit on a tiny microchip? We try to explore some of these curious questions here—but many remain open, inviting you to keep wondering, exploring, and discovering.

The energy (E) of a particle [1] can be related to its temperature using the Boltzmann relation:

$$E = k_B T \tag{1.1}$$

$$k_B \approx 1.38 \times 10^{-23}\,\mathrm{J\,K^{-1}}$$

and 1 eV corresponds to approximately 11 605 K. This implies that particles accelerated to energies in the kiloelectron volt (keV), megaelectron volt (MeV), and gigaelectronvolt (GeV) range are effectively heated to extremely high temperatures. Understanding **particle beam dynamics** is crucial in space physics and accelerator sciences.

When ions are generated in a laboratory as a stream of charged particles, they must be accelerated and transported to a target with minimal intensity loss. These ions are used in various fields of science, such as nuclear physics, materials science, and atomic physics. In discussing heavy ions, which are larger than protons, a key difference lies in their approach to the speed of light as they gain energy. The total energy of a charged particle is given by the following equation:

The relativistic energy–momentum relation is:

$$E^2 = (pc)^2 + (m_0 c^2)^2 \tag{1.2}$$

where E is the total energy of the particle, p is the relativistic momentum, m_0 is the rest mass of the particle, and c is the speed of light.

The relativistic momentum p is related to the velocity v of the particle by:

$$p = \gamma m_0 v \tag{1.3}$$

where γ (the Lorentz factor) is defined as:

$$\gamma = \frac{1}{\sqrt{1-\beta^2}} \tag{1.4}$$

Here $\beta = \frac{v}{c}$. Thus, the total energy of particle is given as follows:

$$E^2 = (\gamma m_0 v c)^2 + (m_0 c^2)^2 = m_0^2 c^4 \left(\gamma^2 \beta^2 + 1\right) = \gamma^2 m_0^2 c^4$$

$$E = \gamma m_0 c^2 \tag{1.5}$$

Finally, the kinetic energy K is given by:

$$K = E - m_0 c^2 = \left(\frac{1}{\sqrt{1-\beta^2}} - 1\right) m_0 c^2 \tag{1.6}$$

$$\beta = \frac{v}{c} = \sqrt{1 - \frac{1}{\gamma^2}} = \sqrt{1 - \left(\frac{1}{1 + \frac{K}{m_0 c^2}}\right)^2}$$

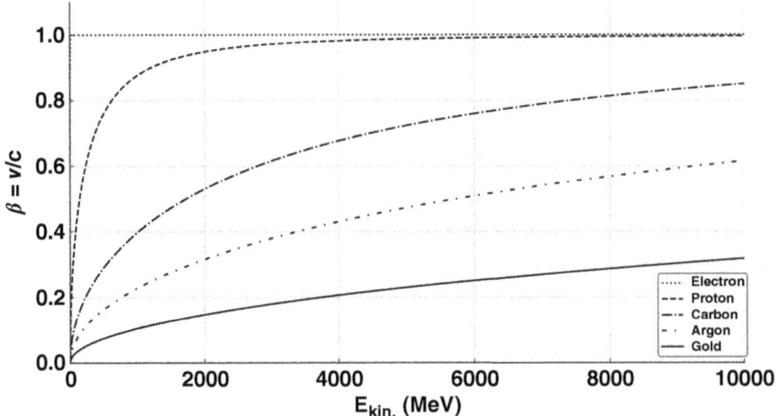

Figure 1.1 Velocity behavior of different charged particles as they approach the speed of light.

If we plot the velocity of different charged particles as per Equation 1.6, then the electrons begin to exhibit significant relativistic behavior at energies around 0.511 MeV, which corresponds to their rest mass energy. However, according to special relativity, they can never reach the speed of light, no matter how much energy they gain. Protons become relativistic at much higher energies, near 938 MeV, their rest mass energy. Similarly, heavy ions, they require even higher energies (in the GeV per nucleon range) to exhibit relativistic effects due to their much greater mass. Regardless of particle type, no material particle can attain the speed of light; they can only asymptotically approach it as their energy increases. This is illustrated in Figure 1.1.

To maintain particles in a stable orbit as a beam, they must be placed in a constant magnetic field, where they spiral around the magnetic lines of force. This leads to the concept of magnetic rigidity, defined alongside the magnetic force, cyclotron frequency, and gyroradius as follows:

Magnetic force on particle:

$$\mathbf{F} = q(\mathbf{v} \times \mathbf{B}) \tag{1.7}$$

Cyclotron frequency of the particle:

$$\omega_c = \frac{qB}{m} \tag{1.8}$$

Larmor radius (gyroradius) of particle:

$$\rho = \frac{mv}{qB} \tag{1.9}$$

Magnetic rigidity:

$$B\rho = \frac{p}{q} = \frac{mv}{q} \tag{1.10}$$

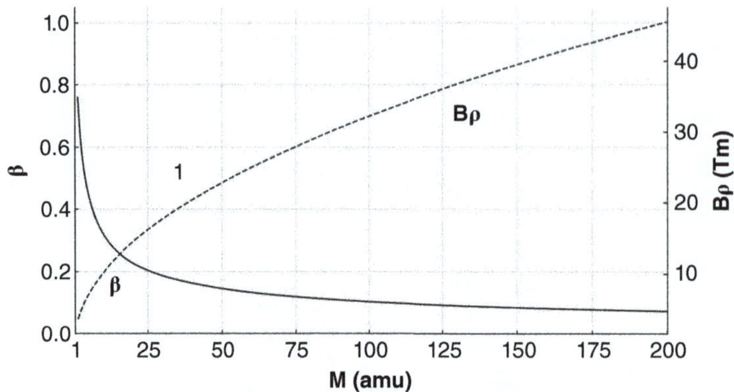

Figure 1.2 Velocity and magnetic rigidity of different ion beams of mass M (amu) at 500 MeV energy.

In terms of practical units:

$$B\rho\ (\mathrm{Tm}) = 3.3356 \cdot \frac{p\ (\mathrm{GeV}/c)}{q} \quad \text{(relativistic)}$$
$$B\rho\ (\mathrm{Tm}) = 0.1439 \cdot \sqrt{\frac{M\ (\mathrm{amu}) \cdot E\ (\mathrm{MeV})}{q^2}} \quad \text{(non-relativistic)} \tag{1.11}$$

Here, E is the total energy of the particle. Using Equation 1.11, if we plot the magnetic rigidity required to bend different ions with a fixed energy of 500 MeV and a unit charge state, we see that it increases with the mass of the ions. This is shown in Figure 1.2, alongside the corresponding normalized particle velocities. Following conclusions can be drawn:

1. Maximum magnetic rigidity is calculated for the heaviest mass, highest energy, and unit positive charge state of the beam desired in the accelerator.
2. Magnetic rigidity increases with mass and energy of charged particles for a given charge state of the beam particles.
3. Magnetic rigidity is maximum for a unit charge state of an ion beam with target energy.
4. Magnetic rigidity depends on the mass-to-charge (m/q) ratio directly, When $E = qV$ where V is the voltage by which particles are accelerated.

In 1909, Ernest Rutherford bombarded alpha particles onto a thin gold foil. To overcome the Coulomb barrier between the alpha particles and the gold nucleus, high energies were necessary to surmount the Coulomb repulsion. The higher the energy imparted to particles, the shorter their de Broglie wavelength (λ). Just as a living cell is observed under an optical microscope using scattered visible light photons, energetic particles can be used to probe matter, depending on their wavelength. The wavelength of energetic particles determines the size of the object to be resolved, so high-energy particles (with mass m, velocity v, and energy E) are required to probe deep into atoms and nuclei. The de Broglie wavelength (λ) is given by the Planck–Einstein relation, the relationship between energy and momentum:

$$E = h\nu = \frac{hc}{\lambda} = pc \tag{1.12}$$

$$\lambda = \frac{h}{p} = \frac{h}{mv} = \frac{h}{\sqrt{2mE}} \tag{1.13}$$

In terms of practical units:

$$\lambda(\mathring{A}) = \frac{12.27}{E(eV)} \tag{1.14}$$

Here, E is the energy of the particle (or photon), h is Planck's constant, ν is the frequency of the de Broglie wave associated with the particle.

Accelerators have evolved over the last two centuries as a result of applying electricity and magnetism to charged particles, driven by the pioneering work of many great scientists and engineers. Initially motivated by Rutherford's famous experiment to explore the nucleus, modern particle accelerators are essential tools for probing the structure of atoms, protons, neutrons, and electrons through high-energy collisions that reveal their internal components. They enable the discovery of new particles, such as quarks, leptons, and the Higgs boson, by recreating the extreme conditions necessary for these short-lived particles to manifest. Studying particle interactions at high energies allows scientists to explore the fundamental forces of nature – gravity, electromagnetism, and the strong and weak nuclear forces – and how they behave and unify. Additionally, accelerators like the Large Hadron Collider (LHC) recreate the energy densities present just after the Big Bang, offering insights into the origins of the universe. They also provide a platform to test and confirm theoretical models of quantum mechanics and relativity, including the Standard Model, by observing particle behavior and interactions with exceptional precision.

Thus, the evolution of accelerators has progressed from natural particle accelerators like cosmic rays and radioactive materials to highly engineered devices capable of producing extreme energetic particles. Cosmic rays, consisting of high-energy particles originating from astrophysical events, represent the earliest and most powerful natural accelerators. Similarly, radioactive sources emit high-energy particles during nuclear decay, naturally accelerating alpha, beta, and gamma particles. The first major technological breakthrough in particle acceleration came with the development of the cathode ray tube (CRT), which used high voltage to accelerate electrons and produce visible images, laying the groundwork for more sophisticated human-made accelerators.

In the early twentieth century, the invention of the Van de Graaff accelerator [2] marked a significant advancement in accelerating charged particles like protons and ions using high voltage generated by a moving belt. This innovation paved the way for circular accelerators like the cyclotron, where charged particles like protons and deuterons are accelerated in a magnetic field and gain energy from an alternating electric field. For electrons, which encounter relativistic effects at higher speeds, the betatron was developed, using a time-varying magnetic field to induce electron acceleration in a circular orbit.

As accelerators grew more powerful, the synchrotron was introduced in the mid-twentieth century, where protons, electrons, and ions are accelerated in a circular path with synchronized magnetic and electric fields, achieving energies in the TeV range. Synchrotrons facilitated the discovery of particles like quarks and leptons and contributed significantly to particle physics. Around the same time, the linear accelerator (LINAC) emerged, allowing particles to be accelerated in a straight path using oscillating electric fields. A variant of the Van de Graaff, the Tandem accelerator, was designed to stabilize high-voltage acceleration using a chain of metal pellets to achieve greater reliability and higher energies.

Following these advancements, the microtron was created to reuse circular orbits for electrons, allowing them to gain energy with each pass. In the mid-twentieth century, the fixed-field alternating gradient (FFAG) accelerator was introduced, accelerating particles in a spiral path while maintaining a constant magnetic field. The nonscaling FFAG improved upon this design by allowing more rapid acceleration with variable particle orbits. Meanwhile, the induction LINAC utilized pulsed magnetic fields to accelerate particles like electrons and ions over shorter distances.

In the era of modern high-energy physics, particle colliders like the LHC emerged, accelerating two beams of particles in opposite directions to create collisions at extreme energies, often in the TeV range. Further innovations included the recirculating LINAC, which allowed particles to pass through the same LINAC multiple times to reach higher energies, as well as the energy recovery LINAC, which focuses on energy efficiency by decelerating electron beams and recovering their energy for reuse.

Most recently, breakthroughs in compact acceleration methods have led to the development of laser plasma accelerators, which use high-intensity lasers to generate plasma wakefields, rapidly accelerating particles like electrons to GeV energies. Beam-driven plasma accelerators follow a similar principle, using a high-energy particle beam to generate a plasma wakefield, offering compact and efficient acceleration over shorter distances.

Plasma wakefield accelerators, driven by either high-intensity lasers or particle beams, can achieve acceleration gradients in the order of gigavolts per meter (GV/m), far exceeding the capabilities of traditional accelerators. This allows for compact designs that can potentially reduce the size and cost of future particle accelerators.

1.1 History of Particle Accelerators and the Chronological Milestones

Before going into modern history accelerators, let's go back to one of the earliest known concepts of the atom in ancient Indian science. Maharishi Kanaad, an ancient Indian sage, was one of the first to propose the idea of atomic theory. He suggested that everything in the universe is made of tiny, indivisible particles called "parmanu," which are similar to what we now think of as atoms. He believed that these particles combine in various ways to form the objects and substances that we see around us. His ideas are recorded in the Vaisheshika Sutras, an important text in Indian philosophy. This was an early understanding that matter is made of basic building blocks.

Particle accelerators began to develop in the late nineteenth century, following important discoveries like X-rays by Wilhelm Röntgen and beta rays by J. J. Thomson. The discovery of radioactivity motivated scientists to find ways to create and control particles artificially. However, it was Ernest Rutherford's groundbreaking experiments that gave scientists the roadmap to study the atomic nucleus in more detail. The following is the timeline of key discoveries and advancements that shaped the history of particle accelerator technology and its pioneer discoveries as shown in Table 1.1. A brief history of accelerators and related discoveries is provided in Table 1.1 chronologically.

Table 1.1 Chronological evolution of particle accelerators worldwide.

Year	Developments
1895	Wilhelm Conrad Röntgen discovered X-rays using CRTs, for which he received the first Nobel Prize in Physics in 1901.
1896	Joseph John Thomson studied the nature of cathode rays, discovering their charge-to-mass ratio and identifying them as electron beams. He was awarded the Nobel Prize in 1906.
1898	Marie Curie discovered groundbreaking work in the study of radioactivity. In 1898, together with her husband Pierre Curie, she discovered two new elements: *polonium* in July and *radium* in December. She also introduced the term "radioactivity" to describe how certain materials give off energy as they break down. Her important discoveries rewarded her two Nobel Prizes: the first in Physics in 1903, shared with Pierre Curie and Henri Becquerel, for their research on radioactivity, and the second in Chemistry in 1911 for discovering radium, polonium, and studying the properties of radium. Marie Curie was the first woman to win a Nobel Prize and is still the only person to have won Nobel Prizes in two different sciences.
1920	John Douglas Cockcroft and Ernest Thomas Sinton Walton conceived the first high-voltage particle accelerator, which used two electrodes inside a vacuum vessel to create a potential drop of around 100 kV.
1923	Rolf Widerøe, a young Norwegian student, first proposed the design of the betatron, incorporating the well-known 2-to-1 rule. Two years later, he refined the design by introducing the condition for radial stability, a crucial factor for successful particle acceleration.
1924	Gustav Ising proposed time-varying fields across drift tubes, introducing the concept of resonant acceleration, which allows for energy levels higher than the system's maximum voltage.
1928–1932	John Cockcroft and Ernest Walton began designing an 800 kV generator under the guidance of Ernest Rutherford in 1928 at the Cavendish Laboratory in Cambridge. In 1932, they achieved a significant milestone by successfully splitting the lithium atom using high-energy protons accelerated by their Cockcroft–Walton generator. This marked the first artificial nuclear reaction, demonstrating the power of particle accelerators in nuclear physics. Their breakthrough experiment laid the foundation for future accelerator technologies and earned them the Nobel Prize in Physics in 1951 for their pioneering work in atomic transmutation.
1928	Rolf Widerøe demonstrated Ising's principle with a 1 MHz, 25 kV oscillator to produce 50 keV potassium ions, suggesting the use of RF voltage between consecutive drift tubes. This was successfully tested by D. H. Sloan and E. O. Lawrence in 1931.
1929	Ernest Lawrence was motivated by Rolf Widerøe's advances in accelerating particles and Gustav Ising's concept of using electric fields for this purpose. In 1929, Lawrence invented the cyclotron, a device that uses magnetic fields to move particles in a circular path while speeding them up with an electric field.
1930	M. Stanley Livingston, a student of Ernest Lawrence, built and tested the first working cyclotron. This machine accelerated hydrogen ions (protons) to an energy of 80 keV, marking a major step forward in particle accelerator technology.

(Continued)

Table 1.1 (Continued)

Year	Developments
1930	Robert Van de Graaff developed the Van de Graaff generator, an important type of electrostatic accelerator that allowed for the creation of extremely high voltages. This innovation was a critical step in early particle acceleration, enabling experiments that required high-energy beams.
1932	Lawrence's cyclotron accelerated protons to 1.25 MeV, successfully achieving atomic splitting shortly after Cockcroft and Walton's breakthrough. For this pioneering achievement, along with his invention and development of the cyclotron and his contributions to artificial radioactive elements, E. O. Lawrence was awarded the Nobel Prize in Physics in 1939.
1940	Donald W. Kerst revolutionized particle acceleration by redesigning the betatron and constructing the first fully functional machine capable of accelerating electrons to 2.2 MeV.
1944	Vladimir Veksler introduced the principle of phase stability, which made it possible to maintain stable orbits while increasing the energy of particles in synchrotrons. This discovery was vital for the development of modern particle accelerators such as synchrocyclotrons and electron synchrotrons.
1946	Luis W. Alvarez built a 32 MeV proton drift tube LINAC at Berkeley, which operated at 200 MHz, marking an important step in the development of LINACs.
1950	Donald W. Kerst constructed the largest betatron in the world, capable of accelerating electrons to 300 MeV, a record at the time.
1950	Nicholas Christofilos solved the challenge of beam stability in synchrotrons by introducing the concept of strong focusing. This advancement allowed synchrotrons to handle higher energy beams and led to more efficient acceleration in circular accelerators.
1951	J. D. Cockcroft and E. T. S. Walton were awarded the Nobel Prize for their pioneering work on the transmutation of atomic nuclei using artificially accelerated particles.
1951	Edwin M. MacMillan, along with Glenn T. Seaborg, was awarded the Nobel Prize for their research on the chemistry of transuranium elements.
1952	E. Courant, M. Livingston, and H. Snyder at Brookhaven, USA, proposed alternating-gradient (AG) magnetic lattices, consisting of bending magnets with AGs or properly spaced quadrupole magnets with alternating polarities. This innovation solved the weak focusing problem in synchrotrons.
1959	Emilio Segrè and Owen Chamberlain received the Nobel Prize in Physics for their discovery of the antiproton, achieved using a particle accelerator at the Lawrence Berkeley National Laboratory.
1960	Bruno Touschek designed the first practical electron–positron collider, which made it possible to study particle collisions in a controlled environment. His work laid the foundation for later developments in colliders like the large electron–positron collider.

Year	Developments
1965	Julian Schwinger, along with Sin-Itiro Tomonaga and Richard P. Feynman, was awarded the Nobel Prize for quantum electrodynamics, which has profound implications for elementary particle physics.
1966	Andrey M. Budker developed the technique of electron cooling, which enhanced beam quality in storage rings and improved the precision of particle accelerator experiments.
1968	Luis W. Alvarez received the Nobel Prize for discovering a large number of resonance states using a hydrogen bubble chamber.
1970	I. M. Kapchinsky and Vladimir Teplyakov introduced the concept of the radio frequency quadrupole in 1970. This innovation improved the acceleration and focusing of low-energy ion beams, leading to more efficient LINACs.
1976	Burton Richter and Samuel Ting were awarded the Nobel Prize in Physics for the independent discovery of the J/Ψ particle, providing evidence for the existence of the charm quark, discovered using high-energy particle accelerators.
1977	Rosalyn Yalow was awarded the Nobel Prize in Physiology or Medicine for developing the radioimmunoassay (RIA), a groundbreaking technique for measuring biological substances using radioactive isotopes. While radioisotopes used in RIA can be produced by particle accelerators, the work was primarily linked to medical diagnostics.
1980	Simon van der Meer invented stochastic cooling, a technique used to refine and manipulate particle beams in high-energy accelerators. This breakthrough was key to achieving the precision needed for particle collisions at the European Organization for Nuclear Research (CERN), which led to the discovery of the W and Z bosons.
1984	Carlo Rubbia and Simon Van der Meer were awarded the Nobel Prize for their contributions to the large project that led to the discovery of the W and Z bosons, the carriers of the weak force.
2013	François Englert and Peter Higgs were awarded the Nobel Prize in Physics for the theoretical discovery of the Higgs mechanism, confirmed by the discovery of the Higgs boson at the LHC at CERN.

1.2 Maxwell's Equations

In accelerator physics, the primary goal is often to solve Maxwell's equations for specific geometries to understand and optimize the electromagnetic fields in an accelerator components, such as radio frequency (RF) cavities, magnets, and electric components. It helps physicists and engineers to design accelerators with the desired beam dynamics, particle focusing, and energy transfer characteristics.

Consider an electromagnetic field moving through free space, which is an area with no electric charges or currents. The electric field, represented by **E**, and the magnetic field, represented by **B**, change over time and space and affect each other. Maxwell's equations describe how these fields behave and interact. The symbol ρ stands for charge density, which is the amount of charge in a given space. The term J is the current density, indicating the flow of electric charge. Now, here are Maxwell's equations in both integral and differential forms:

1.2.1 Maxwell's First Equation: Gauss's Law for Electricity

$$\nabla \cdot \mathbf{E} = \frac{\rho}{\varepsilon_0} \quad \text{(Differential form)} \tag{1.15}$$

$$\oint_S \mathbf{E} \cdot d\mathbf{A} = \frac{Q_{\text{enc}}}{\varepsilon_0} \quad \text{(Integral form)} \tag{1.16}$$

This law states that the divergence of the electric field ($\mathbf{E}$) at any point in space is proportional to the local charge density (ρ). It describes how electric fields emanate from electric charges. In accelerators, electric fields play a key role in accelerating and focusing on charged particle beams. Longitudinal electric fields provide electrostatic acceleration, while transverse electric fields, particularly in electrostatic quadrupoles, focus the beam. Dipole configurations use electric fields for deflecting or steering particles. Gauss's law underpins the theoretical framework for understanding how electric fields interact with charges in accelerator environments.

1.2.2 Maxwell's Second Equation: Gauss's Law for Magnetism

$$\nabla \cdot \mathbf{B} = 0 \quad \text{(Differential form)} \tag{1.17}$$

$$\oint_S \mathbf{B} \cdot d\mathbf{A} = 0 \quad \text{(Integral form)} \tag{1.18}$$

This law states that the divergence of the magnetic field ($\mathbf{B}$) is zero, implying the absence of magnetic monopoles. Magnetic field lines always form closed loops. In accelerators like cyclotrons, synchrotrons, or betatrons, magnetic fields are essential for bending and focusing on charged particles, following this fundamental property.

1.2.3 Maxwell's Third Equation: Faraday's Law of Electromagnetic Induction

$$\nabla \times \mathbf{E} = -\frac{\partial \mathbf{B}}{\partial t} \quad \text{(Differential form)} \tag{1.19}$$

$$\oint_C \mathbf{E} \cdot d\mathbf{l} = -\frac{d}{dt} \int_S \mathbf{B} \cdot d\mathbf{A} \quad \text{(Integral form)} \tag{1.20}$$

Faraday's law describes how a changing magnetic field induces an electric field. In accelerators like the betatron, this principle is employed to induce electric fields that accelerate particles. A time-varying magnetic field generates a circulating electric field, which accelerates electrons or other particles confined within a circular path. This law is fundamental to devices using electromagnetic induction for acceleration.

1.2.4 Maxwell's Fourth Equation: Ampère's Law with Maxwell's Addition

$$\nabla \times \mathbf{B} = \mu_0 \mathbf{J} + \mu_0 \varepsilon_0 \frac{\partial \mathbf{E}}{\partial t} \quad \text{(Differential form)} \tag{1.21}$$

$$\oint_C \mathbf{B} \cdot d\mathbf{l} = \mu_0 I_{\text{enc}} + \mu_0 \varepsilon_0 \frac{d}{dt} \int_S \mathbf{E} \cdot d\mathbf{A} \quad \text{(Integral form)} \tag{1.22}$$

Ampère's law, with Maxwell's correction, incorporates the displacement current ($\frac{\partial \mathbf{E}}{\partial t}$) alongside the conduction current density (**J**). This law describes how electric currents and changing electric fields generate magnetic fields. In accelerator design, this equation is crucial for calculating ampere-turns in dipoles, quadrupoles, sextupoles, and solenoids. It also helps in the design of power supplies for energizing magnets in accelerators.

1.3 Electrostatic Accelerators

As the name suggests, electrostatic accelerators use electric fields to accelerate charged particles. The design of the electric field is crucial, as these accelerators are characterized by a potential gap where a particle experiences a potential difference and accelerates along the direction of the electric field. The voltage can reach very high levels (up to MV) with the use of suitable insulating gases like SF_6. Examples of electrostatic accelerators include Cockcroft–Walton accelerators, Van de Graaff accelerators, and tandem accelerators.

When a charge q moves through an electric field from one point to another, work is done by the electric field. The work done W is given by the change in electric potential energy, which can be expressed as:

$$W = q(V_1 - V_2)$$

where V_1 and V_2 are the electric potentials at the two points.

If we consider the charge moving from a point of zero potential (grounded) to a point with potential V, the work done by the field is:

$$W = qV$$

This work is the energy gained by the charge, so the energy gained by a particle with charge q under a voltage V is simply given by:

$$E = qV$$

The force (F) on a charge particle in an electric field (**E**) is given by:

$$\mathbf{F} = q\mathbf{E}$$

1.3.1 Cockcroft–Walton Accelerator

The Cockcroft–Walton accelerator, first developed in 1932 by John Cockcroft and Ernest Walton [3, 4], played a significant role in nuclear physics. It operates on the principle of cascading voltage multiplication. The system works by charging and discharging capacitors in cascading stages, with the voltage increasing stepwise at each stage. The circuit consists of a

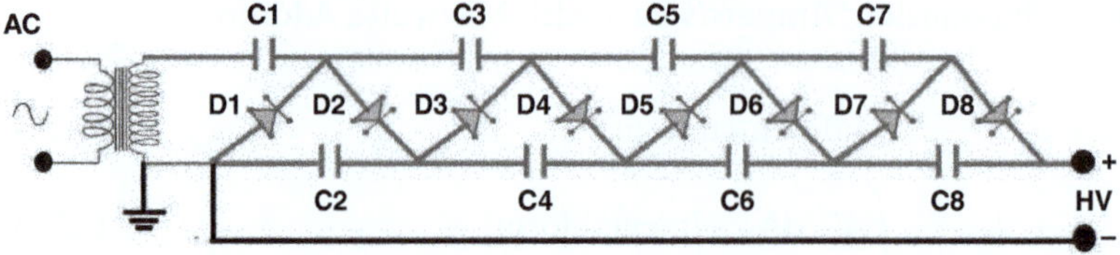

Figure 1.3 Schematic of a Cockcroft–Walton accelerator.

series of capacitors and diodes arranged in a specific configuration to create a voltage multiplier. The output voltage (V_{out}) increases with the number of stages (N), the input voltage (V_{in}), and the capacitance of the capacitors, although size and breakdown limits impose certain constraints. In the setup as shown in Figure 1.3, an AC voltage charges the capacitors in parallel so that each stage generates a potential twice the input voltage. The first capacitor charges in an anticlockwise loop, and the second in a clockwise loop, followed by the discharge of the first capacitor. With more stages, the final voltage increases until it reaches the breakdown voltage, as shown in Figure 1.3, where one can stop further cascading.

In a four-stage Cockcroft–Walton voltage multiplier circuit as shown in Figure 1.3, the circuit consists of two columns of capacitors: the oscillating column and the smoothing column. During the first half-cycle of the AC input, the capacitors in the oscillating column (C_1, C_3, C_5, and C_7) are charged through the odd-numbered diodes (D_1, D_3, D_5, and D_7). During the next half-cycle, the capacitors in the smoothing column (C_2, C_4, C_6, and C_8) are charged through the even-numbered diodes (D_2, D_4, D_6, and D_8). In the steady state, under no-load conditions, each capacitor in the smoothing column is charged to a voltage of $2V_{\mathrm{m}}$, which is twice the peak input voltage.

If the generator supplies any load current I, the output voltage will not reach the theoretical value of $2nV_{\mathrm{m}}$, as depicted in Figure 1.4. Additionally, a ripple is introduced in the voltage, necessitating the consideration of both the voltage drop ΔV_0 and the peak-to-peak ripple $2\delta V$.

The maximum output voltage V_{out} for an n-stage Cockcroft–Walton multiplier is expressed as:

$$V_{\mathrm{out}} = n \cdot 2V_{\mathrm{m}} \tag{1.23}$$

where n represents the number of stages, and V_{m} denotes the peak input voltage.

For a 4-stage multiplier, the maximum output voltage equals eight times the peak input voltage:

$$V_{\mathrm{out}} = 4 \times 2 \cdot V_{\mathrm{m}} = 8V_{\mathrm{m}} \tag{1.24}$$

Let q denote the charge transferred from capacitor C_n to the load in each cycle. The ripple voltage across capacitor C_n is given by:

$$2\delta V = q \sum_{1}^{n} \frac{1}{C_n} \tag{1.25}$$

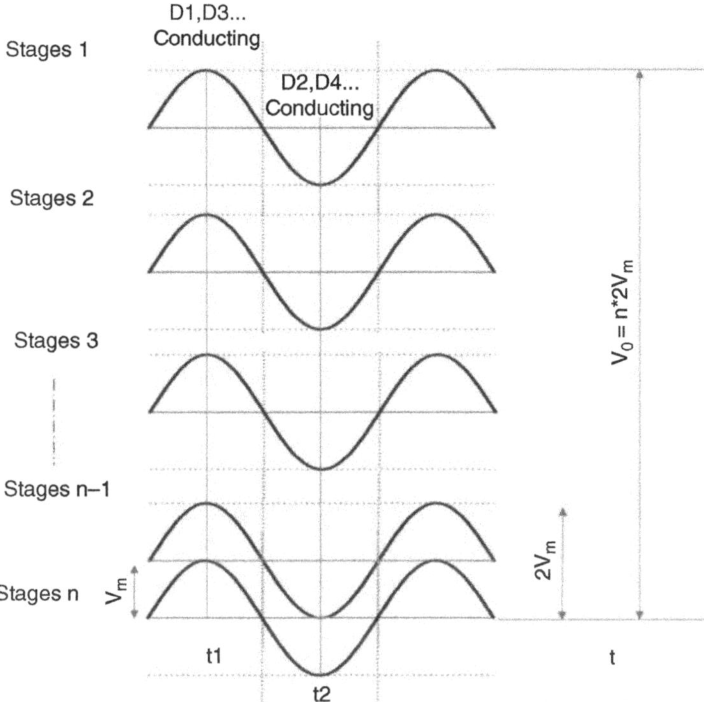

Figure 1.4 Voltage flow in Cockcroft–Walton multiplier circuit.

Assuming that all smoothing capacitors have equal capacitance *C*, and considering that the ripples across all capacitors are in phase, the total ripple voltage on the load is:

$$2\delta V = q\frac{n(n+1)}{2C} = qI\frac{n(n+1)}{2f} \tag{1.26}$$

For equal capacitances in the cascade circuit, the voltage drops across individual stages are:

$$\Delta V_n = \frac{q}{C_n}, \tag{1.27}$$

$$\Delta V_{n-1} = \frac{q}{C}[2n - (n-1)], \tag{1.28}$$

$$\Delta V_1 = \frac{q}{C}\,[2n + 2(n-1) + 2(n-2) + \cdots + 2(2) + 1]. \tag{1.29}$$

By summing and substituting $q = \frac{I}{f}$, the total voltage drop is:

$$\Delta V_0 = \frac{I}{fC}\left(\frac{2n^3}{3} + \frac{n^2}{2} - \frac{n}{6}\right) \tag{1.30}$$

The lower capacitors primarily contribute to ΔV_0, similar to the ripple effect. Doubling C_0 is advantageous as it only needs to withstand half the voltage of other capacitors, i.e. V_{max} Consequently, ΔV_n reduces by, decreasing ΔV_0 of each stage by the same factor, n-times:

$$\Delta V_0 = \frac{I}{fC}\left(\frac{2n^3}{3} - \frac{n}{6}\right). \tag{1.31}$$

For $n \geq 4$, the linear term becomes negligible, and the maximum output voltage is approximated as:

$$V_0^{\max} \approx 2nV_m - \frac{I}{fC} \cdot \frac{2n^3}{3} \tag{1.32}$$

For a fixed number of stages, if ΔV is the voltage drop due to loading effects then the average output voltage is given as follows:

$$V_0 = V_0^{\max} - \Delta V \tag{1.33}$$

This voltage decreases linearly with load current I at constant frequency. Initially, V_0 rises with the number of stages n, but beyond an optimum stage count, it declines. The optimum number of stages that yields the maximum usable output voltage under load is, determined by differentiating Equation (1.32) with respect to n, is:

$$n_{\text{opt}} = \sqrt{\frac{V_m fC}{I}} \tag{1.34}$$

This expression serves as a design criterion for practical CW multipliers to ensure high voltage efficiency without excessive stage losses.

1.3.2 Tandem Accelerators

Tandem accelerators are a type of electrostatic accelerator, similar to the Van de Graaff accelerator, but they utilize the same high-voltage twice. They are also known as Pelletron accelerators [5] because they use metallic pellets to carry charge on a belt, building up a large megavolt potential. A substantial electrostatic potential (V_T) is created at the center of a large insulating tank, usually filled with SF_6 gas, and distributed across various accelerating tubes to ensure linear and smooth acceleration. Tandem accelerators can be configured horizontally or vertically, depending on space constraints. The schematic is shown in Figure 1.5.

Here is the sequence of operation for a non-relativistic particle:

1. The charging chain or belt is given an electrical charge at a lower potential and is moved mechanically towards the high-voltage terminal.
2. The charge is transferred from the chain to the high-voltage terminal, causing the terminal's voltage to increase.
3. The terminal reaches a very high voltage V_T, usually in the range of several million volts, generating a strong electrostatic field. It is utilized linearly by distributing over many accelerating tubes through a set of resistor networks.
4. A negative ion source at the top of the accelerator tank as shown in Figure 1.5 produces negatively charged particles (ions), which are then injected at energy E_{Inj} into the

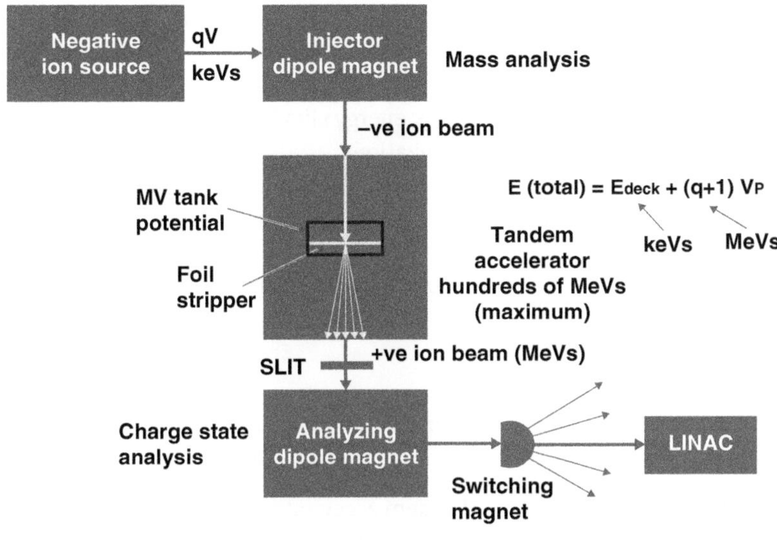

Figure 1.5 Schematic of a Tandem accelerator.

accelerator tank containing many accelerating tubes in series. It is done by an injector dipole magnet, which not only defines the energy of the beam but also selects the particular mass of the beam in one unit negative charge state by the following non-relativistic formula as follows:

$$B_{inj}\rho_{inj}(Tm) = 0.1439\sqrt{\frac{M(amu)E_{inj}(MeV)}{(-1)^2}} \tag{1.35}$$

where ρ_{inj} is the bending radius of the injector magnet. Here, the injector magnet is usually mass selector for the beam under a given acceleration (deck) potential. The main accelerator high voltage (MV) (V_T) is used twice in action, first attracting the negative ion beam from the top of accelerator tank and then stripping the negative ion into a positive ion and hence repelling it by the same high voltage using either a gas or foil stripper.

5. The ions of various charge states after the stripping are accelerated through the tube towards the ground potential, gaining kinetic energy (E_T) by a particular charge state ion (q) as they move as per formula:

$$E_T = E_{Inj} + (q+1)V_T \tag{1.36}$$

6. The accelerated ion beam is then analyzed in terms of energy by an analyzing dipole magnet and the beam is transported to an experimental area with a defined energy and a defined beam current. The corresponding magnetic field (B_{anl}) is given as per non-relativistic formula as follows:

$$B_{anl}\rho_{anl}(Tm) = 0.1439\sqrt{\frac{M(amu)E_T(MeV)}{(q)^2}} \tag{1.37}$$

where ρ_{anl} is the bending radius of analyzer magnet. Here, analyzer magnet is the energy and charge state selector of the beam.

7. In order to stabilize the beam, a dynamic feedback of beam current is taken from analyzer magnet image slits (known as high and low energy slits in bending plane of analyzer magnet), which is fed back for main stabilization.

8. After successful acceleration, beam is transported in beamline. A switching dipole magnet having many output beamlines is used such that for a given magnetic field in the switching magnet, beam can be diverted into one of the beamlines. The corresponding magnetic field (B_{sw}) is given as per non-relativistic formula as follows:

$$B_{sw}\rho_{sw}(Tm) = 0.1439\sqrt{\frac{M(amu)E_T(MeV)}{(q)^2}} \tag{1.38}$$

where ρ_{sw} is the bending radius of the switcher magnet for a particular beamline out of various output ports. Here, a switching magnet is the path selector for beam.

The maximum voltage at the terminal [6] of a Tandem accelerator depends on the size of the terminal electrode, which functions like a spherical capacitor. The capacitance C of this spherical capacitor is given by the formula:

$$C = 4\pi\epsilon_0 r \tag{1.39}$$

which in practical unit becomes:

$$C\,(\mathrm{F}) = 1.11 \times 10^{-10} \cdot r\,(\mathrm{m}) \tag{1.40}$$

where r is the radius of the terminal electrode.

If the spherical terminal with radius r_1 is surrounded by a grounded concentric shell with radius r_2, the capacitance C is calculated by:

$$C = 4\pi\epsilon_0 \frac{r_1 r_2}{r_2 - r_1} = (1.11 \times 10^{-10}\ \mathrm{F/m})\frac{r_1 r_2}{r_2 - r_1} \tag{1.41}$$

The terminal voltage V depends on the charge Q stored on the terminal electrode and is expressed by:

$$V = \frac{Q}{C} \tag{1.42}$$

The maximum voltage that can be achieved at equilibrium, denoted as V_0, is influenced by several currents: the current I_{belt} carried by the belt, the beam current I_{b} in the accelerating tube, the current I_{r} through the resistor chain (which includes the current due to the remaining resistance of the insulating column), and the corona current I_{c}, which rises sharply after a certain point. The time evolution is expressed as follows:

$$\frac{dV}{dt} = \frac{1}{C}\frac{dQ}{dt} = \frac{1}{C}(I_{\mathrm{belt}} - I_{\mathrm{b}} - I_{\mathrm{c}} - I_{\mathrm{res}}) \tag{1.43}$$

When the terminal voltage is low, and the corona current I_{c} is small, this equation can be solved more easily. By introducing the effective resistance R of the resistor chain and the insulating column, the equation simplifies to:

$$\frac{dV}{dt} = \frac{1}{C}\left(I_{\mathrm{belt}} - I_{\mathrm{b}} - \frac{V}{R}\right) \tag{1.44}$$

Here, R represents the effective resistance of the resistor chain and insulating column, measured in ohms (Ω). This resistance plays a crucial role in controlling the voltage drop and charge dissipation in the system. The corona current I_{cor} increases sharply beyond a threshold terminal voltage, typically when the electric field strength exceeds the breakdown strength [7] of the surrounding medium. This threshold depends on factors such as the geometry of the terminal, the type of insulating gas used (e.g. SF_6, and the pressure of the gas. For Tandem accelerators, the terminal radius r_1 typically ranges from tens of centimeters to a few meters, and the equilibrium voltage V_0 can reach several million volts, depending on the application. The exponential term $e^{-t/\tau}$ in the solution indicates how quickly the terminal voltage stabilizes, with the time constant $\tau = \text{RC}$ governing this rate; larger τ results in slower stabilization. The fast feedback control system continuously monitors parameters, such as terminal voltage and beam energy spread. By adjusting the charging current I_{belt} or corona load I_c, it ensures stability within a precision range of typically less than 0.01% energy spread. Maintaining a highly stable terminal voltage is essential for minimizing beam energy fluctuations, which directly affect the resolution and accuracy of experiments relying on the Tandem accelerator.

The time constant $\tau = RC$ and the equilibrium voltage V_0 are defined as:

$$V_0 = R\,(I_{belt} - I_b) \tag{1.45}$$

The solution to the differential equation is:

$$V(t) = V_0 + [V(0) - V_0]\,e^{-t/\tau} \tag{1.46}$$

This solution is valid as long as the currents I_{belt}, I_b, and the resistance R remain constant, and the corona current is negligible. The terminal reaches an equilibrium voltage V_0 when the charge supplied by the belt equals the charge leaving the terminal. In Tandem accelerators, fast feedback control systems usually by analyzing magnet slit currents from beam are used to quickly adjust and stabilize the terminal voltage V. The behavior of V over time is determined by the characteristics of this control system. The terminal voltage should not drift over time and must be highly stabilized to get the beam with a very small energy spread. The terminal voltage can be managed by regulating the charging current I_{belt} and adjusting the corona load current using movable needles.

Regarding strippers, gas and foil strippers are two commonly used methods for changing the charge state of ions in accelerators, each with its advantages and disadvantages. Gas strippers involve passing the ion beam through a low-density gas, where collisions with gas molecules strip electrons from the ions. They generally result in lower charge states and less efficient stripping compared to foil strippers. On the other hand, foil strippers use thin metal foils that the ion beam passes through, where the high-density material more aggressively strips electrons, often leading to higher charge states and more efficient stripping. While foil strippers can produce higher charge states, they tend to cause more scattering and energy loss in the ion beam, which can degrade the beam quality. The choice between gas and foil strippers depends on the specific requirements of the accelerator, such as the desired charge state and beam quality.

The first charge state (either gas or foil) in the Tandem accelerator is usually inside the terminal voltage tank where polarity reversal of ion takes place. Sometimes, one may use two strippers in Tandem accelerator, which makes q1 and q2 charge state of ions so as to get the high charge state q2 of ions. For example, if there is a total n stage of acceleration, the second stripper lies at kth section, then we have total energy from Tandem accelerator as follows:

The total energy E_T of an ion in a Tandem accelerator with a double stripper configuration can be calculated using the following formula:

$$E_T = E_{\text{inj}} + V\left(1 + \frac{q_1 \cdot k}{n} + \frac{q_2 \cdot (n-k)}{n}\right) \tag{1.47}$$

Tandem accelerators, especially Tandem accelerators, are an important advancement in electrostatic acceleration technology. They work by using the same high-voltage potential twice to achieve higher energy for particles. These accelerators are used in many areas, including nuclear physics, material science, and medical treatments. The ability to use both gas and foil strippers allows for flexible control of the particle charge states, helping to improve beam efficiency while reducing energy loss. The tandem design combines sequential electrostatic fields by accelerating tubes to accelerate particles through multiple stages. Magnets such as injector, analyzer, and switching magnets are essential in determining the mass, energy, and direction of the beam. In addition, feedback systems ensure that the terminal voltage remains stable, which helps produce a beam with minimal energy spread.

1.4 Cyclotrons

The concept of the cyclotron [8, 9] was introduced by E. O. Lawrence in 1929. Within a few years, Lawrence, along with Edlefsen, constructed the first circular particle accelerator: a 4-inch cyclotron, and the successful acceleration of particles was demonstrated by M. S. Livingston. This innovation marked a significant advancement, as it was the first accelerator capable of delivering multiple accelerating impulses to a particle beam using a single pair of electrodes. Ernest Lawrence was awarded the Nobel Prize in Physics in 1939 for this invention. Cyclotrons accelerate particles in a spiral trajectory under a homogeneous magnetic field, with acceleration provided by a time-varying electric field.

Lawrence's pioneering work led to significant milestones, such as the successful acceleration of hydrogen ions to 80 keV in 1931 and achieving 1.25 MeV proton acceleration in 1932, shortly after the achievements of Cockcroft and Walton. The cyclotron operates in a high-vacuum environment, typically around 10^{-6} torr, which is essential to minimize beam loss due to collisions with residual gas molecules and to ensure stable high-voltage operation of the dees and deflector. The acceleration process relies on the repeated use of the same electrodes, where charged particles gain energy twice per revolution as they cross the gap between the dees, gradually increasing their spiral orbit radius. The motion of charged particles within the cyclotron occurs in the median plane, where the applied magnetic field guides them in near-circular orbits.

Repeated acceleration requires resonance between the particle's revolution frequency and the applied RF. The resonance condition must be maintained throughout the acceleration process to ensure the particles reach the accelerating gaps at the correct time. Any deviation between the particle revolution frequency and the RF results in a phase shift between the beam particles and the driving RF field, causing the particles to receive less energy. The acceleration process reaches its limit when the phase shift approaches ±90°. Beyond this point, particles experience deceleration and are eventually lost from the system.

The particles are subjected into circular motion using homogeneous magnetic field. As the energy of the particles increases, so does the orbit radius, but if the charge (q), magnetic field (B), and mass (m) remain constant, the frequency stays the same across all energies, velocities, and radii. By positioning two electrodes (known as dees) around an ion source, ions can gain energy twice per revolution, consistently arriving at the same azimuth for each subsequent acceleration. An accelerator that maintains a constant orbital frequency is termed isochronous cyclotron. Positive ions are extracted from the ion source whenever the voltage on the first dee is negative. Each accelerating pulse increases the beam's orbital radius, creating a spiral trajectory. Near the edge of the dipole magnet pole, the beam is extracted using an electrostatic deflector. For a positively charged beam, a negatively charged outer electrode pulls the beam from its orbit, while a thin, grounded inner electrode, known as the septum, separates the exiting beam from the inner orbits.

A cyclotron is not suitable for accelerating electrons primarily because of their low mass, which causes them to quickly reach relativistic speeds even at relatively low energies. As electrons approach the speed of light, their mass effectively increases, leading to a decrease in their angular frequency. This change causes them to fall out of synchronization with the fixed-frequency alternating electric field used in the cyclotron. Thus, the fixed magnetic field in a cyclotron is unable to maintain the proper trajectory for relativistic electrons, making it difficult to keep them on the intended spiral path. These factors together prevent a cyclotron from efficiently accelerating electrons.

The cyclotron operates based on the principle of cyclotron resonance. For a particle of mass (m), charge (q), and velocity (v) moving in a static magnetic field (B) along a circular path of radius (R), the motion is governed by:

$$\frac{mv^2}{R} = qvB \tag{1.48}$$

$$v = \frac{qBR}{m} \tag{1.49}$$

The cyclotron frequency is:

$$f = \frac{v}{2\pi R} = \frac{qB}{2\pi m} \tag{1.50}$$

For relativistic case,

$$f_{\text{cyc, rel}}(\text{MHz}) = \frac{Q}{A} \cdot \frac{15.23 \cdot B(T)}{\gamma} \tag{1.51}$$

$$\gamma = 1 + \frac{E_{\text{kinetic}}}{A \cdot 938.27} \tag{1.52}$$

The kinetic energy of particles is given by:

$$E = \frac{1}{2}mv^2 = \frac{(qBR)^2}{2m} \tag{1.53}$$

Substituting $q = Qe$ and $m = A \cdot m_u$: where m_u is one unit of mass in amu.

Kinetic energy per nucleon:

$$\frac{E}{A} = \frac{(Qe \cdot B \cdot R)^2}{2A^2 \cdot m_u} \tag{1.54}$$

Cyclotron K-factor:

$$K = \frac{(eBR)^2}{2m_u} \tag{1.55}$$

$$\frac{E}{A} = K\left(\frac{Q}{A}\right)^2 \tag{1.56}$$

The K-factor is an important figure of merit for cyclotrons, defining the upper limit of the kinetic energy that can be achieved for proton given the cyclotron's magnetic field and maximum orbital radius. This expression serves as a figure of merit for a cyclotron, also referred to as the bending limit, and can be expressed for protons as:

$$K(\text{MeV}) = 47.9(BR(\text{Tm}))^2 \tag{1.57}$$

The weight of a cyclotron approximately increases with the K value as:

$$W \approx K^{1.5} \approx (BR)^3 \tag{1.58}$$

Finally, the time to complete one revolution is:

$$T = \frac{2\pi R}{v} = \frac{2\pi m}{qB} \tag{1.59}$$

Thus, the period of revolution is independent of the velocity and size of the orbit, depending only on the magnetic field and the charge-to-mass ratio of the particle. When these are constant, the period of rotation is constant. For resonance, the RF period must coincide with this period of revolution. The cyclotron frequency is given by:

$$\omega = \frac{2\pi}{T} = \frac{qB}{m} \tag{1.60}$$

According to the theory of relativity, as the energy or velocity of particles increases, so does their mass as they approach the speed of light. Consequently, the cyclotron frequency of fast particles becomes lower than that of slower particles, leading to a gradual loss of synchronism.

The energy gain (E_k) of a particle under the dee's potential (V) is:

$$E_k = qV \tag{1.61}$$

The particle must pass through the dees n times to reach the required energy (E):

$$E_{\text{max}} = nE_k \tag{1.62}$$

Thus, the maximum radius of the cyclotron orbit is:

$$R_{\text{max}} = \frac{\sqrt{2mE_{\text{max}}}}{qB} \tag{1.63}$$

In summary, the fundamental principle of cyclotron resonance ensures that particles maintain a stable, circular orbit while continuously gaining energy from the electric field as they pass through the dees. The critical relationship between the magnetic field strength, particle charge, mass, and velocity allows the cyclotron to deliver high-energy particles in a compact design. The introduction of the cyclotron K-factor helps quantify the energy limits achievable within the cyclotron based on the magnetic field strength and orbital radius. However, cyclotrons face limitations when dealing with particles that reach relativistic speeds, such as electrons. As particles approach the speed of light, their relativistic mass increases, leading to a loss of synchronization with the RF field, which renders the cyclotron less effective for accelerating particles at such high velocities.

1.5 Synchrotrons

A synchrotron [10] is a circular ring accelerator that functions similarly to a cyclotron, but with a magnetic field that increases with the energy of the particles to maintain a constant circular radius. This variation, known as alternate gradient ensures that the particles remain in resonance with the accelerating RF cavities, which continue to boost energy with each revolution. Synchrotrons are particularly well-suited for accelerating electron beams, making them powerful sources of X-rays and ultraviolet (UV) light. The emitted photons are exceptionally bright, with a narrow emittance cone, and their wavelength is tunable, making them useful in various experiments and applications. Synchrotrons can also accelerate protons and heavy ions. The largest synchrotron-type accelerator in the world is the Large Hadron Collider (LHC), with a circumference of 27 kilometers, located at the European Organization for Nuclear Research (CERN) near Geneva, Switzerland. Other notable synchrotron facilities include the Advanced Photon Source (APS) at Argonne National Laboratory in the United States and the Diamond Light Source in the United Kingdom.

A storage ring is a specialized type of synchrotron where the particles maintain constant energy. The main component of a synchrotron is a ring-shaped vacuum vessel, where charged particles approach nearly the speed of light and are guided by powerful magnets. A synchrotron produces extremely intense beams of light, known as synchrotron radiation. This radiation is used in various disciplines, including physics, chemistry, biology, and materials science.

As particles gain energy in the synchrotron, they emit energy in the form of synchrotron radiation. The energy loss due to synchrotron radiation, especially for high-energy electrons, is significant and scales with the fourth power of the particle's energy. The power radiated due to synchrotron radiation is given by:

$$P_{\text{synch}} = \frac{e^2 c}{6\pi\epsilon_0} \frac{\gamma^4}{R^2} \tag{1.64}$$

where P_{synch} is the synchrotron radiation power, γ is the Lorentz factor, and R is the bending radius of the synchrotron. As the particle's energy increases, the radiated power becomes a limiting factor, requiring careful design considerations to balance energy loss and acceleration.

The alternating gradient (strong focusing) technique is another important feature of synchrotrons. This technique uses quadrupole magnets to focus the particle beam transversely, ensuring that the beam remains stable while being accelerated. The quadrupole magnets create alternating focusing and defocusing forces, increasing beam stability. The focusing strength K of a quadrupole magnet is given by:

$$K = \frac{1}{f} \tag{1.65}$$

where f is the focal length of the quadrupole magnet. This method of strong focusing allows for higher energy beams to be accelerated in more compact synchrotron designs, increasing the efficiency and stability of the accelerator.

For a synchrotron ring containing particles with energy E and N dipole magnets of effective length l, the magnetic rigidity of the bending magnet is given by:

$$B\rho = \frac{\beta E}{q} \tag{1.66}$$

The integrated dipole strength is given by:

$$B.l = B(\rho\theta) = \frac{\beta E}{q}\frac{2\pi}{N} \tag{1.67}$$

The filling factor K_F is defined as the ratio of the total bending trajectory length to the circumference C of the ring:

$$K_F = \frac{Nl}{C} \tag{1.68}$$

The ring circumference is given by:

$$C = \frac{Nl}{K_F} = \frac{2\pi\beta E}{K_F Bq} \tag{1.69}$$

The relationship between the RF $f_{\mathrm{RF}}(t)$ and the revolution frequency $f_{\mathrm{rev}}(t)$ in a synchrotron is given by the harmonic number h, such that:

$$f_{\mathrm{RF}}(t) = h \cdot f_{\mathrm{rev}}(t) \tag{1.70}$$

The revolution frequency $f_{\mathrm{rev}}(t)$ can be expressed as the velocity $v(t)$ of the particle divided by the circumference of the synchrotron $2\pi R_s$:

$$f_{\mathrm{rev}}(t) = \frac{v(t)}{2\pi R_s} \tag{1.71}$$

The velocity $v(t)$ of the particle can be related to its momentum $p(t)$ and total energy $E(t)$ using the equation:

$$v(t) = \frac{p(t) \cdot c^2}{E(t)} \tag{1.72}$$

The momentum $p(t)$ of a charged particle moving in a magnetic field $B(t)$ with a bending radius ρ is given by:

$$p(t) = e \cdot B(t) \cdot \rho \tag{1.73}$$

The total energy $E(t)$ of the particle, considering relativistic effects, is expressed as:

$$E(t) = \sqrt{(m_0c^2)^2 + (p(t) \cdot c)^2} \tag{1.74}$$

By substituting the expression for momentum $p(t)$ into the total energy equation, we obtain:

$$E(t) = \sqrt{(m_0c^2)^2 + (e \cdot B(t) \cdot \rho \cdot c)^2} \tag{1.75}$$

Substituting the expression for velocity $v(t)$ into the equation for the revolution frequency, we have:

$$f_{\text{rev}}(t) = \frac{1}{2\pi R_s} \cdot \frac{e \cdot B(t) \cdot \rho \cdot c^2}{\sqrt{(m_0c^2)^2 + (e \cdot B(t) \cdot \rho \cdot c)^2}} \tag{1.76}$$

Finally, this leads to the expression for the ratio of the RF to the harmonic number, showing how it depends on the magnetic field $B(t)$, the particle's rest mass m_0, the bending radius ρ, and the synchrotron radius R_s:

$$\frac{f_{\text{RF}}(t)}{h} = \frac{c}{2\pi R_s} \cdot \frac{B(t)}{\sqrt{B(t)^2 + \left(\frac{m_0 \cdot c}{e \cdot \rho}\right)^2}} \tag{1.77}$$

The above equation shows how the frequency of the RF system in a synchrotron must be adjusted to stay in synchronization with the particles as they move around the ring. It takes into account the strength of the magnetic field, the mass of the particles, and how fast they are moving. As the particles gain energy and speed up, the magnetic field changes, and this equation helps ensure that the RF system keeps accelerating the particles efficiently by staying perfectly timed with their movement.

1.6 Synchrocyclotron

To overcome the energy limitations of conventional cyclotrons, frequency-modulated cyclotrons, known as synchrocyclotrons, were developed. It is a type of particle accelerator that combines the principles of both a cyclotron and a synchrotron. It operates similarly to a cyclotron but incorporates a technique called *synchrotron resonance* to account for relativistic effects as particles approach the speed of light. As the particles gain energy and their velocity increases, their revolution frequency decreases due to relativistic mass increase. To maintain synchronization, the synchrocyclotron adjusts the frequency of the applied RF electric field to match the changing revolution frequency of the particles. This adjustment allows particles to continue gaining energy efficiently at relativistic speeds. These accelerators utilize the principle of phase stability discovered by Veksler [11], which enables particles to remain in phase with the RF field during acceleration. By gradually lowering the RF to match the decreasing revolution frequency of the accelerating protons, synchrocyclotrons have successfully accelerated protons to energies of several hundred MeV. In synchrocyclotrons, the magnetic field decreases radially to provide axial focusing, ensuring the beam remains stable. However, since the frequency must be continuously reduced during acceleration, synchrocyclotrons produce pulsed beams with a low duty factor and relatively low average beam intensities compared to other accelerators.

The magnetic field B, particle charge q, and mass m determine the motion of the particle in the synchrocyclotron. As the particle accelerates, its mass increases due to relativistic effects, which is represented by the relativistic mass formula:

$$m' = \frac{m}{\sqrt{1-\left(\frac{v}{c}\right)^2}} \tag{1.78}$$

where v is the particle's velocity and c is the speed of light.

The cyclotron frequency f decreases as the particle reaches relativistic speeds and is given by:

$$f = \frac{qB}{2\pi m'} = \frac{qB}{2\pi} \cdot \frac{\sqrt{1-\left(\frac{v}{c}\right)^2}}{m} \tag{1.79}$$

For nonrelativistic particles, the frequency f_0 remains constant and is expressed as:

$$f_0 = \frac{qB}{2\pi m} \tag{1.80}$$

The relationship between the relativistic frequency f and the nonrelativistic frequency f_0 is:

$$\frac{f}{f_0} = \sqrt{1-\left(\frac{v}{c}\right)^2} \tag{1.81}$$

This allows us to find the velocity v as:

$$\frac{v}{c} = \sqrt{1-\left(\frac{f}{f_0}\right)^2} \tag{1.82}$$

The velocity v is also related to the magnetic field B and the radius R of the particle's path:

$$v = \frac{qBR}{m} \tag{1.83}$$

Substituting this into the previous expression gives:

$$\frac{qBR}{mc} = \sqrt{1-\left(\frac{f}{f_0}\right)^2} \tag{1.84}$$

Solving for B gives the expression:

$$B = \frac{mc}{qR}\sqrt{\left(\frac{f}{f_0}\right)^2 - 1} \tag{1.85}$$

This equation shows how the magnetic field adjusts due to the relativistic effects as the particle accelerates. The resonance condition is achieved when the particle's revolution frequency f_0 synchronizes with the frequency f of the RF field, maintaining efficient acceleration:

$$f_0 = nf \tag{1.86}$$

where n is an integer representing the harmonic number.

The synchrocyclotron represents a significant advancement in particle accelerator technology by overcoming the limitations of earlier cyclotrons, particularly with regard to relativistic effects. Through the synchronization of the RF field with the changing frequency of the particles' revolution, synchrocyclotrons efficiently accelerate particles to near-light speeds. This innovation has paved the way for numerous applications, including advancements in medical radiation therapy and nuclear physics research. As particle velocities approach the speed of light, the need to account for relativistic mass increases becomes critical. The formulas governing the relationship between magnetic fields, particle velocity, and energy ensure stable acceleration while maintaining resonance conditions. The insights provided by the relativistic effects on frequency and mass help optimize the design of synchrocyclotrons for high-energy applications.

1.7 Betatron

The betatron, an electron accelerator, was developed in the 1930s by Donald W. Kerst [12] at the University of Illinois, USA. It is designed to accelerate charged particles using rapidly varying magnetic fields, which induce an electric field that accelerates the particles. The particles follow a spiral path within a toroidal, or doughnut-shaped, vacuum chamber. Betatrons are nonrelativistic accelerators and typically operate in the RF range. While betatrons are not as prominent as LINACs today, they were significant in the early development of particle physics, as well as in medical and industrial applications. Betatrons operate on the principle of electromagnetic induction:

$$-e\oint \vec{E}.\vec{dl} = -\frac{d}{dt}\int \vec{B}.\vec{dA} \tag{1.87}$$

The induced electromotive force (emf) is related to the rate of change of magnetic flux Φ, where Φ is the flux through the area enclosed by the orbit:

$$\mathcal{E} = -\frac{d\Phi}{dt}, \quad \Phi = \pi r^2 \langle B \rangle, \quad \frac{d\Phi}{dt} = \pi r^2 \frac{d\langle B \rangle}{dt} \tag{1.88}$$

where $\mathcal{E}$ is the induced emf, Φ is the magnetic flux through the orbit, r is the radius of the orbit, $\langle B \rangle$ is the average magnetic field over the area enclosed by the orbit, and $\frac{d\Phi}{dt}$ is the rate of change of the magnetic flux.

The induced emf $\mathcal{E}$ around the orbit is related to the induced electric field E, which depends on the rate of change of the average magnetic field:

$$\mathcal{E} = E \cdot 2\pi r, \quad E = \frac{r}{2} \cdot \frac{d\langle B \rangle}{dt} \tag{1.89}$$

where E is the induced electric field around the orbit, and $2\pi r$ is the circumference of the orbit.

The energy gained by a particle per revolution is proportional to the induced electric field around the orbit and can be expressed as:

$$\Delta\mathcal{E} = -e \cdot \mathcal{E} = -e \cdot 2\pi r E = -e \cdot \pi r^2 \frac{d\langle B \rangle}{dt} \tag{1.90}$$

This equation shows how the energy gained by the particle in each revolution is tied to the rate of change of the magnetic flux inside the orbit.

The Lorentz force provides the centripetal force necessary to keep the particle in a circular orbit, balanced by the work done by the magnetic field:

$$qvB_{\text{orbit}} = \frac{mv^2}{r}, \quad qvB_{\text{orbit}}r = q\pi r^2 \frac{d\langle B \rangle}{dt} \tag{1.91}$$

where q is the charge of the particle, v is the velocity of the particle, B_{orbit} is the magnetic field at the orbit, m is the mass of the particle, and $\frac{mv^2}{r}$ is the centripetal force acting on the particle.

The **betatron frequency**, or the frequency at which the particles revolve in the circular path due to the magnetic field, is given by:

$$\omega = \frac{qB_{\text{orbit}}}{m} \tag{1.92}$$

where ω is the angular frequency, q is the charge of the particle, and B_{orbit} is the magnetic field at the particle's orbit.

This condition ensures the magnetic field at the orbit is half the average magnetic field within the orbit, allowing stable motion in the betatron:

$$B_{\text{orbit}} = \frac{1}{2}\langle B \rangle \tag{1.93}$$

This condition is crucial for ensuring stable orbits in a betatron. It signifies that the magnetic field at the radius of the particle's orbit must be half the average magnetic field within the area enclosed by the orbit. This balance is necessary to provide the correct centripetal force to keep the particles moving in a stable circular path while they are being accelerated by the induced electric field. If this condition is not met, the particles will either spiral inward or outward, leading to instability. By maintaining this relationship, the betatron can efficiently accelerate particles to high energies while ensuring they remain in a stable orbit.

1.8 Particle Colliders

Particle colliders, as the name suggests, accelerate particles and then collide them to break them apart into more fundamental entities, enabling the exploration of the subatomic world. Particle colliders are specialized types of accelerators used in experimental physics to study the fundamental properties of particles and the forces of nature. Colliders have been instrumental in discovering particles such as the Higgs boson and the top quark, among others. Colliders are designed to achieve high collision rates per unit of time, known as high luminosity, which is essential for increasing the probability of rare particle interactions. Colliders can be categorized into two types:

1. **Circular colliders:** These accelerators guide particles in a circular path using magnetic fields, with collisions occurring at experimental stations along the circumference. Examples include the LHC at CERN, Geneva, Switzerland, and the Relativistic Heavy Ion Collider at Brookhaven National Laboratory, USA.

2. **Linear colliders (LINACs):** These accelerators propel particles in a straight line using RF cavities. Examples include the Stanford Linear Collider and the future International Linear Collider.

The collision energy in a collider is a critical parameter for experiments and is dependent on the energy (E) of the colliding particle. The total energy and momentum of the colliding particles are important parameters for determining the outcome of particle collisions. In symmetric colliders, such as the LHC, the particles in both beams have equal and opposite momentum, leading to maximum center-of-mass energy, given by:

$$E_{\text{collision}} = 2E \tag{1.94}$$

The luminosity of a collider is given by:

$$L = \frac{N_1 N_2 f}{4\pi\sigma_x\sigma_y} \tag{1.95}$$

where N_1 and N_2 are the number of particles per bunch for each beam, f is the bunch collision frequency, σ_x and σ_y are the transverse beam sizes in the horizontal and vertical directions, respectively.

1.8.1 Microtron

A microtron [13, 14] is a type of circular particle accelerator that utilizes a combination of magnetic and electric fields to accelerate charged particles, typically electrons, to high energies. It consists of several key components, including a uniform magnetic field to steer particles in a circular orbit, RF cavities to accelerate particles, and an injection system to introduce particles into the accelerator. The operation of a microtron is based on the principle of phase stability and the synchronization between the particle motion and the applied RF field.

The working of a microtron begins with the injection of charged particles, usually electrons, into the accelerator with an initial kinetic energy K_0. Once inside, the particles are guided in a circular orbit by a uniform magnetic field B, which exerts a Lorentz force on them. The radius of the circular orbit is given by the expression

$$r = \frac{\gamma m v}{qB}$$

where γ is the relativistic factor $\gamma = \frac{1}{\sqrt{1-\frac{v^2}{c^2}}}$, m is the rest mass of the electron, v is its velocity, q is the charge of the particle, and B is the applied magnetic field.

As the particles traverse their circular paths, they pass through an RF cavity at regular intervals, gaining an increment of energy ΔK each time. The energy gain per pass can be expressed as:

$$\Delta K = qV_{\text{RF}} \cos\phi$$

where V_{RF} is the peak RF voltage and ϕ is the phase of the particle relative to the RF field. To ensure efficient acceleration, the phase stability condition must be maintained, allowing the particles to synchronize with the RF field and arrive at the accelerating gaps at the correct time.

A unique feature of the microtron is that, despite the continuous increase in particle energy, the revolution period remains constant. This is because the increase in energy leads to a proportional increase in orbit radius, ensuring that the total time taken per revolution, given by

$$T = \frac{2\pi r_n}{v}$$

remains unchanged. This characteristic simplifies the synchronization process and distinguishes the microtron from other accelerators.

Another important aspect of microtron operation is the gradual expansion of the particle orbit. With each successive acceleration cycle, the orbit radius increases stepwise according to the relation

$$r_n = r_0 + n\Delta r$$

where r_0 is the initial radius and n represents the number of revolutions. The uniform magnetic field ensures constant transverse focusing, which helps maintain beam stability and prevents excessive divergence.

Microtrons are widely used due to their compact design, ease of synchronization, and ability to achieve high energies with relatively low operational costs. These accelerators have found significant applications in the medical field, particularly in radiation therapy for cancer treatment, where electron beams are used to precisely target tumors while minimizing damage to surrounding tissues. In materials science, microtrons are employed to study material properties through electron beam interactions, providing valuable insights into structural composition. Industrial applications also benefit from microtrons, such as in the sterilization of medical equipment, food irradiation for preservation, and polymer modification processes.

In conclusion, the microtron represents an efficient and versatile particle accelerator, capable of providing high-energy electron beams for a range of applications. Its unique properties, such as constant revolution time and phase stability, make it a valuable tool in both scientific research and industrial settings.

1.9 FFAG Accelerators

The acronym FFAG stands for "fixed-field alternating gradient." As the name suggests, these use AG focusing, but with fixed magnetic fields. FFAG accelerators [15] were first conceived in the early 1950s as a method to reduce the accelerator radius compared to traditional circular accelerators. Despite their promising potential, the development of FFAGs faced delays primarily due to the complexity of their magnet designs, which were more intricate compared to those of the rapidly advancing synchrotron technology. Among the various designs, the "radial-sector FFAG" represents the most straightforward implementation. In FFAG accelerators, AG focusing is applied with a series of dipole magnets. This approach replaces a continuous magnetic field with discrete focusing and defocusing elements. This configuration not only provides the necessary beam focusing but also creates gaps to accommodate accelerating cavities and other essential components for beam injection, extraction, and monitoring. FFAG accelerators combine features of synchrotrons and cyclotrons, using a fixed magnetic field and alternating focusing to accelerate particles continuously across a wide range of energies.

FFAG accelerators [16] combine features of both LINACs and circular accelerators. FFAG accelerators use a fixed magnetic field strength with alternating magnetic field gradients to accelerate charged particles. Unlike synchrotron accelerators, where the magnetic field strength increases along the path of the particles, in FFAG accelerators, the magnetic field remains constant while the gradient of the magnetic field changes over time. Particles follow a cycloidal motion within the magnetic field. The equation of motion for a particle's trajectory in an FFAG accelerator can be complex, involving a combination of trigonometric functions and elliptical integrals. More detailed study can be found in Refs. [16–18].

There are two types of FFAGs: scaling and nonscaling. In a scaling FFAG, the focusing of particles remains constant regardless of their energy. The magnetic field must be carefully designed to ensure that particles follow similar orbits at all energies. This design is beneficial for maintaining stable particle paths over a wide energy range, but the strict requirements for the magnetic field make it more challenging to construct.

In contrast, nonscaling FFAGs [18] allow the focusing to change as particle energy increases, meaning that particle orbits can vary. This design offers greater flexibility, as particles can follow different paths at different energies, which is known as chromatic behavior. The magnetic field in a nonscaling FFAG is easier to design, making the overall system simpler and more compact. However, this flexibility can lead to less stability in particle orbits, especially at higher energies, as the precise control over the paths is reduced.

Scaling and nonscaling FFAG accelerators have different designs and characteristics. Scaling FFAG accelerators are designed with magnetic fields that change with energy in a specific way. The magnetic field strength $B(r)$ in a scaling FFAG varies with the radius r such that $B(r) \propto \frac{1}{r}$. This design helps maintain consistent transverse focusing across different energy levels, which leads to stable and predictable beam dynamics over a wide range of energies. However, this approach requires precise control and complex design to ensure that focusing remains uniform.

In contrast, nonscaling FFAG accelerators use magnetic fields that do not necessarily follow this scaling rule. The field profile can vary in a more complex manner, and the transverse focusing also changes with radius and energy. This can result in more complicated beam dynamics. Nonscaling FFAGs offer greater flexibility in energy range and field design but do not provide the same level of predictable stability as scaling FFAGs.

A comparison of various circular accelerators is given in Table 1.2.

Table 1.2 Comparison of circular accelerators.

Accelerator type	Revolution time	Orbit radius	Transverse focusing
Cyclotron	Constant	Variable	Variable
Synchrotron	Variable (except relativistic)	Constant	Constant
Nonscaling FFAG	Variable (small)	Variable (small)	Variable
Scaling FFAG	Variable	Variable	Constant
Betatron	Variable	Constant	Variable
Microtron	Constant	Variable	Constant

Curiosity: *Have you ever watched a swimmer ride a big ocean wave all the way to the shore? Now imagine a tiny particle doing the same thing—not in water, but on a wave made of electric fields inside a special gas called plasma. This is how wakefield acceleration works. When a strong laser pulse or a fast-moving particle beam travels through plasma, it pushes away the electrons and creates a wave behind it—just like a boat makes waves in water. If we place other particles at just the right spot on this wave, they can "surf" it and get pushed forward really fast, gaining a lot of energy in a short distance. Could this help us build smaller and cheaper accelerators that don't need kilometers of space? Why is plasma so good at making strong waves, and how does it help us go beyond what normal machines can do? What happens if a particle ends up in the wrong part of the wave—does it slow down instead of speeding up? How do laser-driven and particle-driven wakefields make these waves differently? And could we one day use this technology for new kinds of medical tools or even space engines? We try to answer some of these exciting questions here, but many more are still waiting for curious minds to explore!*

1.10 Wakefield Accelerators

Conventional accelerators are large and expensive, with maximum achievable fields limited to a few tens of MV/m due to breakdown issues. Plasma waves offer an alternative, enabling more compact and powerful acceleration. Analogous to a boat creating a wake in water, a laser pulse traveling at the speed of light through a plasma generates a wake behind it, termed an electron plasma oscillation. These high-frequency oscillations of electrons in the plasma result in strong electric fields over short distances, reaching the GV/m range. The wakefield amplitude is maximized when the drive duration is shorter than the plasma wavelength.

1.10.1 Laser Wakefield Acceleration

A laser plasma wakefield accelerator (LWFA) [19] uses a powerful, very short laser pulse to push electrons in a plasma, creating wakes. These wakes have strong electric fields that can accelerate charged particles to very high energies in a small space. It is much smaller than traditional particle accelerators. The wakefield can be linear or nonlinear, depending on whether the driver beam density is less than or greater than the plasma electron density, respectively. At high driver densities, a blowout regime occurs, allowing for linear focusing and higher acceleration efficiency.

In LWFA, several fundamental physical quantities are essential to describe the behavior of electrons in intense electromagnetic fields. The normalized vector potential, denoted by a_0, represents the strength of the laser field and determines whether the electron motion becomes relativistic. The resulting electron oscillations in the laser field are described by the quiver velocity, v_q, which increases with a_0. The laser field itself is characterized by its peak electric field E_0, which is directly related to the laser intensity I_0. The laser also has a specific angular frequency ω_0 and wavelength λ_0, which define its temporal and spatial scales. While a free electron in a vacuum plane wave cannot gain net energy due to the Lawson–Woodward theorem, in a plasma, the presence of spatial inhomogeneities allows for net energy gain. The electron experiences a net outward push due to the ponderomotive force, which arises from gradients in the electric field amplitude. The plasma's response is governed by the plasma frequency ω_p, which depends on the electron density n_e, and the associated plasma wavelength λ_p sets the characteristic scale of the wakefield. The longitudinal accelerating field E_p, generated behind the laser pulse, is responsible for electron acceleration. However, electrons eventually outrun the accelerating phase, which defines the dephasing length, denoted as L_d. These quantities together form the foundation for understanding energy transfer from laser pulses to particles in plasma.

The normalized vector potential a_0 describes the behavior of an electron in an intense laser field and is defined as:

$$a_0 = \frac{eE_0}{m_e \omega_0 c}$$

The quiver velocity v_q of the electron in the laser field is:

$$v_q \approx c \frac{a_0}{\sqrt{1 + a_0^2}}$$

In the nonrelativistic regime ($a_0 \ll 1$), the quiver velocity v_q is:

$$v_q = \frac{eE_0}{m_e \omega_0}$$

In the relativistic regime ($a_0 \gtrsim 1$), the quiver velocity approaches the speed of light. The peak electric field E_0 is related to the laser intensity I_0 by:

$$E_0 = \sqrt{\frac{2I_0}{\epsilon_0 c}}$$

Substituting E_0 and $\omega_0 = \frac{2\pi c}{\lambda_0}$ into the expression for a_0, we get:

$$a_0 = \frac{e\lambda_0}{2\pi m_e c^2} \times \sqrt{\frac{2I_0}{\epsilon_0}}$$

For a laser with wavelength $\lambda_0 = 1\,\mu\text{m}$, a_0 is related to the intensity I_0 by:

$$a_0 = 0.855 \times 10^{-9} \times \sqrt{I_0[\text{W/cm}^2]}$$

The Lawson–Woodward theorem states that a free electron cannot gain net energy from a plane electromagnetic wave in vacuum. The force on the electron is described by the Lorentz equation:

$$\frac{d\mathbf{p}}{dt} = q\,(\mathbf{E} + \mathbf{v} \times \mathbf{B})$$

The rate of energy transfer is:

$$\frac{d\mathcal{E}}{dt} = q\mathbf{v} \cdot \mathbf{E}$$

In the case of a plane wave, the work done by the field averages out to zero over a full cycle:

$$\Delta\mathcal{E} = q \int_0^T \mathbf{v}(t) \cdot \mathbf{E}(t)\, dt = 0$$

Thus, no net energy is transferred to the particle. Schemes like laser wakefield acceleration (LWFA) introduce plasma to break the symmetry and enable efficient particle acceleration.

The motion of an electron in an oscillating electric field $\mathbf{E}(t) = \mathbf{E}_s(\mathbf{r})\cos(\omega t)$ is described by the Lorentz force:

$$m_e \frac{d^2\mathbf{r}}{dt^2} = e\mathbf{E}_s(\mathbf{r})\cos(\omega t)$$

The velocity $\mathbf{v}(t)$ is:

$$\mathbf{v}(t) = \frac{e\mathbf{E}_s(\mathbf{r})}{m_e\omega} \sin(\omega t)$$

The time-averaged kinetic energy of the electron is:

$$\langle \mathcal{E}_k \rangle = \frac{e^2 E_s^2(\mathbf{r})}{4m_e\omega^2}$$

The ponderomotive force is the gradient of the time-averaged kinetic energy:

$$\mathbf{F}_\mathrm{p} = -\frac{e^2}{4m_e\omega^2} \nabla E_s^2(\mathbf{r})$$

The plasma frequency ω_p is defined as:

$$\omega_p = \sqrt{\frac{n_e e^2}{\epsilon_0 m_e}}$$

The plasma wavelength λ_p is related to ω_p by:

$$\lambda_p = \frac{2\pi c}{\omega_p}$$

Simplifying:

$$\lambda_p[\mu\mathrm{m}] = 33 \times \left(n_e(10^{18}\,\mathrm{cm}^{-3})\right)^{-1/2}$$

The ponderomotive force creates wakefields that generate an accelerating electric field E_p, proportional to ω_p and the density perturbation $\frac{\delta n_e}{n_e}$. The strength of the field is:

$$E_p[\text{GV/m}] \approx 96 \times \left(n_e(10^{18}\,\text{cm}^{-3})\right)^{1/2} \frac{\delta n_e}{n_e}$$

The laser frequency and wavelength are related similarly. We then express the frequency ratio in terms of wavelengths:

$$\omega_0 = \frac{2\pi c}{\lambda_0}, \qquad \left(\frac{\omega_p}{\omega_0}\right)^2 = \left(\frac{\lambda_0}{\lambda_p}\right)^2 \tag{1.96}$$

In an underdense plasma, the phase velocity of the plasma wave is close to the speed of light. It is approximately given by the laser's group velocity:

$$\beta_p = \frac{v_p}{c} \approx \sqrt{1 - \left(\frac{\omega_p}{\omega_0}\right)^2}, \qquad \text{(from cold plasma dispersion)} \tag{1.97}$$

Applying a binomial expansion for small ω_p/ω_0, we obtain:

$$\beta_p \approx 1 - \frac{1}{2}\left(\frac{\omega_p}{\omega_0}\right)^2, \quad \text{or in terms of wavelengths: } \beta_p \approx 1 - \frac{1}{2}\left(\frac{\lambda_0}{\lambda_p}\right)^2 \tag{1.98}$$

The dephasing length is the distance over which a relativistic electron (moving at nearly c) slips forward by one accelerating half-period of the plasma wave:

$$L_d = \frac{\lambda_p}{2(1 - \beta_p)} \tag{1.99}$$

Substituting the approximation for β_p into the expression:

$$L_d = \frac{\lambda_p}{2 \cdot \frac{1}{2}\left(\frac{\lambda_0}{\lambda_p}\right)^2} = \frac{\lambda_p}{\left(\frac{\lambda_0}{\lambda_p}\right)^2} = \frac{\lambda_p^3}{\lambda_0^2} \tag{1.100}$$

This gives the final 1D expression for dephasing length:

$$\boxed{L_d = \frac{\lambda_p^3}{\lambda_0^2}} \tag{1.101}$$

To rewrite in terms of frequencies, recall:

$$\lambda_p = \frac{2\pi c}{\omega_p}, \quad \lambda_0 = \frac{2\pi c}{\omega_0} \tag{1.102}$$

Substituting these into the wavelength form of L_d, we get:

$$L_d = \frac{\left(\frac{2\pi c}{\omega_p}\right)^3}{\left(\frac{2\pi c}{\omega_0}\right)^2} = \frac{2\pi c \cdot \omega_0^2}{\omega_p^3} \tag{1.103}$$

This gives the dephasing length in terms of plasma and laser frequency:

$$\boxed{L_d = \frac{2\pi c\,\omega_0^2}{\omega_p^3}} \tag{1.104}$$

In higher dimensions, the expression for dephasing length changes due to transverse effects and nonlinear wake evolution.

In the 2D nonlinear regime:

$$L_d^{2D} = \frac{\omega_0^2}{2\omega_p^2}\lambda_p \tag{1.105}$$

In the 3D bubble regime, where the wake forms a near-spherical cavity, the dephasing length becomes:

$$L_d^{3D} = \frac{4}{3} \cdot \frac{\omega_0^2}{\omega_p^2} \cdot \frac{\sqrt{a_0}}{k_p}, \qquad \text{with } k_p = \frac{\omega_p}{c} \tag{1.106}$$

These expressions show how dephasing length depends strongly on the plasma density and laser wavelength, setting a fundamental limit on how far an electron can be accelerated in a single LWFA stage.

1.10.2 Beam Wakefield Acceleration

In *beam-driven plasma wakefield acceleration* (PWFA) [20], a high-energy particle beam, called the *drive beam*, travels through a plasma and excites oscillations in the plasma electrons. These oscillations create a wakefield, which can accelerate another particle beam, called the *witness beam*. The witness beam is placed behind the drive beam to gain energy from the wakefield.

The plasma density, denoted by n_e, is the number of electrons per cubic centimeter. The plasma frequency ω_p, which describes the natural oscillation frequency of the plasma electrons, depends on n_e, the electron charge e, electron mass m_e, and the permittivity of free space ϵ_0. The wake created by the drive beam has a characteristic wavelength, called the *plasma wavelength*, denoted λ_p, which represents the distance between consecutive peaks in the plasma oscillations.

The strength of the electric field created by the wake, represented as E_p, accelerates the witness beam. The wakefield potential, denoted Φ, describes the potential field responsible for generating the wake, and it is directly linked to the charge density of the drive beam, n_b.

The plasma frequency ω_p is given by:

$$\omega_p = \sqrt{\frac{n_e e^2}{m_e \epsilon_0}}$$

This formula shows that the plasma frequency depends on the plasma density. A higher electron density leads to a higher plasma frequency.

The plasma wavelength λ_p, which is the distance between the peaks of the oscillations, is given by:

$$\lambda_p = \frac{2\pi c}{\omega_p}$$

Here, c is the speed of light, and λ_p is inversely proportional to the plasma frequency. The wakefield potential Φ excited by the drive beam is given by:

$$\Phi \approx \frac{en_b}{\epsilon_0 k_p^2}$$

Here, $k_p = \frac{\omega_p}{c}$ is the plasma wavenumber. The longitudinal electric field in the plasma wake, denoted as E_z, is responsible for accelerating particles, and it is proportional to the plasma frequency ω_p. It can be expressed as:

$$E_z \approx \frac{m_e c \omega_p}{e}$$

The energy gained by the witness beam as it travels through the plasma is given by:

$$\Delta\mathcal{E} = m_e c^2 \frac{L}{\lambda_p}$$

Here, L is the length of the plasma, and λ_p is the plasma wavelength. Finally, a practical formula that estimates the peak electric field in beam-driven PWFA is given by:

$$E_p[\mathrm{GV/m}] \approx 96 \times \left(n_e(10^{18}\,\mathrm{cm}^{-3})\right)^{1/2} \frac{\delta n_e}{n_e}$$

This formula shows that the peak electric field E_p scales with the square root of the plasma density n_e and is proportional to the relative plasma density perturbation $\frac{\delta n_e}{n_e}$. These formulae provide an understanding of how beam-driven wakefield acceleration works and how the electric field and energy gain scale with plasma density.

Wakefield accelerators are an exciting new development in the quest for smaller, high-energy particle accelerators. By using plasma waves, these accelerators can potentially overcome the limitations of traditional accelerators, reaching electric fields in the range of GV/m. Both LWFA and PWFA offer unique benefits, including the ability to create very strong acceleration fields over short distances, making them ideal for the next generation of accelerators used in scientific and medical applications. Key concepts such as the ponderomotive force, plasma frequency, and dephasing length are essential to understanding how particle acceleration works in these systems. In advanced regimes, like the bubble regime, wakefield accelerators can achieve higher energy transfer efficiency, which is important for developing practical and scalable devices. Despite their potential, there are still challenges to be solved, such as improving beam quality, managing plasma density, and ensuring efficient energy transfer. As research continues and technology improves, these issues are expected to be resolved, moving wakefield accelerators closer to widespread use. These accelerators have the potential to transform particle physics by providing more accessible and affordable accelerators for a variety of purposes.

Curiosity: *In the late 1890s, a series of amazing discoveries changed science forever. In 1895, Wilhelm Roentgen accidentally found X-rays—an invisible kind of energy that could pass through things like skin and show bones. This inspired Henri Becquerel, who in 1896 discovered that uranium gave off a similar invisible energy all by itself, even in the dark.*

Soon after, Marie and Pierre Curie studied these rays and discovered two new elements—polonium and radium—and called this kind of energy "radioactivity." They all worked hard and made great sacrifices, and between 1901 and 1903, Roentgen, Becquerel, and the Curies won Nobel Prizes. Their discoveries helped start the fields of nuclear science, medical X-rays, and cancer treatment using radiation. Have you ever wondered what radiation really is, and why we take such care to protect ourselves from it in particle accelerators? Radiation isn't always dangerous—it's simply energy that travels through space as particles or waves. Some types, like light and radio waves, are harmless. But when particles move very fast, like in an accelerator, they can create stronger kinds of radiation, like X-rays or neutrons, which can pass through materials and damage living cells. That's why we build thick walls around accelerator areas and use special shielding to block harmful radiation from escaping. But what makes some materials radioactive in the first place? It turns out that certain atoms are naturally unstable—they slowly break apart and give off energy in the form of radiation. This process, called radioactivity, happens in nature (like in rocks and the Sun) but can also happen when accelerator beams hit a target and make new, unstable atoms. How do we measure radiation? And how much is safe? People working near accelerators wear special badges that record how much radiation they're exposed to, and safety rules help limit their dose. Some of the particles created in accelerators disappear quickly, while others stay radioactive for a long time—so how do we safely store or dispose of them? These are just a few of the big questions in radiation physics and protection. We've touched on some of them here, but there's a lot more to discover about how radiation behaves, how it's controlled, and how we stay safe while using it for science and medicine.

1.11 Radiation Physics

Several scientists made major contributions to radiation physics, and their names are used in key units today. Henri Becquerel found radioactivity in 1896, which led to the Becquerel (Bq), a unit measuring the number of decays per second. Marie and Pierre Curie, known for discovering radium and polonium, are honored with the Curie (Ci), used for larger amounts of radioactivity. Louis Harold Gray studied how radiation affects tissues, resulting in the Gray (Gy), a unit for absorbed dose. Rolf Sievert worked on radiation's biological effects, leading to the Sievert (Sv), which considers the type of radiation. Wilhelm Conrad Röntgen, who discovered X-rays, is remembered through the Roentgen (R), which measures radiation exposure in air. These scientists laid the foundation for understanding and measuring radiation. Radiation physics plays a vital role in particle accelerators, where ensuring the safety of radiation workers, especially accelerator staff, is of utmost importance. In this context, radiation refers to both direct and secondary radiation produced in accelerators. The level of radioactivity is commonly measured in Becquerels (Bq), which indicates the number of nuclear decays per second. Specifically, 1 Bq means one decay per second. A higher number of Bq signifies that the substance is undergoing more nuclear decays in a given time, indicating greater radioactivity. However, a higher level of radioactivity does not necessarily equate to higher danger. The type of radiation (such as alpha, beta, or gamma), its energy, and how it interacts with matter are crucial factors that determine its biological effects. For large amounts of radioactivity, the Ci is often used because it represents a much higher level of radioactive decay. **1 Ci =** 3.7×10^{10} **Bq**, making the Ci more convenient for describing high levels of radioactivity.

In radiation physics, different units are used to measure various aspects of radiation, especially in accelerator environments. The following is a list of the common units in radiation physics:

- Absorbed dose (Gray [Gy]): The Gy is the SI unit for absorbed dose, which represents the amount of energy absorbed from radiation per unit mass. One Gray is one joule per kilogram of tissue.
  1 Gy = 100 Rad.
- Absorbed dose (Rad): The Rad (radiation absorbed dose) is an older unit of absorbed dose.
  1 Rad = 0.01 Gy.
- Equivalent dose (Sievert [Sv]): The Sv is the SI unit for equivalent dose, which accounts for the biological impact of radiation. It is calculated by multiplying the absorbed dose (in Gy) by a radiation weighting factor based on the type of radiation.
  1 Sv = 100 Rem.
- Equivalent dose (Rem): The Rem (Roentgen equivalent man) is an older unit of equivalent dose.
  1 Rem = 0.01 Sv.
- Exposure (Roentgen [R]): The R is used to measure radiation exposure in air, particularly for X-rays and gamma rays. **1 R is** 2.58×10^{-4} **Coulombs per kilogram of air.** For X-rays and gamma rays in air, **1 R** $\approx$ **0.009 Gy and 0.009 Sv.**

Let A be the activity of the radioactive source, measured in curies (Ci) or becquerels (Bq). The distance from the source is represented by d in meters. The gamma constant, Γ, is specific to the type of radiation and is measured in $\text{mSv} \cdot \text{m}^2/\text{Ci} \cdot \text{h}$. The dose rate, $\dot{D}$, is the radiation intensity at a given distance, expressed in millisieverts per hour (mSv/h) or rem per hour (rem/h). If shielding is involved, I_0 represents the initial dose rate before the shield, and the shield's thickness is given by x in centimeters. The linear attenuation coefficient μ, which is specific to the shielding material and radiation energy, is measured in cm^{-1}. The reduced dose rate after shielding is H. The total dose received over an exposure time t, measured in hours, is denoted as D. Finally, radioactive decay is described by a decay constant λ, which relates to how fast a substance decays over time, with t representing the elapsed time. The half-value layer (HVL) gives the thickness of the shielding material needed to reduce the radiation intensity by half.

The dose rate at a certain distance d from a radiation source is calculated using the inverse square law:

$$\dot{D} = \frac{\Gamma \cdot A}{d^2} \tag{1.107}$$

When radiation passes through a shield of thickness x, the dose rate after shielding is given as follows:

$$I = I_0 \cdot e^{-\mu x} \tag{1.108}$$

The total dose received over a time t, based on the dose rate after shielding, is given as follows:

$$D = H \cdot t \tag{1.109}$$

The radioactive decay formula, showing how the activity $A(t)$ changes over time, is given as follows:

$$A(t) = A_0 \cdot e^{-\lambda t} \tag{1.110}$$

The formula for the HVL, which gives the thickness required to reduce the radiation intensity by half, is given as follows:

$$HVL = \frac{\ln(2)}{\mu} \tag{1.111}$$

The total radiation exposure, or dose limit, can be calculated is given as follows:

$$D = H \cdot t \tag{1.112}$$

As per exponential radioactive decay law, Let N_0 and N be the number of radioactive atoms of a sample at $t = 0$ and time (t)

$$N = N_0 \exp(-\lambda t)$$

The activity of a radioactive source quantifies the rate at which atom in a radioactive source decays. It is number of disintegration per unit time. Activity of sample also follows same rule:

$$A = A_0 \exp(-\lambda t)$$

Half life of a radionuclide is defined as time in which activity of a sample reduce to half. Thus, One can write:

$$\frac{1}{2} = \exp(-\lambda T_{1/2})$$

$$A = \lambda N = \frac{N \ln 2}{T_{1/2}}$$

where $T_{1/2}$ is the half life of the source.

Next, the specific activity of sample is defined as its activity per unit mass.

$$SA = \frac{N_A \lambda}{M}$$

where NA is the Avogadro's number, M is the atomic weight.

In practical units:

$$SA(Bq/g) = \frac{4.17 \times 10^{23}}{M(amu) T_{1/2}(sec)}$$

Radiation physics is essential for understanding and controlling radiation, especially in places like particle accelerators, where safety is a major concern. The contributions of pioneering scientists such as Henri Becquerel, Marie and Pierre Curie, Louis Harold Gray,

Rolf Sievert, and Wilhelm Conrad Röntgen have provided the foundation for modern radiation measurement systems. We now have key units such as the Becquerel, Curie, Gray, Sievert, and Roentgen to measure different types of radiation and its effects.

Curiosity: *Have you ever thought about how hard it is to keep a group of like-charged particles together? Since they all repel each other, they never really want to stay close. But in particle accelerators, we have to guide these fast-moving particles from their source all the way to a target. How do we do that? We use electric and magnetic fields to steer and control the beam, like invisible hands guiding it along a path. Along the way, we often need to check what kind of particles are in the beam. By passing them through special fields, we can bend them in different ways and learn about their speed and type. To keep the beam sharp and narrow, we bunch the particles closer together in the direction they're moving and also focus them from the sides. This helps control the size and shape of the beam. But first, we need to pull the particles out of the source, and we do that by giving them energy using high voltage. Once the beam is moving through the vacuum inside the accelerator, something interesting happens–if we don't keep focusing it, the beam slowly spreads out in both directions. This shows how carefully we have to control everything to keep the beam stable and useful. We've looked at some of these key ideas here, but there's still much more to explore about how we guide and shape beams of particles with precision.*

1.12 Beam Conceptual Visualization

Finally, from Figure 1.6, one may get an idea of various effects over beams in the simple environment of electromagnetic fields and some basic properties of beams.

a. **Beam Collection of Charges (Coulomb Forces)** Charged particles in the beam interact through Coulomb forces, causing space charge effects that lead to beam expansion.
b. **Beam Transport from Ion Source to Target** Beam transport involves guiding and aligning the beam using electrostatic lenses, magnetic quadrupoles, and drift tubes. This process ensures minimal emittance growth and maintains beam quality.
c. **Beam Bending via Electric or Magnetic Dipole** Electric dipoles deflect beams using electric fields, while magnetic dipoles use the Lorentz force for precise bending. This technique is commonly used in particle accelerators.
d. **Beam Bunching and Focusing** Beam bunching compresses continuous streams into periodic bunches using RF cavities. Focusing counters divergence using magnetic quadrupoles or electrostatic lenses to maintain beam.
e. **Beam Extraction from Ion Source** High voltage applied between the ion source and an extraction electrode creates a stable, well-formed beam with controlled energy and emittance for downstream processes.
f. **Beam in Drift Space** In drift spaces, the beam propagates without external forces, with size increasing in each direction due to space charge effects. Periodic focusing is necessary to preserve beam quality.

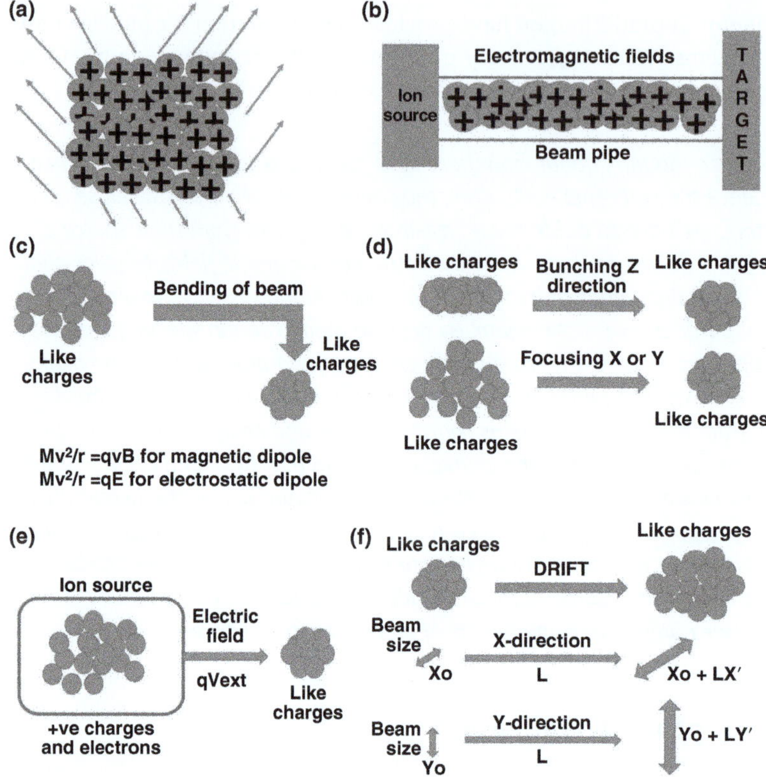

Figure 1.6 (a) Collection of like charges, which never wants to stay together. (b) Beam transport from ion source to target under the effect of electromagnetic fields. (c) Beam analysis using magnetic or electric fields as per the Lorentz force formula due to magnetic or electric fields. (d) Beam bunching and focusing in the longitudinal and transverse directions showing control of beam size by these operations. (e) Beam extraction from source using voltage V_{ext} so that particle energy becomes qV_{ext}. (f) Beam along a drift space showing beam blow up in both transverse planes.

1.13 Numerical Problems

1. Calculate the resolution of an optical microscope based on the wavelength of visible light 500 nm. Then, calculate the resolution of a particle accelerator using the de Broglie wavelength of an electron with a kinetic energy of 10 MeV. Using the electron's rest mass, compare the resolutions of the two devices and determine the ratio of the microscope's resolution to the accelerator's resolution. Finally, explain the significance of this comparison.
2. A Cockcroft–Walton type voltage multiplier has eight stages, each with a capacitance of 0.06 μF. The transformer generates a secondary voltage of 100 kV at 150 Hz frequency. To supply a load current of 10 mA, calculate the percentage ripple, regulation, and the optimal number of stages for minimum voltage drop.
3. Calculate the frequency and radius of a cyclotron for 15 MeV deuterons with a magnetic field of 1 T. The potential difference across the dees is 20 kV. How many turns will the deuteron take to reach a final energy of 15 MeV?

4. A cyclotron accelerates protons by 500 V each time they cross the gap between dees, with an outer radius of 400 mm and a magnetic field of 0.8 T. Calculate the cyclotron frequency of the protons, the maximum kinetic energy, and the total number of revolutions the protons will make.
5. Calculate the **cyclotron frequency** for the following cases:
   (a) A proton is placed in a magnetic field of 1 T (nonrelativistic case).
   (b) A proton with a kinetic energy of 200 MeV is subjected to a magnetic field of 8 T.
   (c) An Ar^{16+} ion, with a kinetic energy of 2 GeV, is placed in a magnetic field of 4 T.
6. Calculate the upper limit for the kinetic energy of protons attainable in a cyclotron with a 1.5 T magnetic field and a 1 m radius.
7. What is the terminal potential required in a Tandem accelerator to obtain 100 MeV proton, carbon, nickel, and gold ion beams? Assume the injection energy to the Tandem accelerator is 200 keV.
8. What is the maximum energy obtained from a Tandem accelerator with a maximum terminal potential of 15 MV for proton, carbon, nickel, and gold ion beams?
9. A synchrotron is designed to accelerate protons to an energy of 2 GeV. It contains 40 dipole magnets, each with an effective length of 2 m, and operates with a magnetic field strength of 1.5 T. The synchrotron has a filling factor K_F of 0.6. Calculate the following: the magnetic rigidity $B\rho$ of the synchrotron, the integrated dipole strength $B \cdot l$ for each dipole magnet, the circumference C of the synchrotron ring, the bending radius ρ of the particle trajectory within the dipole magnets and the required RF for a harmonic number $h = 100$. Assume the particles are moving at nearly the speed of light.
10. A synchrocyclotron is designed to accelerate protons to an energy of 200 MeV. The magnetic field at the proton's orbit is 1.8 T, and the orbit radius is 0.7 m. Calculate the following: the relativistic factor γ of the proton, the final velocity of the proton, the required RF for the synchrocyclotron at this energy, and the time taken for the proton to complete one revolution.
11. A betatron is designed to accelerate electrons to high energies. The average magnetic field $\langle B \rangle$ over the area enclosed by the electron's orbit is changing at a rate of $\frac{d\langle B \rangle}{dt} =$ 50 T/s. The radius of the electron's orbit is 1 m, and the magnetic field at the orbit B_{orbit} is half the average magnetic field $\langle B \rangle$. Calculate the following: the induced electric field E around the electron's orbit, the force F exerted on the electron by the induced electric field, the momentum p of the electron after 1 μs ($t = 10^{-6}$ s), and the magnetic field at the orbit B_{orbit} if the average magnetic field $\langle B \rangle$ is 1 T.
12. An electron is accelerated in a nonscaling FFAG accelerator, where the magnetic field at the reference orbit is 0.5 T with a reference radius of 1.0 m and a magnetic field index of 0.3. The electron is injected with an energy of 1 MeV and accelerated to 5 MeV. Calculate the orbit radius of the electron at 5 MeV and the revolution times at both 1 and 5 MeV, assuming relativistic conditions for the electron's velocity.
13. Calculate the corresponding laser intensities I_0 [W/cm^2] for a laser with a wavelength of 1 μm and normalized vector potential a_0 for the following values: 0.02, 0.1, 0.5, and 1.0.

14. Calculate the plasma wavelength λ_p in micrometers (μm) and the plasma electric field E_p in GV/m for a plasma with an electron density $n_e = 5 \times 10^{18}\,\text{cm}^{-3}$ and a relative density perturbation $\frac{\delta n_e}{n_e} = 0.1$.

15. Calculate the dose rate for a Cesium-137 source with an activity of $A = 6\,\text{Ci}$ at a distance of 1 m using the inverse square law. The gamma constant is given as $\Gamma = 0.33\,\text{mSv} \cdot \text{m}^2/\text{Ci} \cdot \text{h}$. Then, calculate the dose rate after applying 1 cm of lead shielding, where the linear attenuation coefficient for lead is $\mu = 0.55\,\text{cm}^{-1}$, using the exponential attenuation formula. Finally, calculate the total dose received after of exposure by multiplying the dose rate after shielding with the exposure time.

2

Ion Sources

> "We should remember that there was no theory of the earth's magnetic field before the geomagnetic field was discovered, no theory of the solar wind before it was observed, and no theory of magnetospheric electric fields before they were discovered. Similarly, the progress of plasma physics is often determined by unexpected observations rather than by theories."
>
> —*Hannes Alfvén*

After reading this chapter: you should be able to:

- Define how to generate charged particle beam from plasma sources and conceptual understanding of ion sources.
- Define different types of plasma sources.
- Understand different ionization principles to generate gaseous and metallic ions.

Curiosity: *Have you ever wondered how we strip electrons off atoms to make ion beams? From smashing electrons into atoms to heating up surfaces or focusing laser light, each ion source uses a unique trick. Why do ECR sources produce multiple charge states while sputter sources are best for negative ions? What makes laser ionization so selective and what limits its speed? Did you know that the intense electric field at a sharp metal tip can pull atoms into ions just like pulling water from a sponge? And why do plasma-based sources dominate in high-current accelerators? Each mechanism here tells a different physics story. Try asking: which sources are best for light ions like* H^+*, and which ones excel for heavy, highly charged ions? Could a hybrid source combine these techniques for even better performance? We try to answer some questions in this chapter here.*

Various types of ion sources [21] utilize different ionization processes, such as DC discharge, thermal discharge, and radio frequency (RF) discharge etc. Ion sources are classified in Table 2.1 as follows based on their primary process of ionizations.

Charged Particle Beam Physics: An Introduction for Physicists and Engineers, First Edition.
Sarvesh Kumar and Manish K. Kashyap.

Companion Website: https://www.wiley.com/go/Kumar_1e

Table 2.1 Classification of ion sources based on ionization mechanism.

Ionization Mechanism	Examples	Remarks
Electron Impact	Duoplasmatron, Penning Ion Source, Electron Beam Ion Source (EBIS), Colutron Source	Electrons collide with neutral atoms to produce ions; widely used in sources of positive ions.
Surface Ionization	Thermal Ion Source, Source of Negative Ions by Cesium Sputtering (SNICS) (partially)	Hot surfaces ionize atoms depending on the work function; selective for certain elements.
Sputter Ionization	Source of Negative Ions by Cesium Sputtering (SNICS), Cesium-Based Sputter Ion Sources	Ion bombardment ejects and ionizes surface atoms; suitable for generating negative ions.
Photoionization (Laser)	Laser Ion Source, Resonance Ionization Laser Ion Source (RILIS)	Laser photons ionize atoms via resonant or multiphoton absorption; enables element-specific ion production.
Field Ionization	Field Ionization Source, Liquid Metal Ion Source (LMIS)	Extremely strong electric fields near sharp tips ionize atoms; used in focused ion beam (FIB) systems.
Plasma Ionization	Electron Cyclotron Resonance Source (ECR), Radio Frequency Source (RF), Microwave Ion Source, Multicusp Ion Source, Helicon Source, Vacuum Arc Source	Dense plasma generated by electromagnetic fields ionizes atoms; suitable for high-current beams and multiple charge states.

2.1 ECR Ion Sources

Electron cyclotron resonance (ECR) ion source operates based on RF principle and is noted for its superior performance compared to other sources in terms of intensity and high charge state population. The ECR ion source was introduced in the late 1970s [22], and its ongoing development with advancements in RF and magnet technologies has been crucial for research programs. Typically, an ECR ion source extracts an intense, multiply-charged ion beam, which is transported through a low-energy beamline and injected into accelerators, such as radio frequency quadrupole or drift tube linacs, for acceleration to higher energies. Cyclotrons are often equipped with ECR ion sources. In cyclotrons or synchrotrons, the maximum energy achievable is proportional to the square of the charge state (q^2). Therefore, using ions with higher charge states can reduce the size and cost of the accelerator, making the ECR ion source advantageous for producing intense beams of high charge state ions. Furthermore, the absence of cathodes or filaments in the ECR ion source enables the production of stable beams over extended periods using an external gaseous supply. Consequently, ECR ion sources [23, 24] have become a common choice for heavy ion accelerators [25–29].

In an ECR ion source [22–24, 30–32], a plasma consisting of ions and electrons is excited by microwaves. A strong magnetic field is generated throughout the plasma chamber by current-carrying coils or permanent magnets surrounding the chamber. The interaction between microwaves and electrons occurs primarily where the electron cyclotron frequency matches the microwave frequency.

For electrons, a region of ECR is defined by the resonance condition which is relation between ECR field B_{ECR} and cyclotron frequency of electrons f_{cyc} as follows:

$$B_{ECR}(T) = \frac{f_{cyc}(GHz)}{28} \tag{2.1}$$

As electron temperature increases, the resonance broadens due to Doppler shifting of the microwave frequency. Both neutral atoms and ions undergo ionizing collisions with these hot electrons. Provided the magnetic confinement of the plasma is adequate, a species can experience successive ionizing collisions. By applying a positive bias of several kilovolts to the plasma chamber, ions are extracted along magnetic field lines through an aperture toward a grounded or negatively biased puller.

ECR source consists of Hexapole (radial) and Solenoidal confinement (longitudinal). A simple magnetic mirror structure (without a hexapole) has a significant drawback that it cannot effectively prevent plasma from escaping, as the plasma pressure decreases with the radial weakening of the magnetic field. This pressure gradient leads to macroscopic instabilities, such as magnetohydrodynamic (MHD) instabilities, which disrupt plasma confinement and cause substantial radial losses. To address this issue, most ECR ion sources incorporate an additional multipole field, usually a hexapole, which produces a radially increasing magnetic field. This configuration enhances plasma stability and improves confinement by providing a restoring force against radial plasma losses. The schematics of ECR ion source is shown in Figure 2.1.

The incorporation of a hexapole creates what is known as a "minimum-B" (min-B) configuration, where the magnetic field is weakest at the center and increases in all directions. This arrangement helps suppress MHD instabilities by ensuring that plasma remains confined within the trap, as convex magnetic field lines provide a stabilizing effect. In a min-B configuration, regions of equal magnetic field strength, known as magnetic isobars, form closed nested layers, further aiding confinement.

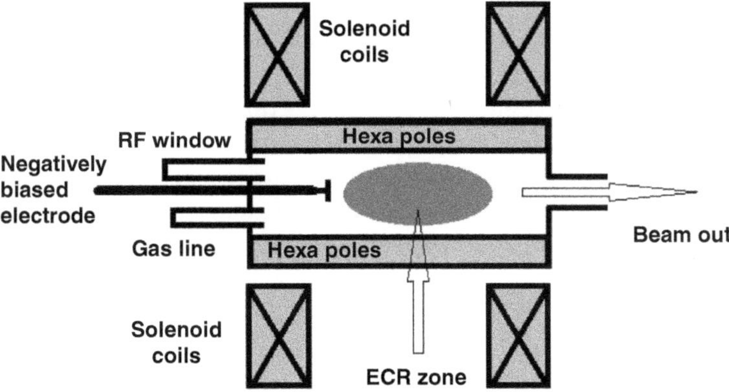

Figure 2.1 Schematic of an ECR ion source.

Thus, the ECR plasma chamber is situated within an axial magnetic field configuration created by two solenoid coils and a radial cusp magnetic field produced by a multipole, typically a sextupole magnet. This combination generates a min-B magnetic field configuration, which has a minimum at the center and increases toward the chamber walls. This configuration provides a stable plasma confinement geometry resistant to MHD instabilities. It is as shown in Figure 2.1.

The min-B configuration is crucial for sustaining plasma stability by utilizing the principles of adiabatic invariants. One such invariant is the magnetic moment, which plays a key role in confining charged particles as they follow helical trajectories along magnetic field lines. The combined effects of the axial mirror field and the radial hexapole field work together to confine the plasma effectively, preventing losses and maintaining high plasma density.

Plasma heating within the ECR ion source primarily occurs through ECR, where microwaves transfer energy to electrons at their gyrofrequency. This heating mechanism facilitates the ionization of neutral atoms and contributes to the production of highly charged ions. Effective heating depends on the careful optimization of microwave frequency to match the local electron gyrofrequency, ensuring efficient energy transfer and sustained plasma operation. Despite these stabilization techniques, plasma loss mechanisms remain a challenge. These include drift instabilities, collisional diffusion, and end losses along the field lines. Drift instabilities arise due to electric and magnetic field interactions, while collisional diffusion results from particle interactions leading to energy loss. End losses occur when particles escape along open magnetic field lines, reducing overall confinement efficiency. The impact of magnetic field gradients on ion trajectories is another critical aspect of ECR ion source performance. Field variations influence ion motion and determine the efficiency of high-charge state production. Understanding and controlling these gradients are essential to optimize plasma confinement and maximize ion beam quality. Furthermore, the plasma within the ECR source consists of different energy populations. Cold electrons contribute to recombination and plasma cooling, while hot electrons are crucial for sustaining high levels of ionization and achieving the desired charge states. Proper management of these populations ensures efficient plasma operation.

A buffer gas is introduced into the chamber. The plasma is created by microwave power injection and confined by the designed magnetic field structure. The processes governing particle confinement in an ECR ion source are complex, with electrons and ions having different temperatures and confinement times. Charged particles and electrons gyrate around magnetic field lines, and plasma electrons, trapped in the magnetic field, follow helical paths along these lines. Electrons gain energy from the RF source, transitioning from cold to warm and then to hot states, contributing to the generation of high-intensity multiple charge states.

The local cyclotron frequency is given by:

$$\omega_{RF} = \omega_{cyc} + k_{||}v_{||} \tag{2.2}$$

where $k_{||}$ is the RF wave vector and $v_{||}$ is the component of the electron velocity parallel to the magnetic field. For $v_{||} = 0$, above equation defines a resonance surface in the min-B magnetic configuration of an ECR ion source.

2.1.1 Electron Energy Distribution Function

In ECR plasma, phenomena such as electromagnetic wave propagation and ECR heating occur. Microwaves accelerate or decelerate electrons depending on their gyration phase relative to the local RF electric field. The energy gain during the accelerating phases typically exceeds the loss during the decelerating phases, leading to stochastic ECR heating and rapid electron energy increases, sufficient for successive ionizations of gaseous atoms or molecules. ECR plasmas are thermodynamically nonequilibrium, and the electron energy distribution function (EEDF) is notably non-Maxwellian. Despite a continuum of electron energies, the EEDF is often represented by three populations related to diffusion, collision, ionization, and confinement processes:

1. Cold electrons (1 eV–100 eV)
2. Warm electrons (100 eV–10 keV)
3. Hot electrons (10 keV to a few hundred keV)

Electrons in the magnetic mirror may be scattered into the loss cone by collisions. The confinement time for electrons is influenced by scattering rates from electron–electron and electron–ion collisions. Energy gain of resonant electrons is affected by the electric field amplitude and the gradient of the external magnetic field at resonance. Adjustments in field gradient impact the thickness of the resonance surface and the efficiency of energy transfer from microwaves to electrons.

Cold electrons, undergoing rapid scattering collisions, are not magnetically confined but are confined by the space charge field of less mobile plasma ions. They contribute minimally to ionization but replenish warm and hot electron populations. Hot electrons, despite having low ionization cross-sections and densities, are well confined due to lower collision rates. Confinement times, proportional to electron temperatures, vary among electron populations. For ions, shorter confinement times may prevent reaching higher charge states, while longer confinement times may lead to charge exchange rather than useful beam current extraction. Increased confinement times enhance the likelihood of multiple consecutive electron impacts, raising the probability of achieving higher charge states. The production of higher charge states rises with microwave power up to an equilibrium charge state, after which it decreases due to RF-induced plasma losses. Optimizing source performance for the highest charge states requires a better understanding of inherent plasma and RF-induced instabilities.

2.1.2 Magnetic Confinement by a Mirror Field

When a charged particle enters a magnetic field, it experiences a force perpendicular to both its velocity and the field direction. This causes the particle to spiral around the field lines while moving forward along them, resulting in a helical path. The velocity of the particle can be split into a component along the magnetic field, $v_{||}$, and a component perpendicular to it, $v_{\perp}$. The perpendicular motion creates a circular orbit known as the Larmor orbit.

To understand magnetic confinement, we first derive the particle's magnetic moment. A circulating charge behaves like a current loop, with magnetic moment given by

$$\mu = IA \tag{2.3}$$

The current is $I = qf = q\omega_c/2\pi = \frac{q^2B}{2\pi m}$, and the area of the orbit is

$$A = \pi r_L^2 = \pi\left(\frac{mv_\perp}{qB}\right)^2 \tag{2.4}$$

Substituting these gives

$$\mu = \frac{q^2B}{2\pi m} \cdot \pi\left(\frac{mv_\perp}{qB}\right)^2 = \frac{mv_\perp^2}{2B} \tag{2.5}$$

This expression shows that the magnetic moment is proportional to the perpendicular kinetic energy and inversely proportional to the magnetic field strength. In a slowly varying magnetic field, this magnetic moment remains constant. Now consider a magnetic field configuration where the field is minimum at the center and increases gradually toward both ends. As a particle moves into the stronger field region, the constancy of μ demands that $v_\perp$ must increase. Since total kinetic energy is conserved,

$$\frac{1}{2}mv^2 = \frac{1}{2}m(v_{||}^2 + v_\perp^2) = \text{constant} \tag{2.6}$$

an increase in $v_\perp^2$ must cause a decrease in $v_{||}^2$. Eventually, $v_{||}$ becomes zero, and the particle is reflected. This is the magnetic mirror effect.

Let us now derive the condition for this reflection. Suppose the particle starts at the mid-plane, where the magnetic field is B_o, with total speed v_o and perpendicular component $v_{\perp o}$. At any other position where the magnetic field is B, energy conservation gives

$$\frac{1}{2}mv_{||}^2 = \frac{1}{2}mv_o^2 - \frac{1}{2}mv_\perp^2 \tag{2.7}$$

$$\Rightarrow v_{||}^2 = v_o^2 - v_\perp^2 \tag{2.8}$$

The magnetic moment is also conserved, so

$$\mu = \frac{mv_\perp^2}{2B} = \frac{mv_{\perp o}^2}{2B_o} \tag{2.9}$$

$$\Rightarrow v_\perp^2 = \frac{v_{\perp o}^2}{B_o}B \tag{2.10}$$

Substituting this into the energy equation gives

$$v_{||}^2 = v_o^2 - \frac{v_{\perp o}^2}{B_o}B \tag{2.11}$$

At the mirror point, $v_{||} = 0$, and the magnetic field reaches its maximum value B_m. So,

$$v_o^2 = \frac{v_{\perp o}^2}{B_o}B_m \tag{2.12}$$

$$\Rightarrow \frac{v_o^2}{v_{\perp o}^2} = \frac{B_m}{B_o} \tag{2.13}$$

Now define the pitch angle θ such that

$$\sin\theta = \frac{v_{\perp o}}{v_o} \Rightarrow \sin^2\theta = \frac{v_{\perp o}^2}{v_o^2} \tag{2.14}$$

Substituting into the reflection condition gives

$$\sin^2\theta > \frac{B_o}{B_m} \tag{2.15}$$

The minimum pitch angle for reflection is defined by equality:

$$\sin^2\theta_m = \frac{B_o}{B_m} \tag{2.16}$$

Defining the mirror ratio

$$R_m = \frac{B_m}{B_o} \tag{2.17}$$

we can also write

$$\sin^2\theta_m = \frac{1}{R_m} \tag{2.18}$$

In addition to axial confinement, the radial mirror ratio is

$$R_r = \frac{B_{\text{min}}}{B_{\text{wall}}} \tag{2.19}$$

and the axial mirror ratio is

$$R_a = \frac{B_{\text{max}}}{B_{\text{min}}} \tag{2.20}$$

The particle distribution becomes anisotropic because only particles with certain pitch angles are reflected. This leads to different temperatures in the parallel and perpendicular directions, resulting in a non-Maxwellian velocity distribution. Magnetic mirrors thus confine particles using conservation of magnetic moment and total energy. Particles with sufficiently large pitch angles reflect between two high-field regions and remain trapped, provided the mirror ratio is high enough.

Due to limitations in superconducting technology, earlier ECR ion sources could achieve mirror ratios only up to around 1.5. Recent advancements allow ECR ion sources to reach higher mirror ratios, improving confinement and shifting the charge state distribution to higher values. Optimized microwave power and field strength enhance ion production, reduce beam emittance, and improve overall source performance by mitigating plasma and RF-induced instabilities.

2.1.3 Modes of ECR Cavity and Plasma Frequency

When microwaves are introduced into the cavity of an ECR ion source, an electromagnetic standing wave can be excited, dependent on the applied frequency and the cavity's geometrical parameters. The electromagnetic field structure within the cavity can be characterized by transverse electric (TE) and transverse magnetic (TM) modes. These modes can be theoretically calculated for a simple cylindrical approximation of the ECR ion source plasma chamber. Each mode represents a specific field distribution within the cavity, characterized by a quality factor (Q), which describes the resonance characteristics by the ratio of energy stored in the system to the energy dissipated:

$$Q = \frac{f}{\Delta f} \tag{2.21}$$

A higher Q value corresponds to a system with lower energy dissipation relative to the energy stored, indicating a more efficient mode. For simple cylindrical cavities, modes can be approximated, providing estimates of mode density. Typically, ECR ion sources operate with a high density of modes around the primary microwave frequency. In GHz-range cavities, mode separation is on the order of a few MHz, leading to overlapping modes. The resulting electromagnetic field structure in real ECR ion source chambers is a superposition of these overlapping modes.

In practice, the real ECR plasma chamber deviates from an ideal cylindrical cavity, incorporating additional components such as pumping ports, extraction apertures, biased discs, microwave input structures, and gas feeding lines. Advanced electromagnetic simulation tools, are necessary for accurate modeling of the resonance behavior in such complex systems. The spatial distribution of confined plasma with varying densities introduces anisotropic permittivity within the cavity, complicating the electromagnetic field structure further. Variations in permittivity and plasma power absorption result in reduced Q values and increased overlap and shifting of modes, adversely affecting the quality of the extracted beam. The plasma acts as a high-pass filter for electromagnetic radiation, with the cutoff frequency determined by the plasma frequency:

$$\omega_p = \sqrt{\frac{n_{ec}e^2}{\epsilon_0 m_e}} = \omega_{\mathrm{RF}} \tag{2.22}$$

where n_{ec} is the cutoff plasma density, e is the electron charge, ϵ_0 is the permittivity of free space, and m_e is the electron mass. Microwaves with frequencies above the plasma frequency can propagate, while those below are reflected. The plasma frequency is dependent on the plasma density, which is controlled by the heating process at a specific microwave frequency. Increasing microwave frequency improves performance by increasing the critical plasma density, necessitating scaled magnetic fields to meet resonance conditions. Higher microwave power provides more energy for electron heating, enabling higher electron energies required for producing high currents of highly charged ions.

2.1.4 Plasma Potential

Electrons, due to their higher mobility in the plasma, have a greater intrinsic loss rate to the chamber walls compared to ions. To maintain quasi-neutrality, the charge lost by escaping electrons is compensated by a positive ion charge, resulting in a positive plasma potential relative to the chamber walls. Extensive experimental studies have determined the plasma potential in ECR ion sources to range from 10 to 100 V. Lower plasma potentials are associated with better ion source performance, yielding enhanced production of highly charged ions.

2.1.5 Geller Scaling Laws

For an optimized charge state q_{op} in an ECR ion source, the scaling laws are considered by the following symbols: $I(q_{op})$ – ion current of charge state q_{op}. It represents the extracted ion beam for a given charge state. B – magnetic field strength. It helps to confine the plasma and increase the production of high-charge-state ions. ω – microwave frequency. It is the frequency of the microwaves used to heat the plasma and ionize the particles. M – atomic mass number of ion. τ – ion confinement time. It indicates how long ions stay trapped in the plasma before escaping. P_{rf} – microwave power. It is the energy supplied to the plasma to increase ionization efficiency and charge state production.

$$I\left(q_{op}\right) \sim \log\left(B\right) \tag{2.23}$$

$$I\left(q_{op}\right) \sim \log\left(\omega^3\right) \tag{2.24}$$

$$I\left(q_{op}\right) \sim \omega^2 M^{-1} \tau^{-1} \tag{2.25}$$

$$I\left(q_{op}\right) \sim P_{rf}^{1/3} \tag{2.26}$$

These laws indicate that the performance of an ECR ion source is highly dependent on the microwave frequency and the confining magnetic field. Increasing the magnetic field improves confinement of ions and electrons, which increases plasma density and shifts the charge state distribution toward higher values. At the extraction region, the magnetic field strength represents a compromise between effective ion confinement and particle loss management for efficient ion extraction. A lower mirror ratio at extraction facilitates particle losses toward the extraction aperture, enhancing extraction efficiency. Loss cones at both ends of the source are typically different in size due to variations in magnetic field maxima. Increasing the axial magnetic field at the injection side reduces the size of the loss cone there, resulting in higher output of ions at the extraction side. The extracted ion current I_q of charge state q is approximately half the ion loss rate and can be expressed as:

$$I_{\text{ext}} \sim \frac{1}{2} \frac{n_q e q V}{\tau_q} \tag{2.27}$$

where n_q is the density of ions with q, τ_q is the ion confinement time for charge state q, and V is the active plasma volume mapping along the magnetic field lines into the extraction area. This volume may vary among different ion species.

The improved performance of ECR ion sources with high magnetic mirror ratios, as described by the scaling laws, is associated with reduced MHD instability limitations. The MHD criterion stipulates that particle pressure in the plasma must be significantly lower than the magnetic pressure induced by the external field. Given that electron temperatures are substantially higher than those of ions and neutrals, the MHD criterion can be expressed as:

$$n_e k T_e \ll \frac{B^2}{2\mu_0} \tag{2.28}$$

where n_e is the electron density, k is the Boltzmann constant, T_e is the electron temperature, B is the magnetic field, and μ_0 is the permeability of free space. Higher magnetic fields increase electron densities and temperatures in the plasma before MHD instabilities occur, thereby enhancing ion source performance.

There are a few review articles [33, 34] also on ECR ion source, which gives a comprehensive understanding about them. Finally, here is the summary of ECR ion source working.

1. Microwaves/RF waves are injected into the ion source cavity through a waveguide, where they interact with electrons. The magnetic field inside the chamber ensures that these electrons move in a circular path, which keeps them within the cavity longer.
2. As the electrons move, they absorb energy from the microwaves, causing them to speed up and gain more energy. Normally, they are termed as cold, warm, and hot electrons depending on their energy range.
3. These electrons then collide with neutral gaseous atoms inside the cavity, knocking out electrons from the atoms and turning them into positively charged ions.
4. The magnetic field also helps to trap the ions and electrons within the chamber, allowing the ionization process to multiply and produce a stable plasma with an inhomogeneous mixture of various charge state ions.
5. Radial confinement of the plasma is achieved by a strong magnetic field created by permanent hexapole magnets, while axial confinement is provided by magnetic mirrors (by solenoid configuration) at the ends of the cavity, ensuring the plasma remains stable and well-contained.
6. A negatively biased electrode is used to repel the electrons and stabilise the plasma.
7. Finally, the positively charged ions are pulled out of the cavity by a set of specially designed electrodes. These electrodes make the plasma into a beam, which can then be used for various scientific and industrial applications.

2.2 SNICS Ion Source

A sputter-type negative ion source (SNICS) was developed in the early 1970s by researchers at both the University of Wisconsin [35] and the University of Pennsylvania [36]. Earlier versions were also reported from Novosibirsk [37]. Commercial versions of SNICS have been produced by companies like National Electrostatic Corporation (NEC) and High Voltage Engineering in Amersfoort. These sources are used to generate heavy-ion beams, which are crucial for applications like ion implantation and nuclear physics research in tandem accelerators.

SNICS is a type of negative ion source that relies on a sputtering mechanism. They can provide one unit negative ion of any atom from a solid pellet. They utilize a combination of immersion lens and Einzel lens. Typically, they are used as injectors into the Tandem Accelerator. Usually, not only the required ion but also the molecular ions in one unit negative form come out of MC-SNICS. Hence, a precise mass analysis using magnetic field is required for the desired ion species. After mass analysis, the negative ion beam is injected into the accelerator tank, where it is accelerated by a Tandem accelerator based upon the Van de Graaff principle. The schematic of SNICS is shown in Figure 2.2. Here is the sequence of operation of MC-SNICS.

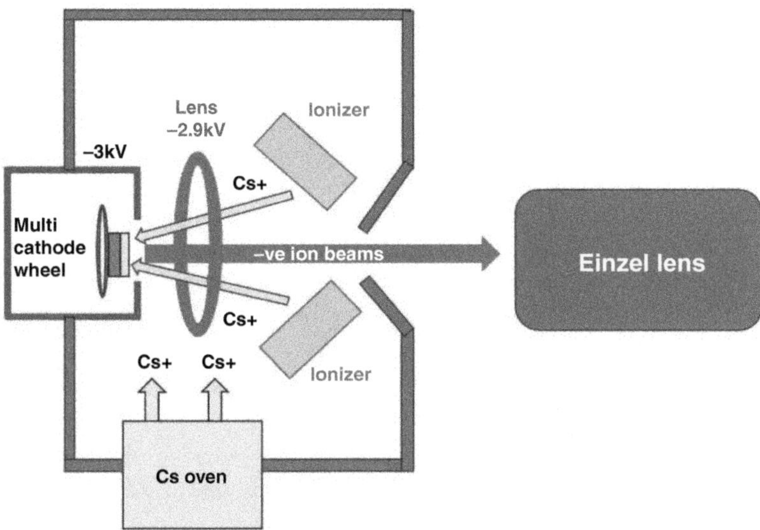

Figure 2.2 Schematics of MC-SNICS.

1. Cesium is heated in an oven to produce vapor, and the oven's temperature is carefully controlled to maintain a consistent supply. The vaporized cesium is then directed to the ionization region through designated pathways.
2. The ionizer filament heats up and emits electrons through a process called thermionic emission. These electrons collide with cesium atoms, turning them into positively charged ions. This ionization is crucial because the cesium ions help in sputtering the cathode material, which then forms negative ions. These negative ions are what the SNICS source ultimately produces and uses for further applications.
3. Once cesium (Cs) is ionized in a SNICS source, the positively charged cesium ions are attracted to the negatively charged cathode. This attraction causes the cesium ions to move toward and deposit on the cathode's surface. This deposition plays a key role in the sputtering process, where the cesium ions assist in knocking atoms off the cathode material, which then leads to the formation of negative ions.
4. These sputtered atoms interact with the cesium layer, gaining electrons and forming negative ions. These negative ions are then extracted from the cathode surface by an electric field applied toward the extraction electrode.
5. In a multi-cathode SNICS source, multiple cathodes made from different materials are available. The system can switch between different cathodes, either mechanically or electrically, exposing the selected cathode to cesium vapor and the sputtering process to generate various types of negative ions. The choice of cathode material depends on the type of negative ions required for the specific application.
6. Once extracted, the negative ion beam is focused and accelerated through an electric field to reach the required energy. The focused ion beam is then directed through the accelerator beamline for experiments or other applications. The beam current is continuously monitored to ensure stability, and the condition of each cathode is regularly

checked, with adjustments made as necessary. The cesium supply and oven temperature are managed to maintain efficient ion production.

Here are some listed properties of using cesium in SNICS sources:

1. Cesium has a low work function, which means it requires less heat to release electrons from its surface. This property makes it easier to produce negative ions.
2. It effectively sputters the material from the cathode, which is necessary for generating a steady supply of ions.
3. It is relatively electropositive, meaning it easily donates electrons. This property enhances the likelihood of neutral atoms gaining an extra electron during sputtering, thereby increasing the production of negative ions.
4. Cesium is chemically reactive and readily forms a thin layer on the cathode surface, facilitating efficient sputtering and ionization processes.

2.3 Duoplasmatron

The duoplasmatron was first developed in 1956 by Manfred von Ardenne to provide a powerful source of gaseous ions. It is a filament-driven ion source [38, 39] and well known for high intensity and high gas efficiency. The cathode is a hot tungsten spiral-shaped filament that emits electrons into a vacuum chamber. A small amount of gas, such as argon, is then introduced into the chamber, where it interacts with these free electrons, leading to the gas ionization and forming a plasma. Initially, this plasma was not very dense. To increase its density, a magnetic field is applied using a solenoid coil that encircles the discharge chamber. This magnetic field compresses the plasma toward the centerline of the plasma chamber, effectively raising the concentration of charged particles. The combined forces of the magnetic and electric fields then confine this denser plasma within a small area. This confinement is crucial for stabilizing the plasma, ensuring it remains dense ($10^{14}\,\text{cm}^{-3}$) and stable enough to produce a focused and high-quality ion beam. Once plasma is compressed, the plasma is pushed through anode to the next stage in the form of a beam. The anode is usually of molybdenum material or similar to withstand the excessive heat power caused by plasma bombardment.

An intermediate electrode (IE) is placed between the cathode and anode, dividing the plasma into two regions: a lower density region between the cathode and IE, and a higher density region between the IE and the anode. To facilitate easier formation and transport of the ion beam, an expansion cup is installed following the anode to reduce ion beam density. The positive potential difference between the IE and the anode accelerates the beam toward the anode. The presence of EC is essential for enhancing the uniformity of plasma distribution across the plasma meniscus, leading to a reduction in beam emittance.

Introducing an additional electrode, known as the reflector electrode (RE), positioned after the anode and maintained at cathode potential, allows the RE to reflect the electrons coming through the anode, thus the ionization efficiency by electrons is increased. This modification of duoplasmatron changes the configuration name to duopigatron. Source needs to satisfy following empirical relation [40].

$$I_{AM} \times B^{1/2} \times P_A^{-1} \times L^{-1} \approx \text{constant} \tag{2.29}$$

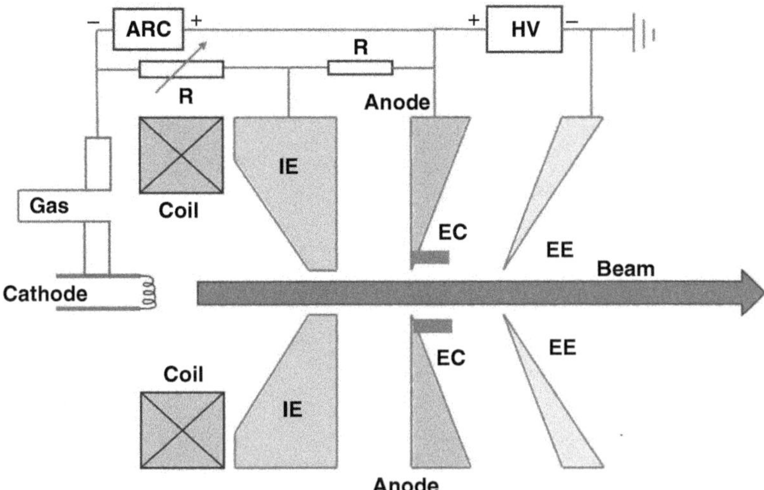

Figure 2.3 Schematics of duoplasmatron. EE: extraction electrode; EC: expansion cup; IE: intermediate electrode.

where:

- I_{AM} is the maximum arc current,
- B is the magnetic field strength,
- P_A is the pressure around the anode,
- L is the length of the anode column.

The equation indicates that the maximum extracted current for normal arc mode increases with an increase in the anode column length L. Finally, here (Figure 2.3) is the stepwise working for duoplasmatron ion source:

1. A neutral gas, often hydrogen, is injected into the ion source chamber.
2. Electrons are emitted from a heated cathode and accelerated toward the anode due to an applied electric field.
3. The electrons collide with gas atoms, ionizing them and creating a dense plasma in the cathode region.
4. The plasma is confined and compressed by a magnetic field, which intensifies the ionization process, increasing the plasma density.
5. The ionized particles move toward an IE called the plasma aperture, where the plasma density is further enhanced by the combination of magnetic and electric fields.
6. Positive ions pass through the plasma aperture and are focused into a beam by an extraction electrode.
7. The ion beam is accelerated to the desired energy level by applying an electric field, forming a highly collimated ion beam.

2.4 Electron Beam Ion Source

Electron beam ion sources (EBIS) [41–45] are usually known for highly charged ions. The concept of EBIS was pioneered by Evgeny Donets and collaborators in the late 1960s at JINR Dubna, USSR.

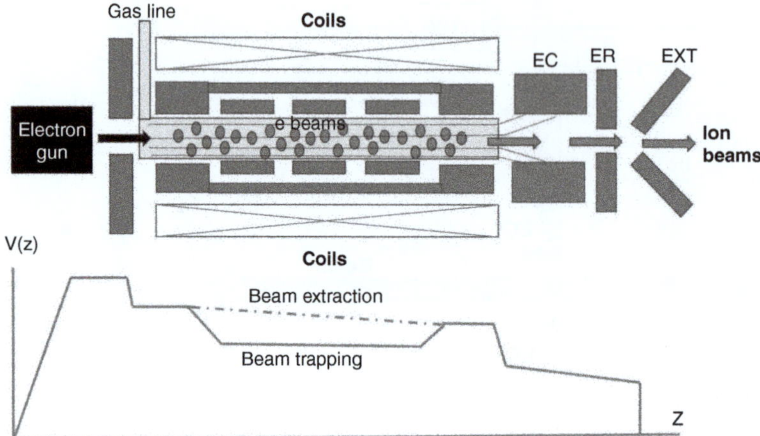

Figure 2.4 Schematics of EBIS. EXT: extraction electrode; ER: electron repeller; EC: electron collector.

They demonstrated the first working EBIS using a high-density electron beam for successive ionization. This innovation enabled controlled production of highly charged ions (HCIs) in a compact laboratory setup. Following this, the first compact Electron Beam Ion Trap (EBIT) was developed by Levine et al. [46]. EBIS are not as efficient as ECR ion sources but can provide high charge states of ions more comparatively. The schematic of the source is shown in Figure 2.4.

Electron beam of particular energy (keV) and particular current (mA) gets generated by an electron gun, which interacts with gaseous molecules and strips off their electrons. Thus, the ions are made. The ions are confined by magnetic field radially as well as by electric field axially. They are raised to high charge states by multiple impacts of electron beams. The series of drift tubes creates a potential well for confining ions over a longer period of time, and the trap potential is lowered whenever the ion beam is extracted. The beam does not remain confined for enough time so that the fully stripped ions are maximized as the instabilities gets developed, which overheat the ions and subsequently escape from the trap. The formation of higher charge states depends on the trapping conditions and the ionization time. The time taken [47] by an ion to go from charge state Z to Z+1 is as follows:

$$\tau_z(s) = \frac{1}{\sigma_z v_b n_b} = \frac{1.6 \times 10^{-19}}{\sigma_z J} \tag{2.30}$$

where σ_z is the appropriate cross-section for the particular charge state, n_b is the density of the electron beam, v_b is the electron beam velocity and J is the beam current density. Finally, here is the stepwise summary of the electron beam ion source (EBIS):

1. A high-density, monoenergetic electron beam is generated and magnetically compressed into a narrow path using a strong axial magnetic field, typically provided by superconducting solenoids.
2. An electrostatic axial potential well is established along the beam axis using a set of drift tube electrodes, forming an ion trap for longitudinal confinement.

3. Low-charged or neutral atoms of the working gas (injected either continuously or in a pulsed mode) are introduced into the trap region where they interact with the passing electron beam.
4. Through successive electron-impact ionization events, electrons are stripped from the injected atoms or low-charge-state ions, gradually increasing their charge state over time.
5. During the ionization process, the ions are confined radially by the strong magnetic field and axially by the electrostatic potential well, allowing sufficient time for them to reach the desired high charge states.
6. Once the required charge state is achieved, the axial potential barrier on one side of the trap is lowered, and the highly charged ions are extracted along the direction of the electron beam.
7. The extracted ions can then be post-accelerated as needed, forming a beam of highly charged ions suitable for downstream applications. The system is then re-initialized for the next ion production cycle.

2.5 Penning Ion Sources

Penning-type ion sources [48–53] generate ions through electrostatic discharges, offering a simpler mechanism compared to discharges initiated by RF waves or microwaves. The typical configuration of a cold cathode Penning source is cylindrically symmetric around its central axis at R = 0, as shown in Figure 2.5. High DC voltages (ranging from 500 to 5000 V) are applied between the anode and cathode. This symmetrical electrode arrangement enhances discharge efficiency by extending the path length that ionizing electrons travel within the discharge region. The electric field generated by the applied voltage causes the electrons to oscillate between the two cathodes. Additionally, the presence of a magnetic field along the axis of the discharge chamber further lengthens the electron path. This magnetic field causes the electrons to undergo Larmor gyration along the magnetic field lines at a characteristic frequency as follows:

$$\omega = \frac{eB}{m_e} \tag{2.31}$$

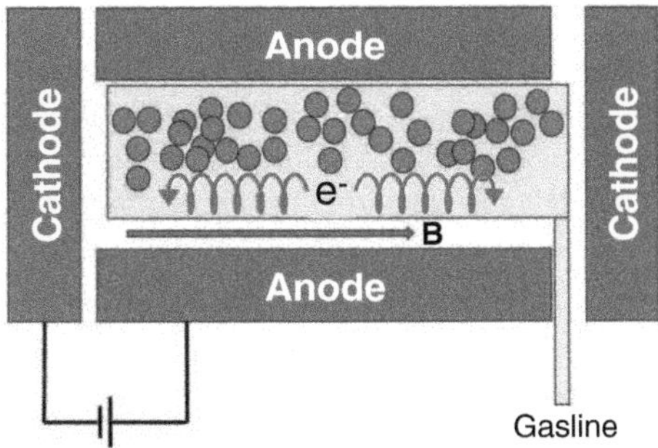

Figure 2.5 Schematics of the Penning ion source, with two cathodes at same potential.

Figure 2.5 illustrates a schematic of the Penning source electrode configuration. The two cathodes are maintained at the same potential, while the anode voltage VA is applied across the anode and cathodes. The magnetic field B, aligned along the axial Z direction, causes electrons in the discharge region to spiral around the magnetic field lines. During this process, the electrons oscillate along the Z-axis, colliding with neutral particles and ionizing them.

The resulting gyration radius of electrons and ions due to the magnetic field is typically much smaller than the discharge chamber dimensions. This confinement, coupled with the symmetric electrode configuration, increases the path length of the ionizing electrons, allowing the discharge to be initiated and sustained at much lower gas pressures than would otherwise be required. It should be noted that these descriptions most accurately reflect electron behavior prior to discharge ignition; once plasma is formed, the local electric and magnetic fields create more complex electron trajectories.

In the neutral gas, free electrons are accelerated by the applied electric field, leading to collisions with neutral gas particles. If an electron has sufficient energy, these collisions may result in the removal of one or more electrons from the neutral particles, creating ions and secondary electrons. The primary electron loses energy with each collision but continues to ionize neutrals until it either runs out of energy or is absorbed by the chamber walls. The secondary electrons, in turn, are also accelerated by the electric field and can cause further ionization. This chain of events ignites and sustains the plasma.

Below, we outline the various collision processes and products for electrons in hydrogen gas, with electron impact ionization of hydrogen molecules being the dominant process. The ions produced in these reactions are much more massive than the electrons and are not significantly confined by the axial magnetic field. These ions are drawn toward the cathodes by the symmetric potential in the discharge region and are further accelerated by the applied electric field. As the ions strike the cathodes, secondary electrons are emitted, which then accelerate back into the discharge region, playing a critical role in maintaining the Penning discharge. Here are stepwise and different electron impact collision processes in details:

1. $\mathrm{H_2} + e^- \rightarrow \mathrm{H} + \mathrm{H} + e^-$ (8.0 eV)
2. $\mathrm{H} + e^- \rightarrow \mathrm{H^+} + 2e^-$ (13.6 eV)
3. $\mathrm{H_2} + e^- \rightarrow \mathrm{H_2^+} + 2e^-$ (15.4 eV)
4. $\mathrm{H_2} + e^- \rightarrow \mathrm{H^+} + \mathrm{H} + 2e^-$ (18.0 eV)
5. $\mathrm{H_2^+} + e^- \rightarrow 2\mathrm{H^+} + 2e^-$ (27.0 eV)
6. $\mathrm{H_2} + e^- \rightarrow 2\mathrm{H^+} + 3e^-$ (30.0 eV)
7. $\mathrm{H_2} + e^- \rightarrow e^- + e^- + e^- + \mathrm{H^+} + \mathrm{H^+}$ (46 eV)

The behavior of a Penning discharge [53] is influenced by the pressure within the chamber, especially when the magnetic field and voltage are constant. As pressure increases, so does the discharge current, which is why early experiments used Penning discharges as pressure gauges. Higher pressure means more gas particles, which shortens the distance between ionizing collisions. Pressures are typically measured at the gas inlet. Moreover, as the axial magnetic field increases, plasma density and discharge current generally rise. However, after a certain point, further increases in the magnetic field cause the discharge current to drop. This may happen because stronger magnetic fields reduce electron losses to the anode, thereby reducing ion losses to the cathode. This balancing act maintains plasma neutrality, which might suggest that plasma density actually increases with a stronger magnetic field, despite the drop in current. Finally, for a set operating pressure and axial magnetic field, both the discharge current and the extracted ion current tend to increase as the discharge voltage rises. Higher applied voltages between the anode and cathodes result in greater acceleration of electrons by the electric field within the discharge region. Consequently, electrons gain more energy, enabling them to ionize a greater number of neutral particles before their energy drops below the ionization threshold.

Here is the stepwise summary of the PIG ion sources:

1. A neutral gas is introduced into the ion source chamber.
2. Electrons are emitted from a cathode and are accelerated toward an anode due to an applied electric field.
3. A perpendicular magnetic field causes the electrons to spiral within the chamber, which increases their path length and the likelihood of ionizing gas molecules.
4. Electron collisions with gas atoms or molecules result in ionization, generating positive ions.
5. The magnetic field confines the electrons inside the chamber, allowing for more ionization events and the creation of a dense plasma.
6. Positive ions are extracted from the plasma through a slit in the anode or by an extraction electrode.
7. The ions are then accelerated by an electric field, forming an ion beam.

2.6 Laser Ion Sources

In 1969, Peacock et al. [54] as well as Byckovsky et al. [55] independently discovered that laser-produced plasma could be utilized as ion sources. This technique involves focusing a laser beam onto a solid, movable target, which results in the evaporation of the target material. During this process, electrons are heated to hundreds of electron volts (eV), which subsequently ionize the target material through collisions. The energy absorption mechanism primarily involves inverse bremsstrahlung, wherein plasma electrons, accelerated by the laser, scatter off plasma ions. In the dense, high-temperature plasma, ions are ionized progressively due to collisions with electrons. A schematic of laser ion source is shown in Figure 2.6.

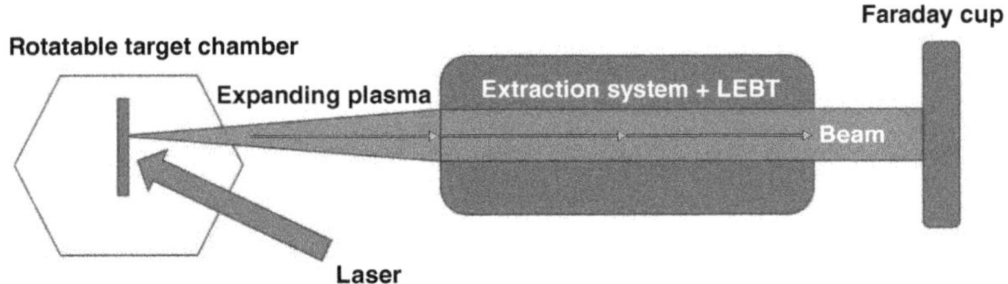

Figure 2.6 Schematics of laser ion source. LEBT: low energy beam transport.

The temperature of the plasma – and consequently, the charge state of the ions – depends on the laser power applied. Higher laser power results in a higher plasma temperature and the generation of more highly charged ions. The duration of the ion pulse is influenced by the distance between the target and the extraction plane; this distance is constrained by the maximum plasma density tolerable without causing arcing and by the minimum current required, as the extracted current density diminishes with increasing distance.

The system does not require additional power supplies except a stepper motor for target rotation. Typically, a CO_2 laser is used, which can vary in pulse energy from 1 to 50 J, has a pulse length of approximately 1 μs, a radiation wavelength of 10.6 μm, a divergence of a few milliradians, and a repetition rate of 1 Hz. The laser gas mixture usually comprises CO_2, N_2, and He. With a low power density (around 10^9 W/cm^2), high currents of low-charge state ions can be achieved, while a high power density ($\geq 10^{12}$ W/cm^2) yields an intense beam of highly charged ions.

A longitudinal magnetic field helps in confining the plasma expansion, shaping it into a cylindrical form along the source axis. When the laser pulse interacts with the target, energy is initially transferred from photons to free electrons in the metal, and then from these electrons to the lattice. The coupling timescales for free-electron heating and thermalization range from a few to tens of femtoseconds, while energy transfer from electrons to the lattice occurs in timescales of tens of picoseconds. Given that nanosecond pulses are much longer than these timescales, laser energy continues to transfer to the system even after ion emission begins, resulting in a quasi-steady-state system where ion energy is largely governed by plasma hydrodynamics.

For picosecond and femtosecond laser pulses, which are comparable to or shorter than the electron-phonon coupling times, the system may not achieve thermal equilibrium before the pulse ends, complicating theoretical descriptions of the ion emission. While charge state-resolved energy spectra of ions from long-nanosecond laser-produced plasmas have been extensively studied, such measurements for plasmas created by ultrashort laser pulses are less well-documented. There is ongoing debate regarding whether steady-state assumptions, valid for long nanosecond pulses, can be applied to shorter pulses. These assumptions are frequently used to infer charge state-specific energies from charge state-insensitive results, such as those obtained with Faraday cups.

Finally, here is the stepwise summary of processes in the laser ion source:

1. A solid target, usually metallic, is placed inside a vacuum chamber.
2. A pulsed, high-energy laser is directed toward the target, with parameters such as pulse duration and energy controlled precisely.
3. When the laser pulse strikes the target, material from the surface is evaporated, resulting in a plume of particles.
4. The evaporated material forms a plasma, consisting of free electrons and ions.
5. The laser energy is absorbed by the plasma, increasing the electron temperature and enhancing the ionization process.
6. High-energy electrons cause the atoms in the plasma to undergo ionization, producing singly or multiply charged ions.
7. The plasma expands rapidly away from the target.
8. An electric field is applied to extract the ions from the expanding plasma.
9. The extracted ions are then collimated and focused into a beam using ion optics.

2.7 Vacuum Arc Ion Source

The metal vapor vacuum arc (MEVVA) ion source [21, 56], developed at Berkeley in the early 1980s, represented a major advancement in ion source technology. This device produces ion beams by creating a vacuum arc between a heated cathode and a cold anode, which vaporizes the cathode material. The vapor is then ionized to generate a plasma, which can be extracted as a metal ion beam. MEVVA sources are known for their ability to generate high-current metal ion beams, making them valuable for applications, such as materials processing and ion implantation.

1. The process begins with the ignition of a vacuum arc between hot cathode spots and a cold anode using a high voltage spark.
2. The arc plasma is sustained between the cathode spots and the anode, primarily consisting of metal vapor from the cathode.
3. This plasma expands through a hole in the anode into an expansion area, which is connected to the anode via a resistor to manage the current.
4. An extractor grid, located at the end of the expansion area, draws out the ion beam from the plasma.
5. The ion current produced remains stable throughout the cathode's life and shows little variation between different cathodes.

2.8 Numerical Problems

1. In an electron source operating under thermionic emission and space charge-limited conditions, calculate the current density from thermionic emission using the Richardson–Dushman law and the space charge-limited current using the Child–Langmuir law. The work function of the cathode material is $\phi = 4.5\,\text{eV}$, the cathode temperature is $T = 2000\,\text{K}$, and the cathode area is $A_c = 1\,\text{cm}^2$. A potential difference of $V = 5\,\text{kV}$ is applied between the cathode and anode, with a separation distance of $d = 1\,\text{mm}$. The Richardson constant is $A_R = 1.2 \times 10^6\,\text{A/m}^2\text{K}^2$. Finally, determine whether the emission is limited by thermionic effects or space charge effects.
2. Calculate the magnetic mirror loss cone condition, ionization rate, and plasma confinement time for an ECR ion source using the following parameters: magnetic field at the midplane $B_{min} = 0.1\,\text{T}$, maximum magnetic field at the mirror points $B_{max} = 0.4\,\text{T}$, electron density $n_e = 5 \times 10^{17}\,\text{m}^{-3}$, neutral gas density $n_g = 1 \times 10^{19}\,\text{m}^{-3}$, ionization cross-section $\sigma_i = 2 \times 10^{-19}\,\text{m}^2$, and electron velocity $v_e = 3 \times 10^7\,\text{m/s}$, along with a plasma radius of 0.05 m and particle perpendicular velocity of $v_\perp = 1 \times 10^6\,\text{m/s}$.
3. For an electron beam ion source (EBIS), find the electron density, ionization time, ionization efficiency, trap capacity, and total ions produced under the following conditions: electron beam current $I_e = 2\,\text{A}$, electron beam energy $E_e = 5\,\text{keV}$, beam radius $r_b = 2\,\text{mm}$, trap length $L = 0.3\,\text{m}$, trap potential $V_t = 3\,\text{kV}$, ionization cross-section $\sigma_i = 2{\times}10^{-19}\,\text{m}^2$, neutral gas density $n_g = 1{\times}10^{20}\,\text{m}^{-3}$, average ion charge state $z = 5$, and confinement time $t_c = 0.5$ seconds.
4. Using the given parameters for a duoplasmatron ion source, including electron temperature $T_e = 5\,\text{eV}$, plasma density $n_p = 1 \times 10^{19}\,\text{m}^{-3}$, extraction potential $V_{ext} = 15\,\text{kV}$,

ion charge $z = 1$, extraction aperture radius $r_a = 1.5$ mm, plasma area $A_p = 1\,\text{cm}^2$, and plasma length $L_p = 5$ cm, determine the extracted ion beam current, ion energy after extraction, ion density at the extraction point, total ions in the plasma volume, and the space charge limit for the beam current.

5. In a Penning ion source, calculate the extracted ion beam current, the energy of the deuteron beam after extraction, plasma density in the source, and magnetic field required for confinement with the provided parameters: extraction potential $V_{ext} =$ 10 kV, magnetic field $B = 0.3$ T, plasma density $n_p = 5 \times 10^{17}\,\text{m}^{-3}$, plasma length $L_p = 10$ cm, and electron temperature $T_e = 5$ eV.
6. For a SNICS ion source generating negative oxygen ions (O^-), determine the negative ion beam current, the energy of the extracted ions, cesium vapor pressure in pascals, the total number of negative oxygen ions produced during sputtering over 1 second, and the number of cesium atoms sputtered per second. The parameters provided are: sputtering voltage $V_s = 20$ kV, extraction potential $V_{ext} = 10$ kV, sputtering area $A_s = 2\,\text{cm}^2$, ionization efficiency $\eta = 5\%$, cesium vapor pressure $P_{Cs} = 1\times10^{-4}$ Torr, electron temperature $T_e = 2$ eV, and sputter yield $Y = 0.1$ ions/atom.
7. For a laser ion source, find the laser intensity on the target, ion beam current for a carbon ion ($z = +1$), energy of the extracted ion beam, plasma density in the laser-induced plasma, and total number of ions produced per laser pulse. The parameters given are: laser wavelength $\lambda = 1064$ nm, laser pulse energy $E_{pulse} = 1$ J, pulse duration $t_{pulse} = 10$ ns, spot size $A_{spot} = 0.1\,\text{mm}^2$, plasma density $n_p = 1 \times 10^{18}\,\text{m}^{-3}$, and extraction potential $V_{ext} = 20$ kV.
8. A beam dump with a surface area of $0.5\,\text{m}^2$ and an emissivity of 0.9 is heated to a temperature of 1500 K. Calculate the power radiated by the beam dump.
9. A cavity wall with a surface area of $1.5\,\text{m}^2$ and an emissivity of 0.8 is heated by electromagnetic fields to a temperature of 2000 K. Determine the power radiated by the cavity walls.

3

Beam Optics

"Particles are the ultimate reality, and fields are their shadows."

—Richard Feynman, *The Feynman Lectures on Physics*

After reading this chapter, you should be able to:

- To learn the basic terminology of beam optics.
- To understand the beam's evolution in phase space dynamics.
- To understand the significance of transfer matrix approach and Twiss parameter formalism.

Beam optics focuses on the processes of ion beam extraction, acceleration, and transportation from the source to a designated target. In the context of accelerator physics, an ion beam refers to a stream of charged particles, all sharing the same charge polarity. These beams are typically modeled as ellipses within phase space, a representation based on conjugate variables such as the particle's position and momenta. The beam characteristics are often described using Twiss parameters, which will be covered in this chapter. When the forces acting on the beam remain linear within the aperture of the optical elements, and the beam current is relatively low, typically in the microampere range, the beam's behavior can be analyzed using linear optics. However, as the beam current increases to the milliampere level, and higher-order optical elements like sextupoles are introduced, nonlinear effects become more prominent. Under these conditions, higher-order corrections must be considered, and the beam dynamics shift to a nonlinear regime, potentially leading to phase space filamentation. In linear optics, however, the phase space ellipse of the beam rotates, with its area remaining constant.

As the beam converges, it reaches a focal point known as the beam waist, after which it begins to diverge again, as illustrated in Figure 3.1. The line labeled AB represents the central axis of the beampipe. The upright ellipse indicates the beam waist at the focal point. During focusing, the ellipse rotates counterclockwise, while during defocusing, it rotates clockwise, maintaining a constant area.

Charged Particle Beam Physics: An Introduction for Physicists and Engineers, First Edition.
Sarvesh Kumar and Manish K. Kashyap.

Companion Website: https://www.wiley.com/go/Kumar_1e

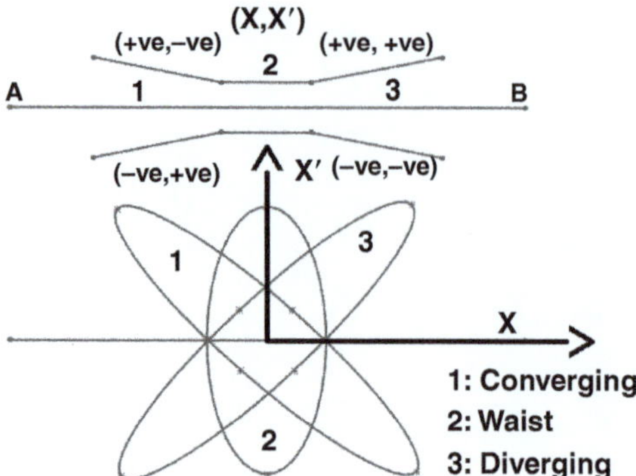

Figure 3.1 Phase space ellipse in various forms.

3.1 Phase Space and Liouville's Theorem

Phase space is a representation where the position and momentum coordinates of beam particles are plotted. Each particle corresponds to a point in phase space, providing a snapshot of the beam's profile at a given position and time within the accelerator. The particle positions, and momenta evolve during beam transport. For decoupled horizontal and vertical planes, horizontal phase space (x vs. p_x) can be used to calculate the two-dimensional (2D) horizontal emittance by measuring the area covered by the beam in phase space. Likewise, vertical phase space (y vs. p_y) provides the 2D vertical emittance, while longitudinal phase space ($\Delta\phi$ vs. ΔE) captures phase and energy spread, defining longitudinal emittance. A single measurement, such as from a beam profile monitor, provides particle position information but not momentum. To infer momentum, measurement of beam size at two or more locations are required, enabling the calculation of beam divergence.

Liouville's theorem serves as a fundamentals in transverse beam dynamics, stating that the area enclosed by the points representing beam particles in transverse phase space remains constant over time. Typically, this contour forms an ellipse, driven by linear restoring forces, similar to the harmonic motion seen in simple pendulums. While the shape of this ellipse may change as the beam is transported, its area remains preserved.

In most accelerators, particle bunches contain huge number of particles. Tracking the motion of such large numbers of particles over accelerator lengths ranging from a few meters to tens of kilometers is computationally demanding. Fortunately, statistical mechanics provides tools to handle these large ensembles efficiently, allowing the beam's collective properties to be determined.

Liouville's theorem states that if there is a flow of matter inside a given volume, the density (ρ) inside the volume must change to conserve the mass (M), given by:

$$M = \int \rho(x, y, z, t)\, dx\, dy\, dz = \int \rho(x, y, z, t)\, dV \tag{3.1}$$

The sign of the normal component of the velocity of matter determines whether the matter flows in or out of the control volume. To find the mass flux through the control surface,

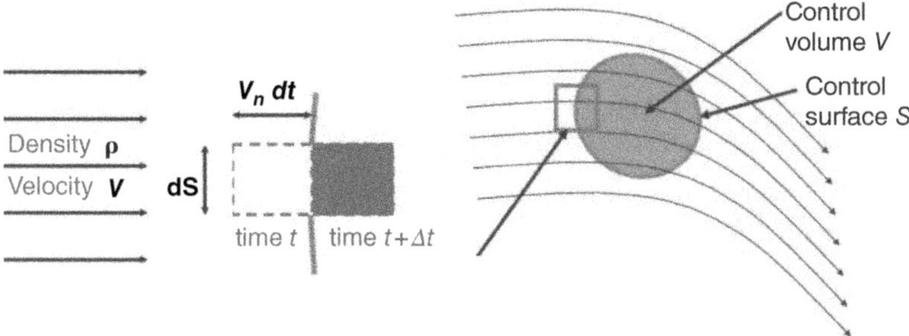

Figure 3.2 Mass flux into a given volume.

consider a small part of the surface where the velocity is normal to the surface as shown in Figure 3.2.

$$\frac{dM}{dt} = -\rho \mathbf{v} \cdot d\mathbf{S} \tag{3.2}$$

The net inflow through the boundary of the control volume is:

$$\frac{\partial M}{\partial t} = -\int_S \rho \mathbf{v} \cdot d\mathbf{S} \tag{3.3}$$

Using Gauss theorem and Equation (3.1):

$$\frac{\partial}{\partial t}\int_V \rho \, dV = -\int_V \nabla \cdot (\rho \mathbf{v}) \, dV \tag{3.4}$$

This gives the continuity equation:

$$\frac{\partial \rho}{\partial t} + \nabla \cdot (\rho \mathbf{v}) = 0 \tag{3.5}$$

For a function $f(x, p_x)$ and velocity $v(\dot{x}, \dot{p_x})$, the equation becomes:

$$\frac{\partial f}{\partial t} + \nabla \cdot (f\mathbf{v}) = \frac{\partial f}{\partial t} + \frac{\partial (\dot{x} f)}{\partial x} + \frac{\partial (\dot{p_x} f)}{\partial p_x} \tag{3.6}$$

For Hamiltonian systems:

$$\dot{x} = \frac{\partial H}{\partial p_x} \quad \text{and} \quad \dot{p_x} = -\frac{\partial H}{\partial x} \tag{3.7}$$

This leads to:

$$\frac{\partial \dot{x}}{\partial x} = \frac{\partial^2 H}{\partial x \partial p_x} \quad \text{and} \quad \frac{\partial \dot{p_x}}{\partial p_x} = -\frac{\partial^2 H}{\partial x \partial p_x} \tag{3.8}$$

Hence, we get:

$$\frac{\partial f}{\partial t} + \nabla \cdot (f\mathbf{v}) = \frac{df}{dt} \tag{3.9}$$

Thus, from the continuity equation:

$$\frac{df}{dt} = 0 \quad \Rightarrow \quad \frac{d\rho}{dt} = 0 \tag{3.10}$$

The phase space density in a Hamiltonian system remains constant during motion, meaning the volume occupied by the system in phase space is preserved. An ion beam, which is a collection of charged particles surrounding a reference particle, can be described by spatial coordinates (x, y, z) and momentum coordinates (p_x, p_y, p_z), forming a six-dimensional (6D) phase space. Under linear restoring forces, this can be decomposed into three independent 2D phase spaces.

However, processes like beam cooling challenge this conservation principle. Beam cooling reduces the phase space volume, effectively increasing phase space density. This leads to decreased emittance and increased beam brightness, despite Liouville's theorem.

The Heisenberg uncertainty principle imposes a quantum limit on how small the phase space area (or beam emittance) can be:

$$\Delta x \Delta p \geq \frac{\hbar}{2} \tag{3.11}$$

In high-energy accelerators, particularly for light particles like electrons, this quantum limit becomes significant. No matter how much emittance is reduced via cooling, it cannot be reduced below this limit.

In accelerator physics, controlling beam emittance is crucial for achieving desired beam properties. The Heisenberg uncertainty principle sets the lower bound on emittance, while Liouville's theorem ensures that the phase space volume remains constant unless external forces like damping are applied.

3.2 Emittance and Acceptance

The beam's characteristics in both transverse and longitudinal phase space are described by emittance, [57] with the corresponding terms being transverse and longitudinal emittance. An ideal beam exhibits minimal energy spread and travels with minimal divergence. According to Liouville's theorem, when a system is governed by conservative forces, the phase space area occupied by particles remains constant, known as emittance.

Liouville's theorem further implies that in conservative fields, with no particle interaction, the phase space density and the phase space volume of a particle ensemble are conserved. This conservation means that it is impossible to alter the phase space volume using static electric or magnetic fields, although this is not necessarily true for time-varying fields. While the volume remains constant, its shape can change significantly, akin to an incompressible fluid. For independent motions in the x, y, and z directions, the theorem holds for the subspaces (x, p_x), (y, p_y), and (z, p_z).

The divergence angles are represented as x' and y'. Considering particle velocity (v_x, v_y, v_z) and momentum (p_x, p_y, p_z), the beam divergence is defined as:

$$x' = \tan\theta_x = \frac{v_x}{v_z} = \frac{p_x}{p_z} = \frac{p_x}{p} \tag{3.12}$$

$$y' = \tan\theta_y = \frac{v_y}{v_z} = \frac{p_y}{p_z} = \frac{p_y}{p} \tag{3.13}$$

Emittance is then defined as:

$$\epsilon_x = \pi x x' \tag{3.14}$$

This geometrical emittance remains constant in the absence of acceleration or deceleration of charged particles. To account for such effects, normalized emittance is introduced as:

$$\epsilon_n = \beta\gamma\pi x x' \tag{3.15}$$

Normalized emittance is a conserved quantity and serves as a figure of merit for ion sources or accelerators. It is typically expressed in units of πmm-mrad $\sqrt{\text{MeV}}$ and is inversely proportional to beam momentum, an effect known as adiabatic damping.

Emittance is a property of the ion beam, while acceptance pertains to the beam optical component. Acceptance is defined as the phase space area encompassing all points that can be transmitted by a beam optical device. For effective beam transport:

Acceptance of beam optical device ≥ emittance of ion beam

The emittance of a beam emerging from an accelerator is typically constrained by the space available for axial and transverse oscillations during acceleration, in addition to the characteristics of the beam at injection. The equation of motion for the beam can often be modeled as simple harmonic motion:

$$x'' = -kx \tag{3.16}$$

where k represents the spring constant associated with the restoring force. The solution to this equation takes a simple sinusoidal form:

$$x = a\sin(kz + \phi) \tag{3.17}$$

The trajectory of an individual particle can be described as a sine wave with an amplitude a and a phase ϕ determined by the initial conditions:

$$x' = ka\cos(kz + \phi) \tag{3.18}$$

By combining Equations 3.17 and 3.18, we obtain:

$$\left(\frac{x}{a}\right)^2 + \left(\frac{x'}{ka}\right)^2 = 1 \tag{3.19}$$

This is the equation of an upright ellipse in phase space. As the particle undergoes sinusoidal oscillations in real space, the phase space ellipse is traced out by the representative points. In Figure 3.3, numbered particles are shown traversing a beam pipe under the influence of restoring forces, and the corresponding phase space ellipses are depicted at various z positions. The particle rotates counterclockwise within the phase space ellipse.

To reduce transverse emittance while keeping the longitudinal momentum, several techniques are employed. Synchrotron radiation damping reduces transverse energy as particles move in a circular path, with energy in the forward direction restored. Electron cooling

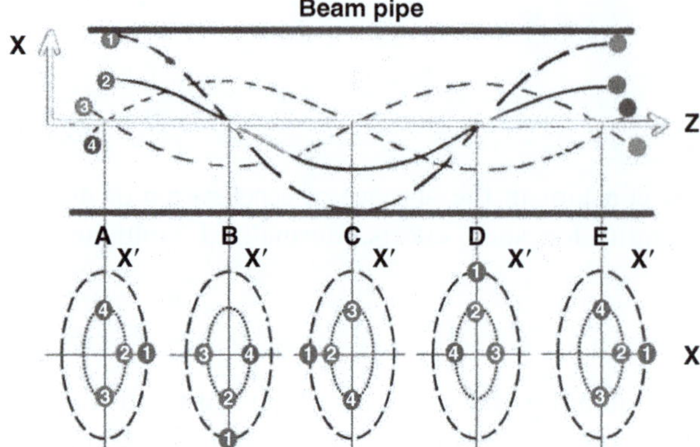

Figure 3.3 Sinusoidal oscillations of particles and their elliptical phase representations and rotation of particles in phase spaces.

involves using a beam of electrons to reduce the transverse motion of heavier particles like protons. Stochastic cooling detects particle path deviations and corrects them, reducing transverse emittance. Laser cooling targets particles with higher transverse energy, reducing their sideways motion. Adiabatic damping naturally reduces sideways motion as particles gain forward energy during acceleration. Bunch compression focuses on shortening the particle bunch, reducing transverse spread while maintaining forward momentum. These methods improve beam quality and focus in accelerators.

3.3 Brightness

In particle accelerators, brightness refers to a measure of the beam's quality, taking into account both its intensity and focus. It is a key parameter that quantifies how many particles are concentrated within a given phase space volume, which includes both position and momentum. Higher brightness is crucial for various applications, such as in particle colliders and light sources, where focused, intense beams are necessary for precise experiments and higher collision rates.

Mathematically, brightness B is given by:

$$B = \frac{I}{\epsilon_x \epsilon_y}$$

where I is the beam current (intensity). ϵ_x and ϵ_y are the beam *emittances* in the horizontal and vertical directions, respectively.

Brightness in photon beams, such as those produced by synchrotron sources or free-electron lasers (FELs), measures how many photons are focused on a specific area and energy range. It includes the intensity of the photon beam, how well it is focused, and the spread of its energy (or wavelength). High brightness is important for experiments that require precise and intense light, such as X-ray imaging and spectroscopy.

The brightness B_{photon} for photon beams is given by:

$$B_{\text{photon}} = \frac{\Phi}{\epsilon_x \epsilon_y \left(\frac{\Delta\lambda}{\lambda}\right)}$$

where Φ = photon flux is the number of photons emitted per second. $\frac{\Delta\lambda}{\lambda}$ is the relative bandwidth, showing the energy spread of the beam.

Photon beams with high brightness are often produced in particle accelerators, especially in synchrotron radiation facilities and FELs. In these accelerators, high-energy electron beams are guided through magnetic fields, causing them to emit photons. The brightness of the photon beam is influenced by the quality of the electron beam, particularly its emittance and current. By controlling the electron beam's properties, accelerators can produce highly focused and intense photon beams for scientific research and industrial applications.

3.4 Luminosity

Luminosity for head-on Gaussian bunches represents the rate at which particles in two colliding beams collide and interact, crucial in high-energy physics experiments. It depends on the frequency of beam crossings, the number of particles in each beam, and the concentration of these particles at the interaction point. When beams are more focused, with particles packed closely together, the collision rate increases, leading to higher luminosity. Thus, luminosity rises with more particles in each beam and frequent crossings, while it decreases if the beams are more spread out. This concept is essential for the effective operation of particle colliders. Consider two bunches in a particle collider going head on with the following profiles:

The particle density of bunch 1 is represented by:

$$\rho_1(x,y) = \frac{N_1}{2\pi\sigma_x\sigma_y} \exp\left(-\frac{x^2}{2\sigma_x^2} - \frac{y^2}{2\sigma_y^2}\right) \tag{3.20}$$

where N_1 is the number of particles in the bunch, and σ_x and σ_y are the horizontal and vertical beam sizes, respectively.

The particle density of bunch 2 is represented by:

$$\rho_2(x,y) = \frac{N_2}{2\pi\sigma_x\sigma_y} \exp\left(-\frac{x^2}{2\sigma_x^2} - \frac{y^2}{2\sigma_y^2}\right) \tag{3.21}$$

where N_2 is the number of particles in the bunch, and σ_x and σ_y are the horizontal and vertical beam sizes, respectively.

The interaction rate R is given by:

$$R = f \int_{-\infty}^{\infty} \int_{-\infty}^{\infty} \rho_1(x,y)\rho_2(x,y)\, dx\, dy \tag{3.22}$$

where f is the bunch crossing frequency, and the integral represents the overlap of the two Gaussian distributions over all space.

After substituting the Gaussian distributions and simplifying, the interaction rate becomes:

$$R = \frac{fN_1N_2}{(2\pi\sigma_x\sigma_y)^2}\int_{-\infty}^{\infty}\exp\left(-\frac{x^2}{\sigma_x^2}\right)dx\int_{-\infty}^{\infty}\exp\left(-\frac{y^2}{\sigma_y^2}\right)dy \tag{3.23}$$

The integrals are standard Gaussian integrals, which evaluate to:

$$\int_{-\infty}^{\infty}\exp\left(-\frac{x^2}{\sigma_x^2}\right)dx = \sigma_x\sqrt{\pi}, \quad \int_{-\infty}^{\infty}\exp\left(-\frac{y^2}{\sigma_y^2}\right)dy = \sigma_y\sqrt{\pi} \tag{3.24}$$

The luminosity L is defined as the interaction rate R per unit cross-sectional area σ_{int}:

$$L = \frac{R}{\sigma_{\text{int}}} \tag{3.25}$$

For the head-on collision of two Gaussian beams, assuming that all interactions are effective, we take $\sigma_{\text{int}} = 1$, simplifying the expression for luminosity:

$$L = R \tag{3.26}$$

Finally, the luminosity L is given by:

$$L = \frac{fN_1N_2}{4\pi\sigma_x\sigma_y} \tag{3.27}$$

which shows that luminosity is proportional to the bunch crossing frequency f, the number of particles in each bunch N_1 and N_2, and inversely proportional to the product of the beam sizes σ_x and σ_y.

The generalized luminosity expression for head-on Gaussian bunches with unequal beam sizes is given by:

$$L = \frac{fN_1N_2}{2\pi\sqrt{\sigma_{1x}^2+\sigma_{2x}^2}\sqrt{\sigma_{1y}^2+\sigma_{2y}^2}} \tag{3.28}$$

Here, f is the frequency of bunch crossings, N_1 and N_2 represent the number of particles in each bunch, σ_{1x} and σ_{2x} are the horizontal beam sizes for bunch 1 and bunch 2, respectively, and σ_{1y} and σ_{2y} are the vertical beam sizes for bunch 1 and bunch 2, respectively. This expression accounts for the differing horizontal and vertical beam sizes between the two bunches, providing a comprehensive formula for calculating luminosity when the beams are not identical in size.

For two beams colliding head-on, the luminosity L_0 is given by:

$$L_0 = \frac{fN_1N_2}{2\pi\sqrt{\sigma_{1x}^2+\sigma_{2x}^2}\sqrt{\sigma_{1y}^2+\sigma_{2y}^2}} \tag{3.29}$$

However, when the beams collide at an angle, the overlap in the transverse plane is reduced. The effective horizontal beam size σ_x' in the presence of a crossing angle (ϕ) is given by:

$$\sigma_x' = \frac{\sigma_x}{\sqrt{1+\left(\frac{\phi\sigma_z}{\sigma_x}\right)^2}} \tag{3.30}$$

where σ_z is the bunch length. The luminosity with a crossing angle is then:

$$L_{\text{cross}} = L_0 \cdot \frac{1}{\sqrt{1 + \left(\frac{\phi\sigma_z}{\sigma_x}\right)^2}} \tag{3.31}$$

Substituting the expression for L_0 gives:

$$L_{\text{cross}} = \frac{fN_1N_2}{2\pi\sqrt{\sigma_{1x}^2 + \sigma_{2x}^2}\sqrt{\sigma_{1y}^2 + \sigma_{2y}^2}} \cdot \frac{1}{\sqrt{1 + \left(\frac{\phi\sigma_z}{\sigma_x}\right)^2}} \tag{3.32}$$

The crab crossing scheme is a technique used to counteract the reduction in luminosity that occurs when beams collide at a crossing angle. In this scheme, special RF cavities tilt the particle bunches just before they collide, effectively aligning them for a head-on collision despite the crossing angle. The correction factors account for this by adjusting the luminosity calculation to include the effects of beam offset and crossing angle.

The formula (3.27) is corrected for beam offset using the factor W, which is expressed as:

$$W = \exp\left(-\frac{(d_2 - d_1)^2}{4\sigma_x^2}\right) \tag{3.33}$$

where d_1 and d_2 are the displacements of the two beams, and σ_x is the horizontal beam size.

To account for the crossing angle ϕ, an additional correction factor S is introduced:

$$S = \frac{1}{\sqrt{1 + \left(\frac{\phi\sigma_z}{\sigma_x}\right)^2}} \tag{3.34}$$

The combined effect of both beam offset and crossing angle is corrected using an exponential factor, where B and A are defined as:

$$B = \frac{(d_2 - d_1)\sin\left(\frac{\phi}{2}\right)}{2\sigma_x^2} \tag{3.35}$$

$$A = \sin^2\left(\frac{\phi}{2}\right)\sigma_x^2 + \cos^2\left(\frac{\phi}{2}\right)\sigma_z^2 \tag{3.36}$$

The final luminosity expression incorporating these corrections is:

$$L = L_0 \cdot W \cdot S \cdot \exp\left(\frac{B^2}{A}\right) \tag{3.37}$$

These correction factors allow the luminosity to accurately reflect the conditions in the crab crossing scheme, ensuring that the effects of beam offset and crossing angle are properly accounted for in the calculation.

3.5 Matrix Formalism

In accelerator physics, when a charged particle beam moves through an optical element such as a drift space or a quadrupole, its individual particle's positions and angles change according to linear equations. Instead of tracking each particle separately, we describe the beam statistically using its second-order moments, forming what is called the sigma matrix or beam matrix. For one transverse plane, say the horizontal x-plane, the sigma matrix at any location is defined as

$$\sigma = \begin{pmatrix} \langle x^2 \rangle & \langle xx' \rangle \\ \langle xx' \rangle & \langle x'^2 \rangle \end{pmatrix} \quad \text{where} \quad \langle x^2 \rangle, \langle x'^2 \rangle, \langle xx' \rangle \tag{3.38}$$

are second-order moments.

When a beam passes through any optical element, the phase space coordinates (x, x') transform linearly according to

$$\begin{pmatrix} x_1 \\ x_1' \end{pmatrix} = \begin{pmatrix} R_{11} & R_{12} \\ R_{21} & R_{22} \end{pmatrix} \begin{pmatrix} x_0 \\ x_0' \end{pmatrix} \quad \text{with} \quad R = \begin{pmatrix} R_{11} & R_{12} \\ R_{21} & R_{22} \end{pmatrix}. \tag{3.39}$$

This particle-level transformation leads to the beam matrix transformation

$$\sigma_1 = R\sigma_0 R^T \tag{3.40}$$

where R^T is the transpose of R. Suppose initially

$$\sigma_0 = \begin{pmatrix} \sigma_{11} & \sigma_{12} \\ \sigma_{12} & \sigma_{22} \end{pmatrix}. \tag{3.41}$$

Then multiplying $R\sigma_0$ and then by R^T gives

$$\begin{aligned} \sigma_{11}(1) &= R_{11}^2\sigma_{11} + 2R_{11}R_{12}\sigma_{12} + R_{12}^2\sigma_{22}, \\ \sigma_{12}(1) &= R_{11}R_{21}\sigma_{11} + (R_{11}R_{22} + R_{12}R_{21})\sigma_{12} + R_{12}R_{22}\sigma_{22}, \\ \sigma_{22}(1) &= R_{21}^2\sigma_{11} + 2R_{21}R_{22}\sigma_{12} + R_{22}^2\sigma_{22}. \end{aligned} \tag{3.42}$$

In the special case of a drift space of length L, the transfer matrix becomes

$$R_{\text{drift}} = \begin{pmatrix} 1 & L \\ 0 & 1 \end{pmatrix} \tag{3.43}$$

which leads to

$$\begin{aligned} \sigma_{11}(1) &= \sigma_{11}(0) + 2L\sigma_{12}(0) + L^2\sigma_{22}(0), \\ \sigma_{12}(1) &= \sigma_{12}(0) + L\sigma_{22}(0), \\ \sigma_{22}(1) &= \sigma_{22}(0). \end{aligned} \tag{3.44}$$

Thus, in a drift, the beam size grows quadratically with L, the position-angle correlation grows linearly, and the divergence remains unchanged.

The phase space distribution of a beam is described by

$$X^T\sigma^{-1}X = 1 \quad \text{with} \quad X = \begin{pmatrix} x \\ x' \end{pmatrix}, \quad \sigma^{-1} = \frac{1}{\det(\sigma)} \begin{pmatrix} \sigma_{22} & -\sigma_{12} \\ -\sigma_{12} & \sigma_{11} \end{pmatrix}. \tag{3.45}$$

Expanding gives

$$\sigma_{22}x^2 - 2\sigma_{12}xx' + \sigma_{11}x'^2 = \det(\sigma). \tag{3.46}$$

The maximum extents of the ellipse are

$$x_{\text{max}} = \sqrt{\sigma_{11}}, \quad x'_{\text{max}} = \sqrt{\sigma_{22}}, \tag{3.47}$$

the normalized correlation coefficient is

$$r_{12} = \frac{\sigma_{12}}{\sqrt{\sigma_{11}\sigma_{22}}}, \tag{3.48}$$

and the slopes of the principal axes are

$$\text{slope} = r_{12}\sqrt{\frac{\sigma_{22}}{\sigma_{11}}}, \quad \text{or} \quad \text{slope} = \frac{1}{r_{12}}\sqrt{\frac{\sigma_{22}}{\sigma_{11}}}. \tag{3.49}$$

The intercepts at $x' = 0$ and $x = 0$ are

$$x_{\text{int}} = \sqrt{\sigma_{11}(1 - r_{12}^2)}, \quad x'_{\text{int}} = \sqrt{\sigma_{22}(1 - r_{12}^2)}. \tag{3.50}$$

The beam ellipse can also be expressed using Twiss parameters α, β, and γ as

$$\sigma = \epsilon \begin{pmatrix} \beta & -\alpha \\ -\alpha & \gamma \end{pmatrix} \quad \text{with} \quad \beta\gamma - \alpha^2 = 1, \tag{3.51}$$

giving

$$\langle x^2 \rangle = \epsilon\beta, \quad \langle xx' \rangle = -\epsilon\alpha, \quad \langle x'^2 \rangle = \epsilon\gamma. \tag{3.52}$$

An important property of beam transport is that the emittance remains constant if the transport matrix R is symplectic, that is

$$R^TJR = J \quad \text{with} \quad J = \begin{pmatrix} 0 & 1 \\ -1 & 0 \end{pmatrix}. \tag{3.53}$$

Symplecticity implies that $\det(R) = 1$, and as a result, the beam emittance, being proportional to $\sqrt{\det(\sigma)}$, is preserved under the transformation. Thus, in any linear, symplectic system, the phase space area enclosed by the beam, and hence the beam quality, is conserved.

Now coming to full 3D, each beam optical component is represented mathematically by 6×6 transfer matrix and a particle is defined by six variables, which define its position and momentum in the 6D phase space: $P = (x, x', y, y', z, z')$. To first order, the evolution of particle coordinates is governed by the matrix equation:

$$x_i(t) = \sum_{j=1}^{6} R_{ij} x_j(0) \tag{3.54}$$

Here, R_{ij} are the elements of the transfer matrix R. The determinant $|R| = 1$, which is a direct consequence of the equations of motion for a charged particle reflects Liouville's theorem, which states the conservation of phase space area.

The coordinates after transformation by the matrix can be expressed as follows:

$$\begin{bmatrix} x_1 \\ x_1' \\ y_1 \\ y_1' \\ l_1 \\ \delta_1 \end{bmatrix} = \begin{pmatrix} R_{11} & R_{12} & R_{13} & R_{14} & R_{15} & R_{16} \\ R_{21} & R_{22} & R_{23} & R_{24} & R_{25} & R_{26} \\ R_{31} & R_{32} & R_{33} & R_{34} & R_{35} & R_{36} \\ R_{41} & R_{42} & R_{43} & R_{44} & R_{45} & R_{46} \\ R_{51} & R_{52} & R_{53} & R_{54} & R_{55} & R_{56} \\ R_{61} & R_{62} & R_{63} & R_{64} & R_{65} & R_{66} \end{pmatrix} \begin{bmatrix} x_0 \\ x_0' \\ y_0 \\ y_0' \\ l_0 \\ \delta_0 \end{bmatrix}$$

In other notations, the transfer matrix may be represented as:

$$\begin{pmatrix} (x|x_0) & (x|x_0') & (x|y_0) & (x|y_0') & (x|l) & (x|\delta) \\ (x'|x_0) & (x'|x_0') & (x'|y_0) & (x'|y_0') & (x'|l) & (x'|\delta) \\ (y|x_0) & (y|x_0') & (y|y_0) & (y|y_0') & (y|l) & (y|\delta) \\ (y'|x_0) & (y'|x_0') & (y'|y_0) & (y'|y_0') & (y'|l) & (y'|\delta) \\ (l|x_0) & (l|x_0') & (l|y_0) & (l|y_0') & (l|l) & (l|\delta) \\ (\delta|x_0) & (\delta|x_0') & (\delta|y_0) & (\delta|y_0') & (\delta|l) & (\delta|\delta) \end{pmatrix}$$

In general:

1. x is the horizontal displacement of the arbitrary ray with respect to the assumed central trajectory.
2. x' is the horizontal angle of the ray with respect to the assumed central trajectory.
3. y is the vertical displacement of the ray with respect to the assumed central trajectory.
4. y' is the vertical angle of the ray with respect to the assumed central trajectory.
5. l is the path length difference between the arbitrary ray and the central trajectory.
6. δ is the fractional momentum deviation of the ray from the assumed central trajectory.

Due to mid-plane symmetry and the decoupling of x and y motions, certain elements of the matrix are zero:

$$R_{13} = R_{14} = R_{23} = R_{24} = R_{31} = R_{32} = R_{41} = R_{42} = R_{36} = R_{46} = 0$$

Additionally, the zero elements in the fifth column arise because all variables are independent of the path length difference:

$$R_{15} = R_{25} = R_{35} = R_{45} = R_{65} = 0$$

The simplified form of the matrix is then:

$$R = \begin{pmatrix} R_{11} & R_{12} & 0 & 0 & 0 & R_{16} \\ R_{21} & R_{22} & 0 & 0 & 0 & R_{26} \\ 0 & 0 & R_{33} & R_{34} & 0 & 0 \\ 0 & 0 & R_{43} & R_{44} & 0 & 0 \\ R_{51} & R_{52} & 0 & 0 & 1 & R_{56} \\ 0 & 0 & 0 & 0 & 0 & 1 \end{pmatrix}$$

The final coordinates can be expressed as:

$$x_1 = R_{11}x_0 + R_{12}x'_0 + R_{16}\delta_0, \quad x'_1 = R_{21}x_0 + R_{22}x'_0 + R_{26}\delta_0$$

$$l_1 = R_{51}x_0 + R_{52}x'_0 + l_0 + R_{56}\delta_0, \quad y_1 = R_{33}y_0 + R_{34}y'_0$$

$$y'_1 = R_{43}y_0 + R_{44}y'_0, \quad \delta_1 = \delta_0$$

Several specific cases arise depending on the nature of the transfer matrix:

1. **Parallel-to-point focusing:** $R_{11} = R_{33} = 0$ implies that the final image size is independent of the initial beam size.
2. **Point-to-point focusing:** $R_{12} = R_{34} = 0$ implies that the final image size is independent of the initial divergence.
3. **Parallel-to-parallel focusing:** $R_{21} = R_{43} = 0$ implies that the final beam divergence is independent of the initial size.
4. **Point-to-parallel focusing:** $R_{22} = R_{44} = 0$ implies that the final beam divergence is independent of the initial divergence.
5. **Achromatic condition:** $R_{16} = R_{26} = 0$ ensures that particles with different momenta have the same path length and focus to the same point.
6. **Isochronous condition:** $R_{51} = R_{52} = R_{56} = 0$ ensures that particles with different momenta have the same time of flight, independent of their initial momentum, and energy.

3.5.1 Significance of Transfer Matrix Elements

1. $R_{11} = \frac{x_1}{x_0}$ and $R_{33} = \frac{y_1}{y_0}$ represent beam size magnification in the x and y planes, respectively, for point-to-point imaging, where unit magnification is preferred for one-to-one beam transport.

2. $R_{22} = \frac{x_1'}{x_0'}$ and $R_{44} = \frac{y_1'}{y_0'}$ represent angular magnification of the beam for parallel-to-point imaging.
3. $R_{21} = -\frac{1}{f_x}$ and $R_{43} = -\frac{1}{f_y}$ represent the inverse of the focal length in the x and y planes for all types of imaging.
4. The first-order momentum resolution or momentumresolving power is given by:

$$R = \frac{p}{\Delta p} = \frac{R_{16}}{x_1} = \frac{R_{16}}{2x_0 R_{11}}$$

5. By symmetry considerations, $R_{16} = -R_{52}$ and $R_{26} = -R_{51}$.

For systems with rotational symmetry, such as electrostatic lenses, the dispersive elements of the transfer matrix are zero, and the radial and axial submatrices are identical:

$$R = \begin{pmatrix} R_{11} & R_{12} & 0 & 0 & 0 & 0 \\ R_{21} & R_{22} & 0 & 0 & 0 & 0 \\ 0 & 0 & R_{11} & R_{12} & 0 & 0 \\ 0 & 0 & R_{21} & R_{22} & 0 & 0 \\ 0 & 0 & 0 & 0 & R_{55} & R_{56} \\ 0 & 0 & 0 & 0 & 0 & R_{66} \end{pmatrix}$$

3.6 Equation of Motion in a Co-moving Coordinate System

Consider a beam moving along the designed reference orbit parameterized by s. The particle's position is described using a co-moving coordinate system (x, y, s), where x and y are small transverse displacements from the orbit. Since the beam transverse size is small compared to the accelerator size, we expand about the reference orbit keeping only first-order terms. In the co-moving frame, steering primarily affects the $x-s$ plane. Such particle dynamics are shown in Figure 3.4.

Initially, the coordinate system has unit vectors $\boldsymbol{x}_{0A}$ and $\boldsymbol{s}_{0A}$ along x and s directions respectively. After rotation by angle φ, they transform to:

$$\boldsymbol{x}_0 = \boldsymbol{x}_{0A} \cos\varphi + \boldsymbol{s}_{0A} \sin\varphi, \quad \boldsymbol{s}_0 = -\boldsymbol{x}_{0A} \sin\varphi + \boldsymbol{s}_{0A} \cos\varphi. \tag{3.55}$$

Differentiating with respect to φ gives:

$$\frac{d\boldsymbol{x}_0}{d\varphi} = \boldsymbol{s}_0, \quad \frac{d\boldsymbol{s}_0}{d\varphi} = -\boldsymbol{x}_0. \tag{3.56}$$

Since the path element is $ds = Rd\varphi$, where R is bending radius, we have:

$$\frac{d\varphi}{dt} = \frac{1}{R}\frac{ds}{dt} = \frac{\dot{s}}{R}. \tag{3.57}$$

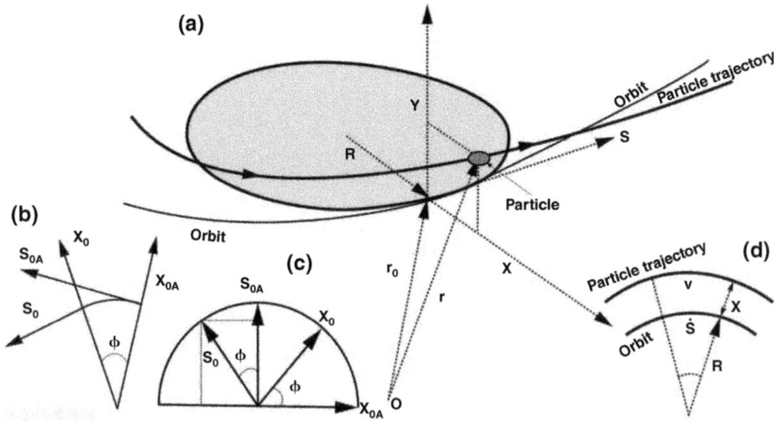

Figure 3.4 (a) Rotated co-moving coordinate system describing the beam motion relative to the orbit [58]. (b, c) Rotation of a Cartesian coordinate system (x, z, s) around the z-axis. (d) Comparison between the velocity v of an individual particle and the velocity $\dot{s}$ of the co-moving coordinate system along the orbit.

Thus, time derivatives of the unit vectors are:

$$\dot{\boldsymbol{x}}_0 = \frac{d\boldsymbol{x}_0}{dt} = \frac{1}{R}\dot{s}\boldsymbol{s}_0, \quad \dot{\boldsymbol{s}}_0 = \frac{d\boldsymbol{s}_0}{dt} = -\frac{1}{R}\dot{s}\boldsymbol{x}_0, \quad \dot{\boldsymbol{y}}_0 = 0. \tag{3.58}$$

The particle's position vector is:

$$\boldsymbol{r} = \boldsymbol{r}_0 + x\boldsymbol{x}_0 + y\boldsymbol{y}_0, \tag{3.59}$$

where $\boldsymbol{r}_0$ is the reference orbit position, moving with $\dot{\boldsymbol{r}}_0 = \dot{s}\boldsymbol{s}_0$.
Taking time derivative:

$$\begin{aligned}
\dot{\boldsymbol{r}} &= \frac{d}{dt}(\boldsymbol{r}_0 + x\boldsymbol{x}_0 + y\boldsymbol{y}_0) \\
&= \dot{\boldsymbol{r}}_0 + \dot{x}\boldsymbol{x}_0 + x\dot{\boldsymbol{x}}_0 + \dot{y}\boldsymbol{y}_0 + y\dot{\boldsymbol{y}}_0 \\
&= \dot{s}\boldsymbol{s}_0 + \dot{x}\boldsymbol{x}_0 + x\frac{1}{R}\dot{s}\boldsymbol{s}_0 + \dot{y}\boldsymbol{y}_0 \quad (\text{since } \dot{\boldsymbol{y}}_0 = 0) \\
&= \dot{x}\boldsymbol{x}_0 + \dot{y}\boldsymbol{y}_0 + \left(1 + \frac{x}{R}\right)\dot{s}\boldsymbol{s}_0.
\end{aligned} \tag{3.60}$$

Taking second time derivative and using $\dot{\boldsymbol{s}}_0 = -\frac{1}{R}\dot{s}\boldsymbol{x}_0$:

$$\begin{aligned}
\ddot{\boldsymbol{r}} &= \frac{d}{dt}\left(\dot{x}\boldsymbol{x}_0 + \dot{y}\boldsymbol{y}_0 + \left(1 + \frac{x}{R}\right)\dot{s}\boldsymbol{s}_0\right) \\
&= \ddot{x}\boldsymbol{x}_0 + \dot{x}\dot{\boldsymbol{x}}_0 + \ddot{y}\boldsymbol{y}_0 + \dot{y}\dot{\boldsymbol{y}}_0 + \left(\frac{\dot{x}}{R}\dot{s} + \left(1 + \frac{x}{R}\right)\ddot{s}\right)\boldsymbol{s}_0 + \left(1 + \frac{x}{R}\right)\dot{s}\dot{\boldsymbol{s}}_0 \\
&= \ddot{x}\boldsymbol{x}_0 + \frac{1}{R}\dot{x}\dot{s}\boldsymbol{s}_0 + \ddot{y}\boldsymbol{y}_0 + \left(\frac{\dot{x}}{R}\dot{s} + \left(1 + \frac{x}{R}\right)\ddot{s}\right)\boldsymbol{s}_0 - \left(1 + \frac{x}{R}\right)\frac{1}{R}\dot{s}^2\boldsymbol{x}_0
\end{aligned} \tag{3.61}$$

Grouping similar terms:

$$\ddot{\boldsymbol{r}} = \left(\ddot{x} - \left(1 + \frac{x}{R}\right)\frac{\dot{s}^2}{R}\right)\boldsymbol{x}_0 + \ddot{y}\boldsymbol{y}_0 + \left(\frac{2}{R}\dot{x}\dot{s} + \left(1 + \frac{x}{R}\right)\ddot{s}\right)\boldsymbol{s}_0. \tag{3.62}$$

The Lorentz force equation for a particle of charge q is:

$$m\ddot{\boldsymbol{r}} = q(\dot{\boldsymbol{r}} \times \boldsymbol{B}), \tag{3.63}$$

where $\boldsymbol{B}$ is the magnetic field.

Assuming only vertical magnetic field B_z, so $\boldsymbol{B} = (0, 0, B_z)$, expanding the cross product:

$$\dot{\boldsymbol{r}} \times \boldsymbol{B} = \begin{vmatrix} \boldsymbol{i} & \boldsymbol{j} & \boldsymbol{k} \\ \dot{x} & \dot{y} & \left(1 + \frac{x}{R}\right)\dot{s} \\ 0 & 0 & B_z \end{vmatrix} \tag{3.64}$$
$$= (\dot{y}B_z)\,\boldsymbol{i} - (\dot{x}B_z)\,\boldsymbol{j} + 0\boldsymbol{k}.$$

Thus:

$$\dot{\boldsymbol{r}} \times \boldsymbol{B} = (\dot{y}B_z, -\dot{x}B_z, 0)\,. \tag{3.65}$$

Substituting into the Lorentz force:

$$m\ddot{\boldsymbol{r}} = qB_z(\dot{y}\boldsymbol{x}_0 - \dot{x}\boldsymbol{y}_0). \tag{3.66}$$

Matching components:

From $\boldsymbol{x}_0$ direction:

$$m\left(\ddot{x} - \left(1 + \frac{x}{R}\right)\frac{\dot{s}^2}{R}\right) = qB_z\dot{y}, \tag{3.67}$$

From $\boldsymbol{y}_0$ direction:

$$m\ddot{y} = -qB_z\dot{x}. \tag{3.68}$$

Now, expressing everything in terms of derivatives with respect to s:

Since:

$$\dot{x} = x'\dot{s}, \quad \ddot{x} = x''\dot{s}^2 + x'\ddot{s}$$

$$\dot{y} = y'\dot{s}, \quad \ddot{y} = y''$$

Substituting:

$\boldsymbol{x}_0$ direction gives:

$$m\left(x''\dot{s}^2 + x'\ddot{s} - \left(1 + \frac{x}{R}\right)\frac{\dot{s}^2}{R}\right) = qB_z y'\dot{s}, \tag{3.69}$$

$\boldsymbol{y}_0$ direction gives:

$$m(y''\dot{s}^2 + y'\ddot{s}) = -qB_z x'\dot{s}. \tag{3.70}$$

Assuming $\ddot{s} \approx 0$ (nearly constant longitudinal velocity), simplifies to:

$$x'' - \left(1 + \frac{x}{R}\right)\frac{1}{R} = \frac{qB_z}{m\dot{s}}y', \tag{3.71}$$

$$y'' = -\frac{qB_z}{m\dot{s}}x'. \tag{3.72}$$

Next, using that the full particle speed is:

$$v = \dot{s}\left(1 + \frac{x}{R}\right), \tag{3.73}$$

thus:

$$\dot{s} = v\left(1 - \frac{x}{R}\right) \quad \text{(first-order expansion).}$$

Substituting v, neglecting second-order small terms, final simplified transverse equations are:

$$x''(s) + \left(\frac{1}{R^2(s)} - k(s)\right)x(s) = \frac{1}{R(s)}\frac{\Delta p}{p}, \tag{3.74}$$

$$y''(s) + k(s)y(s) = 0, \tag{3.75}$$

where $k(s)$ is the normalized quadrupole gradient. These equations form the basis for calculations in linear beam optics.

3.7 Hill's Equation and Twiss Parameters Formalism

The introduction of alternating gradient focusing, as described by E. Courant in [59], marked a significant advancement in synchrotron design, enabling the use of much stronger focusing forces in both the horizontal and vertical planes. This concept was discovered somewhat serendipitously when Courant investigated the effects of reversing the gradient in some magnets of the weak-focusing Cosmotron. This modification proved highly beneficial, leading to the development of more powerful focusing systems. In general, when both the bending strength $1/\rho(s)$ and the focusing strength $k(s)$ vary along the reference orbit, the resulting motion is described by an oscillatory equation with a variable restoring force. The general form of the equation of motion is:

$$u''(s) + k(s)u(s) = \frac{1}{\rho(s)}\frac{\Delta p}{p_0} \tag{3.76}$$

The general solution to this equation can be expressed as:

$$u(s) = C(s)u_0 + S(s)u_0' + D(s)\frac{\Delta p}{p_0} \tag{3.77}$$

$$u'(s) = C'(s)u_0 + S'(s)u_0' + D'(s)\frac{\Delta p}{p_0} \tag{3.78}$$

Here, $C(s)$ and $S(s)$ are cosine-like and sine-like solutions of the homogeneous equation, with initial conditions:

$$\begin{bmatrix} C_0 & S_0 \\ C_0' & S_0' \end{bmatrix} = \begin{bmatrix} 1 & 0 \\ 0 & 1 \end{bmatrix}$$

$D(s)$, known as the dispersion trajectory, is a particular solution to the inhomogeneous equation when $\frac{\Delta p}{p_0} = 1$, representing a small momentum deviation, with initial conditions:

$$\begin{bmatrix} D_0 \\ D_0' \end{bmatrix} = \begin{bmatrix} 0 \\ 0 \end{bmatrix}$$

Thus, we can express the system in matrix form as:

$$\begin{bmatrix} u_2 \\ u_2' \\ \left(\frac{\Delta p}{p_0}\right)_2 \end{bmatrix} = \begin{bmatrix} C(s) & S(s) & D(s) \\ C'(s) & S'(s) & D'(s) \\ 0 & 0 & 1 \end{bmatrix} \begin{bmatrix} u_1 \\ u_1' \\ \left(\frac{\Delta p}{p_0}\right)_1 \end{bmatrix}$$

The determinant of this transformation matrix is zero, as shown by:

$$(CS' - SC')' = CS'' - SC'' = -k(s)(CS - SC) = 0$$

This determinant equals unity at $s = s_0$ for very small momentum deviations, i.e. $\frac{\Delta p}{p_0} = 1$. Equation 3.76 can be rewritten as:

$$u''(s) + k(s)u(s) = \frac{1}{\rho(s)} \tag{3.79}$$

The general solution to this equation is:

$$u(s) = aC(s) + bS(s) + D(s) \tag{3.80}$$

Using Green's function, we can express $D(s)$ as:

$$D(s) = \int_0^{s_0} G(s_0, z)\frac{1}{\rho(z)}dz \tag{3.81}$$

where the Green's function is:

$$G(s_0, z) = S(s_0)C(z) - C(s_0)S(z) \tag{3.82}$$

From these equations, we obtain:

$$D(s) = S(s)\int_0^s \frac{1}{\rho(z)}C(z)dz - C(s)\int_0^s \frac{1}{\rho(z)}S(z)dz \tag{3.83}$$

Using a general approach, the equation can be rewritten in the form of Hill's equation:

$$u''(s) + k(s)u(s) = 0 \tag{3.84}$$

Let us assume an ansatz for the solution of this equation:

$$u(s) = \sqrt{\beta(s)\epsilon}\,\cos(\psi(s) + \psi_0) \tag{3.85}$$

where $\beta(s)$ is the amplitude function at position s and $\psi(s)$ is the phase function.

Substituting into the equation leads to:

$$f_1(s)\cos(\psi(s) + \psi_0) + f_2(s)\sin(\psi(s) + \psi_0) = 0 \tag{3.86}$$

Here, $f_1(s)$ and $f_2(s)$ are coefficients that must vanish independently, leading to the condition:

$$\beta'(s)\psi'(s) + \beta(s)\psi''(s) = 0 \quad \text{or} \quad (\beta(s)\psi'(s))' = 0 \tag{3.87}$$

$$\beta(s)\psi'(s) = 1 \tag{3.88}$$

Thus, the phase advance $\psi(s)$ is given by:

$$\psi(s) = \int_0^s \frac{1}{\beta(z)} dz \tag{3.89}$$

Next, using the condition from $f_1(s)$, we find:

$$\frac{1}{2}\beta(s)\beta''(s) - \frac{1}{4}(\beta'(s))^2 + \beta(s)^2 k(s) = 1 \tag{3.90}$$

We now define the Twiss parameters:

$$\alpha(s) = -\frac{1}{2}\beta'(s) \quad \text{and} \quad \gamma(s) = \frac{1+\alpha(s)^2}{\beta(s)} \tag{3.91}$$

By differentiating, we have:

$$u'(s) = -\sqrt{\frac{\epsilon}{\beta(s)}}\alpha(s)\cos(\psi(s)+\psi_0) - \sqrt{\frac{\epsilon}{\beta(s)}}\sin(\psi(s)+\psi_0) \tag{3.92}$$

Using this, we form the combination:

$$\beta(s)u'(s) + \alpha(s)u(s) = -\sqrt{\beta(s)\epsilon}\,\sin(\psi(s)+\psi_0) \tag{3.93}$$

From this, we obtain the final equation for the phase space ellipse and its characteristic dimensions as follows:

$$\gamma(s)u^2 + 2\alpha(s)uu' + \beta(s)(u')^2 = \epsilon \tag{3.94}$$

$$a = -\alpha\sqrt{\frac{\epsilon}{\gamma}}, \quad b = -\alpha\sqrt{\frac{\epsilon}{\beta}}, \quad c = \sqrt{\frac{\epsilon}{\gamma}},$$

$$d = \sqrt{\epsilon\beta}, \quad e = \sqrt{\frac{\epsilon}{\beta}}, \quad f = \sqrt{\epsilon\gamma}$$

$$R1 = \sqrt{\frac{\epsilon}{2}}(\sqrt{H+1} + \sqrt{H-1}),$$

$$R2 = \sqrt{\frac{\epsilon}{2}}(\sqrt{H+1} - \sqrt{H-1}), \quad H = \frac{\beta+\gamma}{2}$$

The phase space ellipse in Figure 3.5 is characterized by the Twiss parameters. The orientation of the ellipse depends on the sign of α: positive α corresponds to a converging beam, negative α indicates a diverging beam, and $\alpha = 0$ represents a beam waist. The parameter β is related to the beam size, with larger values indicating a larger beam. The parameter γ

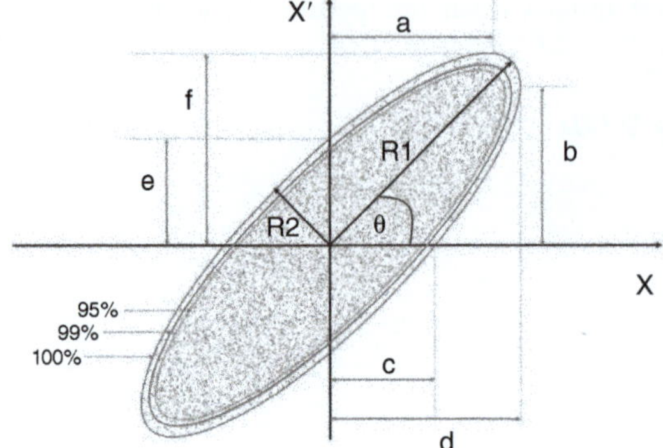

Figure 3.5 Phase space ellipse characterized by Twiss parameters and emittances.

depends on both α and β. These Twiss parameters, also known as Courant–Snyder parameters, describe the area, shape, and orientation of the ellipse, providing a good approximation of the beam shape in phase space.

The angle θ is given by:

$$\tan(2\theta) = \frac{\alpha}{\gamma - \beta} \tag{3.95}$$

Considering the equation of an ellipse:

$$ax^2 + 2bxy + cy^2 = d \tag{3.96}$$

The area of the ellipse, according to analytic geometry, is:

$$A = \frac{\pi d}{\sqrt{ac - b^2}} \tag{3.97}$$

Thus, according to Liouville's theorem:

$$\frac{\pi \epsilon}{\sqrt{\beta\gamma - \alpha^2}} = \pi \epsilon \tag{3.98}$$

This simplifies to:

$$\beta\gamma - \alpha^2 = 1 \tag{3.99}$$

Substituting into the matrix form solution:

$$\begin{bmatrix} u_1 \\ u_1' \end{bmatrix} = \begin{bmatrix} C(s) & S(s) \\ C'(s) & S'(s) \end{bmatrix} \begin{bmatrix} u_0 \\ u_0' \end{bmatrix} \tag{3.100}$$

For $K > 0$:

$$C(s) = \cos(\sqrt{K}s), \quad S(s) = \frac{\sin(\sqrt{K}s)}{\sqrt{K}} \tag{3.101}$$

For $K < 0$:

$$C(s) = \cosh(\sqrt{|K|}s), \quad S(s) = \frac{\sinh(\sqrt{|K|}s)}{\sqrt{|K|}} \tag{3.102}$$

Finally, the Twiss parameters in matrix form become:

$$\begin{bmatrix} \beta(s) \\ \alpha(s) \\ \gamma(s) \end{bmatrix} = \begin{bmatrix} C^2(s) & -2SC(s) & S^2(s) \\ -CC'(s) & S'C(s) + SC'(s) & -SS'(s) \\ C'^2(s) & -2S'C'(s) & S'^2(s) \end{bmatrix} \begin{bmatrix} \beta_0 \\ \alpha_0 \\ \gamma_0 \end{bmatrix} \tag{3.103}$$

For example, a simple drift space of length l, the transformation becomes:

$$\begin{bmatrix} \beta(s) \\ \alpha(s) \\ \gamma(s) \end{bmatrix} = \begin{bmatrix} 1 & -2l & l^2 \\ 0 & 1 & -l \\ 0 & 0 & 1 \end{bmatrix} \begin{bmatrix} \beta_0 \\ \alpha_0 \\ \gamma_0 \end{bmatrix} \tag{3.104}$$

In transverse beam dynamics, a charged particle beam is often described statistically, since it consists of many particles with slightly different transverse positions and angles. For a Gaussian beam in one transverse plane, such as the horizontal x-plane, the distribution of particles in phase space (x, x') typically forms an elliptical shape.

To represent this distribution, we define the beam matrix, or sigma matrix, denoted by σ. This is a symmetric 2×2 matrix that contains the second-order statistical moments of the beam:

$$\sigma = \begin{pmatrix} \sigma_{x^2} & \sigma_{xx'} \\ \sigma_{xx'} & \sigma_{x'^2} \end{pmatrix} \tag{3.105}$$

Here, $\sigma_{x^2} = \langle x^2 \rangle$, $\sigma_{x'^2} = \langle x'^2 \rangle$, and $\sigma_{xx'} = \langle xx' \rangle$ are the variances and covariance of the beam distribution.

From these moments, we define the root-mean-square (rms) emittance as

$$\epsilon_{\text{rms}} = \sqrt{\langle x^2 \rangle \langle x'^2 \rangle - \langle xx' \rangle^2} = \sqrt{\det \sigma}$$

This quantity characterizes the statistical spread of the beam in phase space. It is invariant under linear transformations and provides a measure of beam quality.

For a Gaussian beam, the distribution of particles in phase space is given by

$$f(x, x') = \frac{1}{2\pi\sqrt{\det \sigma}} \exp\left[-\frac{1}{2}\left(\frac{x^2\,\sigma_{x'^2} - 2xx'\,\sigma_{xx'} + x'^2\,\sigma_{x^2}}{\det \sigma}\right)\right]$$

This function defines a two-dimensional normal distribution centered at the origin. The argument of the exponential determines the phase space ellipse that bounds the density contour.

The ellipse equation corresponding to a contour of constant density is

$$x^2\sigma_{x'^2} - 2xx'\sigma_{xx'} + x'^2\sigma_{x^2} = \epsilon^2$$

This represents the shape and orientation of the beam in phase space and encloses a fraction of the beam depending on the value of ϵ.

The sigma matrix can also be expressed in terms of the Twiss parameters α, β, and γ, which are related by

$$\beta\gamma - \alpha^2 = 1$$

Using these parameters, the beam matrix takes the form

$$\sigma = \epsilon \begin{pmatrix} \beta & -\alpha \\ -\alpha & \gamma \end{pmatrix}$$

Here, β is associated with the beam size, γ with the angular spread, and α with the correlation or tilt in phase space.

The Gaussian distribution function can now be written equivalently as

$$f(x,x') = \frac{1}{2\pi\epsilon} \exp\left[-\frac{1}{2\epsilon}\left(\gamma x^2 + 2\alpha xx' + \beta x'^2\right)\right]$$

and the corresponding ellipse equation becomes

$$\gamma x^2 + 2\alpha xx' + \beta x'^2 = \epsilon$$

This equation defines the phase space boundary of the beam for a given emittance. For a Gaussian beam, this ellipse encloses approximately thirty-nine percent of the particles. The rms emittance quantifies the intrinsic phase space area of the beam, while the sigma matrix and Twiss parameters describe its geometry and orientation. These concepts are fundamental for the analysis, matching, and transport of beams in accelerator and beamline systems.

3.8 Horizontal and Vertical Root Mean Square Beam Sizes in Accelerators

It is quite important to know horizontal and vertical root mean square (RMS) beam size at any position of accelerators, especially at the beam waist. Let us consider both betatron motion and dispersion effects, which are critical in understanding the spatial extent of the particle beam.

Let $x(s)$ represent the horizontal displacement of the particle at position s along the accelerator, where s is the path length. The focusing function $K_x(s)$ depends on the magnetic fields in the accelerator. The amplitude of the oscillation is denoted by A, and the horizontal beta function, which describes the amplitude of the betatron oscillation, is denoted by $\beta_x(s)$. The phase advance is represented by $\psi(s)$, while the horizontal emittance, which measures the phase space area occupied by the beam, is denoted by ϵ_x. The Courant–Snyder parameters, α_x and γ_x, describe the shape of the phase space ellipse. The horizontal dispersion function, $\eta_x(s)$, indicates the sensitivity of the particle position to momentum deviation, with

$\delta = \frac{\Delta p}{p_0}$ representing the relative momentum deviation and σ_δ its standard deviation. The RMS beam size contributions from betatron motion and dispersion are denoted by σ_{betatron} and $\sigma_{\text{dispersion}}$, respectively. The total horizontal RMS beam size is denoted by σ_x, while the vertical RMS beam size is denoted by σ_y, with the vertical beta function $\beta_y(s)$ and vertical emittance ϵ_y defined analogously to their horizontal counterparts.

The horizontal motion of a particle in an accelerator can be described by the equation of betatron oscillations:

$$\frac{d^2x}{ds^2} + K_x(s)x(s) = 0 \tag{3.106}$$

where $x(s)$ is the horizontal displacement and $K_x(s)$ is the focusing function.

The general solution to this equation is a sinusoidal function with a varying amplitude:

$$x(s) = A \cdot \sqrt{\beta_x(s)} \cdot \cos(\psi(s) + \psi_0) \tag{3.107}$$

where A is the oscillation amplitude, $\beta_x(s)$ is the beta function, and $\psi(s)$ is the phase advance.

The emittance ϵ_x is related to the phase space area and is given by:

$$\epsilon_x = \frac{x^2}{\beta_x} + 2\alpha_x x\theta_x + \gamma_x\theta_x^2 \tag{3.108}$$

where α_x and γ_x are the Courant–Snyder parameters.

The maximum amplitude A of the oscillation is related to the emittance by:

$$A^2 = \epsilon_x \tag{3.109}$$

Thus, the horizontal displacement $x(s)$ can be expressed as:

$$x(s) = \sqrt{\beta_x(s) \cdot \epsilon_x} \cdot \cos(\psi(s)) \tag{3.110}$$

The RMS beam size due to betatron motion is:

$$\sigma_{\text{betatron}} = \sqrt{\langle x^2 \rangle} = \sqrt{\beta_x(s)\epsilon_x} \tag{3.111}$$

The dispersion function $\eta_x(s)$ relates the position of a particle to its momentum deviation δ:

$$x(s) = \eta_x(s)\delta \tag{3.112}$$

The RMS beam size due to dispersion is:

$$\sigma_{\text{dispersion}} = \eta_x\sigma_\delta \tag{3.113}$$

where σ_δ is the standard deviation of the relative momentum spread.

The total horizontal RMS beam size σ_x is the quadrature sum of the betatron and dispersion contributions:

$$\sigma_x = \sqrt{\sigma_{\text{betatron}}^2 + \sigma_{\text{dispersion}}^2} \tag{3.114}$$

Substituting the individual contributions:

$$\sigma_x = \sqrt{\beta_x\epsilon_x + (\eta_x\sigma_\delta)^2} \tag{3.115}$$

In the absence of significant vertical dispersion, the vertical RMS beam size σ_y is determined primarily by the betatron motion:

$$\sigma_y = \sqrt{\langle y^2 \rangle} = \sqrt{\beta_y \epsilon_y} \tag{3.116}$$

In accelerator physics, understanding both the horizontal and vertical RMS beam sizes is essential for accurately describing the beam's spatial distribution. The horizontal RMS beam size σ_x is influenced by both betatron motion and dispersion, with the total beam size being the quadrature sum of these effects. Betatron motion contributes through the beta function $\beta_x(s)$ and the beam emittance ϵ_x, while dispersion effects are linked to the relative momentum spread σ_δ and the dispersion function $\eta_x(s)$.

Similarly, in the vertical plane, where dispersion is often negligible, the beam size σ_y is primarily governed by betatron oscillations, determined by the vertical beta function $\beta_y(s)$ and emittance ϵ_y.

Overall, these formulations allow for precise calculations of beam sizes at any point in the accelerator, providing critical insights for optimizing beam transport and focusing. Understanding the interplay between betatron oscillations and dispersion is crucial for designing efficient accelerator systems, especially at beam waists where accurate control of the beam size is necessary.

3.9 Betatron Phase Advance and Tune

Betatron phase advance and tune [60, 61] are more important in circular accelerators because the particles travel around the same path repeatedly. This repetition requires careful control of the phase and resonance to keep the beam stable over many turns. The beta function, denoted as $\beta(s)$, describes the envelope of oscillation amplitude of a particle's path in an accelerator. It plays a critical role in determining the beam size and focusing characteristics at different locations along the accelerator. The value of $\beta(s)$ changes with the position s along the beamline, making it essential for understanding the behavior of the beam. The betatron phase advance, $\Delta\psi$, refers to the change in phase that a particle undergoes as it moves through a segment of the accelerator. It is mathematically defined as:

$$\Delta\psi(s) = \int_{s_1}^{s_2} \frac{ds}{\beta(s)} \tag{3.117}$$

where $\beta(s)$ is the beta function, and the integral is taken over the distance from s_1 to s_2.

The betatron tune, Q, represents the number of betatron oscillations a particle completes in one full turn around the accelerator. It is defined as:

$$Q = \frac{1}{2\pi} \oint \frac{ds}{\beta(s)} \tag{3.118}$$

where the integral is taken over the entire circumference of the accelerator. It is of two types known as integer tune and fractional tune:

The integer tune, Q_{int}, is the whole number part of the betatron tune Q. When the tune is an integer, the beam can resonate with imperfections in the accelerator, which may lead to instability. Integer tunes can make the beam sensitive to errors in the magnetic fields or imperfections in the accelerator lattice. Accelerator designs typically avoid integer tunes to prevent large oscillations and potential beam losses.

The fractional tune, Q_{frac}, is the fractional part of the betatron tune Q. It is adjusted to optimize the stability of the beam and avoid resonance conditions. It helps to steer clear of resonances with simple fractional values, which could cause instabilities, although these are generally less severe than integer resonances. By selecting an appropriate fractional tune, the beam's focus and stability are maintained, preventing the growth of small oscillations due to resonances.

In circular accelerators, particles do not always follow the exact circular path. Instead, they oscillate around the ideal orbit due to magnetic fields that focus the beam. These oscillations are called betatron oscillations and occur in both horizontal and vertical planes. The oscillation results from the focusing elements, such as quadrupole magnets, which guide the particles back to the ideal path. The betatron phase advance refers to how much the phase of these oscillations changes as a particle moves through a section of the accelerator. It measures the progression of the oscillation along the beam path. The total phase advance over one complete turn in the accelerator is related to the betatron tune, Q, which is the number of oscillations the particle completes in one revolution. The phase advance over a segment of the accelerator is given by $\Delta\mu = 2\pi \times \frac{1}{n}$, where n is the number of segments.

The **synchrotron tune** refers to the number of oscillations a particle makes in the longitudinal direction during one full revolution around a particle accelerator. These oscillations happen because particles experience slight variations in energy when they pass through the accelerator's radio frequency (RF) cavities. The cavities provide energy to the particles, and depending on how much energy a particle has, it may speed up, or slow down, causing it to oscillate around the reference energy.

Let E be the total energy of a particle and E_s the synchronous energy. The energy deviation is defined as:

$$\Delta E = E - E_s$$

Since the revolution time T depends on the energy, small deviations lead to a change in revolution time given by the phase slip factor η:

$$\frac{\Delta T}{T} = -\eta \frac{\Delta E}{E_0}$$

where $E_0 = \gamma mc^2$ is the total reference energy of the particle, and the slip factor is:

$$\eta = \frac{1}{\gamma_t^2} - \frac{1}{\gamma^2}$$

where γ_t is the transition gamma, and γ is the Lorentz factor.

The RF phase ϕ of a particle evolves with time due to its revolution time deviation. This is given by:

$$\frac{d\phi}{dt} = h\omega_0 \frac{\Delta T}{T}$$

where h is the harmonic number, and $\omega_0 = 2\pi f_0$ is the revolution angular frequency. Using the time deviation relation:

$$\frac{d\phi}{dt} = -h\omega_0 \eta \frac{\Delta E}{E_0}$$

This equation shows that energy deviations cause changes in the phase.
The energy of a particle changes due to the accelerating RF voltage V_{RF}, following:

$$\frac{dE}{dt} = eV_{\mathrm{RF}} \sin\phi$$

For small phase deviations around the synchronous phase ϕ_s, we approximate:

$$\sin\phi \approx \sin\phi_s + (\phi - \phi_s)\cos\phi_s$$

Since $eV_{\mathrm{RF}} \sin\phi_s$ corresponds to the synchronous particle, we obtain:

$$\frac{d\Delta E}{dt} = eV_{\mathrm{RF}}(\phi - \phi_s)\cos\phi_s$$

From the phase evolution equation, expressing $\phi - \phi_s$ in terms of ΔE:

$$\phi - \phi_s = -\frac{E_0}{h\omega_0\eta}\frac{d\Delta E}{dt}$$

Substituting this into the energy equation:

$$\frac{d\Delta E}{dt} = eV_{\mathrm{RF}}\cos\phi_s\left(-\frac{E_0}{h\omega_0\eta}\frac{d\Delta E}{dt}\right)$$

Taking the time derivative of both sides gives a second-order differential equation:

$$\frac{d^2\Delta E}{dt^2} + \frac{heV_{\mathrm{RF}}\cos\phi_s\eta}{E_0}\Delta E = 0$$

which describes simple harmonic motion with frequency:

$$\omega_s = \sqrt{\frac{heV_{\mathrm{RF}}\cos\phi_s\eta}{E_0}}$$

Since the RF system affects momentum evolution rather than direct energy changes, the correct normalization factor involves β^2, where:

$$p_0 = \beta\gamma mc$$

Replacing E_0 with $\beta^2 E_0$ in the denominator, we get:

$$\omega_s = \sqrt{\frac{heV_{\mathrm{RF}}\cos\phi_s\eta}{\beta^2 E_0}}$$

Finally, the **synchrotron tune** is defined as:

$$Q_s = \frac{\omega_s}{\omega_0}$$

Substituting for ω_s:

$$Q_s = \sqrt{\frac{heV_{\mathrm{RF}}\cos\phi_s\eta}{2\pi\beta^2 E_0}}$$

Thus, the synchrotron tune measures how many times a particle oscillates in energy and phase during one full revolution in a circular accelerator. It depends on the RF voltage, particle energy, synchronous phase, and the phase slip factor. A higher RF voltage strengthens the restoring force, leading to faster oscillations, while a higher particle energy makes the oscillations slower. The synchronous phase and phase slip factor control the stability of these oscillations. Overall, the synchrotron tune indicates the strength of longitudinal focusing inside the accelerator.

3.10 Chromaticity

The concepts of tune and chromaticity in particle accelerators are similar to ideas in other fields like acoustics, where vibrations of a musical string depend on its tension, or optics, where light bends differently depending on its wavelength (like how a prism separates light into colors). In particle accelerators, chromaticity [60, 61] describes how the tune changes with the particle's momentum. There is only a small range of tunes that keep the beam stable, and a wide chromaticity range can lead to unwanted spread in the tune due to variations in particle momentum. Now, we establish this concept mathematically as follows:

The phase space coordinates at position s and after one complete turn at $s + C$ are related by the transfer matrix M_0:

$$\mathbf{X}(s + C) = M_0\mathbf{X}(s) \tag{3.119}$$

where C is the circumference of the accelerator and $\mathbf{X}(s) = \begin{pmatrix} x(s) \\ x'(s) \end{pmatrix}$ is the phase space vector.

The transfer matrix M_0 has the general form:

$$M_0 = \begin{pmatrix} M_{11} & M_{12} \\ M_{21} & M_{22} \end{pmatrix} \tag{3.120}$$

One complete oscillation corresponds to a phase advance of 2π radians. If the particle completes Q_x oscillations in one turn, the total phase advance is $2\pi \times Q_x$ radians. The matrix elements $M_{11}, M_{12}, M_{21}, M_{22}$ are related to the phase advance $\mu = 2\pi Q_x$, where Q_x is the betatron tune, and the Twiss parameters α_x, β_x, and γ_x. The transverse motion of a particle in an accelerator can be described by the general solution:

$$x(s) = \sqrt{\epsilon_x \beta_x(s)} \cos(\psi(s) + \psi_0) \tag{3.121}$$

Here $x(s)$ is the transverse displacement at position s. ϵ_x is the emittance (a constant). $\beta_x(s)$ is the beta function. $\psi(s)$ is the phase advance. ψ_0 is the initial phase.

The derivative with respect to s gives the slope $x'(s)$:

$$x'(s) = -\sqrt{\epsilon_x}\left(\frac{\sin(\psi(s) + \psi_0)}{\sqrt{\beta_x(s)}} + \frac{\alpha_x(s)}{\sqrt{\beta_x(s)}} \cos(\psi(s) + \psi_0)\right) \tag{3.122}$$

where $\alpha_x(s) = -\frac{1}{2}\frac{d\beta_x(s)}{ds}$.

After one complete turn, the phase advance is $\psi(s+C) = \psi(s) + 2\pi Q_x$, where Q_x is the betatron tune. The phase space coordinates after one turn are:

$$\mathbf{X}(s+C) = \begin{pmatrix} x(s+C) \\ x'(s+C) \end{pmatrix} = \begin{pmatrix} \sqrt{\epsilon_x \beta_x(s+C)} \cos(\psi(s) + 2\pi Q_x + \psi_0) \\ -\sqrt{\epsilon_x}\left(\frac{1}{\sqrt{\beta_x(s+C)}} \sin(\psi(s) + 2\pi Q_x + \psi_0) \right. \\ \left. + \frac{\alpha_x(s+C)}{\sqrt{\beta_x(s+C)}} \cos(\psi(s) + 2\pi Q_x + \psi_0) \right) \end{pmatrix} \tag{3.123}$$

Using the trigonometric identities:

$$\begin{aligned} \cos(\psi(s) + 2\pi Q_x + \psi_0) = {} & \cos(\psi(s) + \psi_0) \cos(2\pi Q_x) \\ & - \sin(\psi(s) + \psi_0) \sin(2\pi Q_x) \end{aligned}$$

$$\begin{aligned} \sin(\psi(s) + 2\pi Q_x + \psi_0) = {} & \sin(\psi(s) + \psi_0) \cos(2\pi Q_x) \\ & + \cos(\psi(s) + \psi_0) \sin(2\pi Q_x) \end{aligned}$$

Substituting these back into the expressions for $x(s+C)$ and $x'(s+C)$, we expand the terms:

$$\begin{aligned} x(s+C) = {} & \cos(2\pi Q_x) \cdot \sqrt{\epsilon_x \beta_x(s+C)} \cos(\psi(s) + \psi_0) \\ & - \sin(2\pi Q_x) \cdot \sqrt{\epsilon_x \beta_x(s+C)} \sin(\psi(s) + \psi_0) \end{aligned} \tag{3.124}$$

$$\begin{aligned} x'(s+C) = {} & -\sqrt{\epsilon_x} \left[\frac{\sin(\psi(s) + \psi_0)}{\sqrt{\beta_x(s+C)}} \cos(2\pi Q_x) \right. \\ & \left. + \frac{\cos(\psi(s) + \psi_0)}{\sqrt{\beta_x(s+C)}} \sin(2\pi Q_x) \right] \\ & - \alpha_x(s+C) \sin(2\pi Q_x) \sqrt{\epsilon_x} \cos(\psi(s) + \psi_0) \end{aligned} \tag{3.125}$$

To express $\mathbf{X}(s+C)$ in terms of $\mathbf{X}(s)$, we use the matrix form:

$$\mathbf{X}(s+C) = M_0 \mathbf{X}(s) \tag{3.126}$$

where the matrix elements are:

$$M_{11} = \cos(2\pi Q_x) + \alpha_x(s+C) \sin(2\pi Q_x) \tag{3.127}$$

$$M_{12} = \beta_x(s+C) \sin(2\pi Q_x) \tag{3.128}$$

$$M_{21} = -\gamma_x(s+C) \sin(2\pi Q_x) \tag{3.129}$$

$$M_{22} = \cos(2\pi Q_x) - \alpha_x(s+C)\sin(2\pi Q_x) \tag{3.130}$$

with $\gamma_x(s+C) = \frac{1+\alpha_x^2(s+C)}{\beta_x(s+C)}$.

Thus, the one-turn transfer matrix M_0 is:

$$M_0 = \begin{pmatrix} \cos(2\pi Q_x) + \alpha_x \sin(2\pi Q_x) & \beta_x \sin(2\pi Q_x) \\ -\gamma_x \sin(2\pi Q_x) & \cos(2\pi Q_x) - \alpha_x \sin(2\pi Q_x) \end{pmatrix} \tag{3.131}$$

When a particle has a momentum deviation δ, the tune Q_x is shifted by an amount proportional to the chromaticity ξ_x:

$$Q_x(\delta) = Q_x + \delta \cdot \xi_x. \tag{3.132}$$

The transfer matrix $M(\delta)$ incorporating the chromaticity is given by:

$$M(\delta) = \begin{pmatrix} \cos(2\pi(Q_x + \delta \cdot \xi_x)) + \alpha_x \sin(2\pi Q_x \delta \cdot \xi_x) & \beta_x \sin(2\pi(Q_x + \delta \cdot \xi_x)) \\ -\gamma_x \sin(2\pi(Q_x + \delta \cdot \xi_x)) & \cos(2\pi(Q_x + \delta \cdot \xi_x)) - \alpha_x \sin(2\pi Q_x \delta \cdot \xi_x) \end{pmatrix} \tag{3.133}$$

The equation represents how the particle's motion changes in a circular accelerator when there is a deviation in momentum. The matrix modifies the betatron phase advance to account for the tune shift caused by chromaticity, and it adjusts the coupling between position and momentum accordingly. This is crucial for understanding and correcting the behavior of a beam in an accelerator, especially to avoid resonances that could destabilize the beam.

3.11 RMS Emittance

The RMS emittance [57, 62, 63] provides a statistical measure of the beam's phase space spread, reflecting the average particle behavior rather than extreme values. This makes RMS emittance a more stable and practical representation of the beam's size and divergence compared to the full 100% emittance.

Typically, the normalized emittance remains constant throughout an accelerator, though it can decrease due to collimation. However, emittance often increases due to factors such as system mismatches, coupled systems, or nonconservative forces like stripper foils. In theory, the emittance of 100% of a Gaussian beam is infinite, as the distribution extends to infinity. In practice, beam emittance is mostly influenced by particles on the outer edges of the beam, which are more likely to encounter nonlinear fields and be lost. Therefore, emittance is usually defined for a fraction of the beam, commonly around 90%. Since emittance varies significantly depending on the fraction of the beam being considered, it is important to specify the percentage of the beam included when reporting emittance values.

Emittance is often discussed in terms of the RMS transverse beam size. For a Gaussian distribution in the transverse coordinate x, the normalized particle density function is:

$$n(x)dx = \frac{1}{\sqrt{2\pi}\sigma} \exp\left(-\frac{x^2}{2\sigma^2}\right) dx \tag{3.134}$$

This distribution is stationary in time at a specific location. The trajectories in the $u-(\alpha u+\beta u')$ phase space form circles. In equilibrium, the distribution in the coordinate $\alpha u + \beta u'$ will also be Gaussian with standard deviation σ. All particles on a given circle rotate turn by turn, so the equilibrium distribution depends only on the radius and is independent of the position along the circle. The 2D phase space distribution is:

$$n(u, \alpha u + \beta u')\, du\, d(\alpha u + \beta u') = \frac{1}{2\pi\sigma^2} \exp\left(-\frac{u^2 + (\alpha u + \beta u')^2}{2\sigma^2}\right) du\, d(\alpha u + \beta u') \tag{3.135}$$

Switching to polar coordinates where $r^2 = u^2 + (\alpha u + \beta u')^2$, the distribution becomes:

$$n(r, \theta) r\, dr\, d\theta = \frac{1}{2\pi\sigma^2} \exp\left(-\frac{r^2}{2\sigma^2}\right) r\, dr\, d\theta \tag{3.136}$$

Next, we define a radius a within which a fraction F of the particles are contained:

$$F = \int_0^{2\pi} \int_0^a n(r, \theta) r\, dr\, d\theta = \int_0^a \exp\left(-\frac{r^2}{2\sigma^2}\right) \frac{r\, dr}{\sigma^2}$$

Solving for a, we get:

$$a^2 = -2\sigma^2 \ln(1 - F)$$

Since the area is conserved as per the phase space description:

$$\beta\epsilon = \pi a^2 = \pi\left(-2\sigma^2 \ln(1 - F)\right)$$

This leads to:

$$\epsilon = -\frac{2\pi\sigma^2}{\beta} \ln(1 - F) \tag{3.137}$$

It gives the geometric emittance ϵ required to enclose a fraction F of a beam with a two-dimensional Gaussian distribution. In this formula, σ is the root mean square transverse beam size, and β is the Twiss beta function at the observation point. This equation provides a direct connection between the physical size of the beam, its optical focusing properties, and the percentage of beam content enclosed.

The rms emittance is related to the beam size and optics by

$$\epsilon_{\text{rms}} = \frac{\sigma^2}{\beta} \tag{3.138}$$

Substituting into the original formula yields

$$\epsilon = -2\pi\epsilon_{\text{rms}} \ln(1 - F) \tag{3.139}$$

This version makes clear how the emittance required to contain a certain beam fraction scales with the rms emittance. As F approaches 1, the logarithmic term diverges, reflecting the infinitely long tails of the Gaussian distribution.

The table below shows the fraction of the beam enclosed for selected values of total emittance expressed as multiples of the rms emittance:

Emittance ϵ (as a multiple of ϵ_{rms})	Fraction of Beam Enclosed %
$1 \cdot \epsilon_{rms}$	14.71
$\pi \cdot \epsilon_{rms}$	39.35
$2\pi \cdot \epsilon_{rms}$	63.21
$4\pi \cdot \epsilon_{rms}$	86.47
$6\pi \cdot \epsilon_{rms}$	95.02
$8\pi \cdot \epsilon_{rms}$	98.17

These values demonstrate that the rms emittance corresponds only to the beam core, while practical beamline components often need to accommodate much larger emittances to ensure efficient transport. For instance, an acceptance of $6\pi\,\epsilon_{rms}$ is required to transmit 95 percent of the beam.

This formulation is essential in accelerator design, allowing engineers and physicists to match beamline apertures and acceptance regions to the expected beam distribution, based on statistical beam quality.

In conclusion, the fraction of the beam enclosed increases as the emittance increases. For the RMS emittance, about 39.35% of the beam is enclosed. As the emittance increases to four and six times the RMS emittance, the fractions of the beam enclosed rise to 86.47 and 95.02%, respectively. To enclose the entire beam, theoretically, an infinite emittance would be required. In reality, the tail of the beam will always be cut off and the total emittance will always be finite and the tail particles are lots in the beam pipe.

For standard beam distribution such as Kapchinskij–Vladimirskij (KV), Waterbag, Parabolic, and Gaussian type distributions, It can be shown as total emittance is four times, six times, six times, and n^2 times the RMS emittance, respectively where n is the standard deviation while considering Gaussian distribution.

3.12 Space Charge Effects on Ion Beam

Let the beam radius, represented by R, define the size of the beam at any given point along its path. The external focusing forces that act on the beam, such as those provided by magnetic lenses, are represented by κ_0^2, which refers to the zero-current focusing term. This term indicates the focusing strength when no space charge effects are present. However, when the beam carries current, the mutual repulsion between charged particles results in space charge forces [64], quantified by the generalized perveance K. This parameter measures the degree of space charge influence, which tends to expand the beam. The emittance, ϵ, measures the beam's intrinsic properties, including its spatial spread and angular divergence, and remains independent of current while resisting expansion caused by space charge.

The envelope equation [65] describes how the beam radius evolves as it travels through a focusing system. It accounts for external forces, emittance, and space charge effects. The equation governing the beam radius in the presence of space charge is:

$$R'' + \kappa_0^2 R - \frac{K}{R^2} - \frac{\epsilon^2}{R^3} = 0 \tag{3.140}$$

Here, R'' describes the change in beam radius, $\kappa_0^2 R$ represents external focusing forces, $\frac{K}{R^2}$ represents space charge forces, and $\frac{\epsilon^2}{R^3}$ reflects the emittance resisting expansion.

Space charge forces become critical in low-energy, intense ion beams, where the charge density is high and velocities are relatively low. These forces tend to increase emittance, leading to beam blow-up. In higher-energy sections of the accelerator, the magnetic forces generated by the beam particles can counteract this blow-up, but the effect remains small.

Electromagnetic fields generated by the beam itself act back on the beam. Consider a continuous beam with cylindrical symmetry moving with a constant velocity $v = \beta c$ in the z direction, with a uniform charge density. Let q be the enclosed charge and L the bunch length. Consider a Gaussian surface of radius r and length l. Applying Gauss's law to this surface inside the cylinder, the Coulomb repulsion pushes the test particle outward. The induced force is zero at the beam center and increases toward the beam's edge. The magnetic force is radial and attractive for the test particle due to the current running in parallel. This is depicted in Figure 3.6.

For symmetry reasons, the magnetic field has only an azimuthal component B_θ. Using the integral form of Ampere's law over a cylinder centered on the beam axis, we have:

$$B_\theta(r) = \frac{\mu_0 \beta c}{r} \int_0^r \rho(r') r' dr' \tag{3.141}$$

Similarly, for symmetry reasons, the electric field has only a radial component E_r. Using Gauss's law over the same cylindrical surface, we obtain:

$$E_r(r) = \frac{1}{\epsilon_0 r} \int_0^r \rho(r') r' dr' \tag{3.142}$$

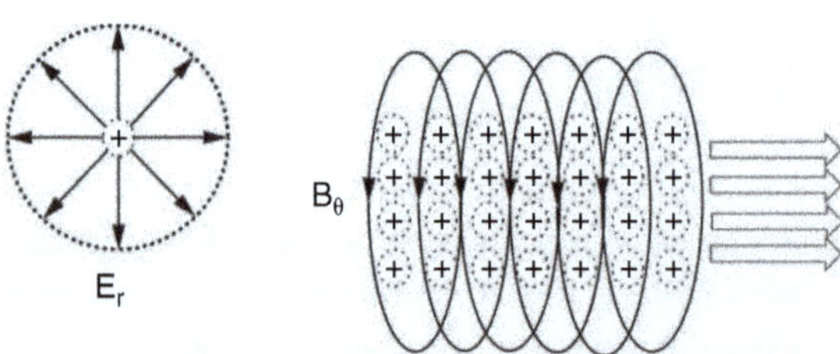

Figure 3.6 Parallel beam current formed by positively charged particles.

By comparing Equations 3.141 and 3.142, the relationship between the electric and magnetic fields for a relativistic beam can be written as:

$$B_\theta(r) = \frac{\beta}{c}E_r(r) \tag{3.143}$$

The Lorentz force on a test particle within the beam is:

$$F = q(E + \mathbf{v} \times \mathbf{B}) \tag{3.144}$$

where q is the charge of the test particle, E is the electric field, and $\mathbf{v} \times \mathbf{B}$ is the magnetic force. Substituting $E_r(r)$ and $B_\theta(r)$ from Equations 3.141 and 3.142, the radial force is:

$$F_r = qE_r(1 - \beta^2) = \frac{qE_r}{\gamma^2} \tag{3.145}$$

Here, $\gamma = \frac{1}{\sqrt{1-\beta^2}}$ is the Lorentz factor. This equation shows that the radial force due to space charge is reduced by the factor $\frac{1}{\gamma^2}$, meaning that as the beam velocity approaches the speed of light ($\beta \to 1$), space charge effects diminish.

The total radial force on the particle can also be derived from the Lorentz force expression:

$$F = q(E_r\hat{r} + \mathbf{v} \times \mathbf{B}) \tag{3.146}$$

For a particle moving along the z-axis, with velocity $\mathbf{v} = v\hat{z}$ and the magnetic field B_θ having an azimuthal component $\hat{\theta}$, the magnetic force is:

$$\mathbf{v} \times \mathbf{B} = v\hat{z} \times B_\theta\hat{\theta} = vB_\theta\hat{r} \tag{3.147}$$

Thus, the total radial force on the particle is:

$$F_r = q(E_r - vB_\theta) \tag{3.148}$$

Substituting B_θ from Equation 3.141, we get:

$$F_r = qE_r\left(1 - \frac{v^2}{c^2}\right) = \frac{qE_r}{\gamma^2} \tag{3.149}$$

This shows that at low velocities (small β), space charge forces are large, leading to beam expansion. As the beam velocity increases and approaches c, space charge effects become negligible as $\frac{1}{\gamma^2}$ approaches zero. This explains why space charge effects are more pronounced in low-energy beams and diminish in relativistic beams.

In summary, space charge forces significantly impact low-energy, highly charged ion beams by causing beam expansion and increasing emittance. For Gaussian beams, the total emittance is equal to the RMS emittance when the distribution is truncated at $1 - \sigma$. In the absence of space charge, the beam's radius R_0 is determined by external focusing forces alone. However, with space charge present, the radius increases due to particle repulsion.

3.13 Numerical Problems

1. What is the normalized emittance of your beam of 100 keV with a beam size of 5 mm and a divergence of 10 mrad?
2. Draw the phase space ellipses of your beam at 100 keV in converging, diverging, and waist positions. Assume the emittance of the beam is 100 π mm-mrad.
3. Calculate the velocity and beta, gamma of an ion beam with an energy of 1 MeV/u, and 1 GeV/u.
4. Calculate the horizontal and vertical RMS beam sizes given a location in an accelerator lattice with a horizontal beta function of 10 m, a vertical beta function of 5 m, a peak horizontal dispersion of 4 m, a horizontal and vertical emittance of 2 mm-mrad, and a relative momentum spread of 0.001.
5. A 2 GeV storage ring operates with six electron bunches and six positron bunches circulating in opposite directions. The current per bunch is initially 5 mA, and the ring has a circumference of 600 m. The horizontal beam size is 6.8×10^{-4} m, the vertical beam size is 4.0×10^{-5} m, and the longitudinal beam size is 2.5 cm. You are asked to calculate the initial luminosity of this experiment. Additionally, the beam lifetimes for both the electrons and positrons are 1.5 hours, and you need to determine the integrated luminosity for a 2-hour run, assuming the beam currents decay exponentially over time.
6. Consider a FODO cell with a focusing and defocusing quadrupole of focal length $f =$ 10 m, length $L = 1$ m, and drift spaces of lengths $d_1 = 2$ m, $d_2 = 4$ m, and $d_3 = 2$ m. Derive the total transfer matrix M_{FODO} and compute the Twiss parameters α, β, and γ for this system.
7. Consider a circular accelerator with a circumference of $C = 600$ m, consisting of 12 identical FODO cells, where the particles undergo horizontal and vertical betatron oscillations due to focusing elements. The quadrupole magnets provide focusing with an average focal length of $f = 10$ m. First, calculate the betatron tune Q for both horizontal and vertical oscillations if the average phase advance per FODO cell is $\Delta\mu = 60°$, and determine the total phase advance for one complete revolution of the ring. Next, if the chromaticity of the system is $\xi_x = -5$ in the horizontal plane and $\xi_y = -3$ in the vertical plane, calculate the change in tune when the relative momentum deviation is $\Delta p/p = 0.001$. Finally, given a momentum compaction factor $\alpha_c = 0.01$, an RF harmonic number $h = 400$, and a slip factor $\eta = -0.001$, compute the synchrotron tune Q_s.
8. Prove that for a beam with for KV distribution: total emittance is four times the RMS emittance, for a beam with non-stationary waterbag or parabolic distribution: total emittance is six times the RMS emittance, for a beam with nonstationary Gaussian distribution with a cutoff at nth standard deviations: total emittance is n^2 times the RMS emittance.
9. Prove that for a phase space ellipse characterized by Twiss parameters ($\alpha, \beta,$ and γ) and emittance (ϵ) as shown in Figure 3.5.

$$a = -\alpha\sqrt{\frac{\epsilon}{\gamma}}, \qquad b = -\alpha\sqrt{\frac{\epsilon}{\beta}}, \qquad c = \sqrt{\frac{\epsilon}{\gamma}},$$

$$d = \sqrt{\epsilon\beta}, \qquad e = \sqrt{\frac{\epsilon}{\beta}}, \qquad f = \sqrt{\epsilon\gamma}$$

$$R1 = \sqrt{\frac{\epsilon}{2}}(\sqrt{H+1} + \sqrt{H-1}),$$

$$R2 = \sqrt{\frac{\epsilon}{2}}(\sqrt{H+1} - \sqrt{H-1}), \qquad H = \frac{\beta+\gamma}{2}$$

10. What are units for longitudinal emittance that are normalized throughout the accelerator and hence the figure of merit?

4

Motion in Magnetostatic Devices

> "I had, also, studied the laws of electromagnetic phenomena, and knew that the magnetic field of force rotates, as solid bodies do, and that the energy of motion is available through its agencies."
>
> —Richard Feynman, *The Feynman Lectures on Physics*

Magnetic fields are crucial for controlling charged particles in various configurations. They offer several advantages over electrostatic methods, such as the ability to handle high-energy particles (starting from MeV energies) more effectively in beamlines. Additionally, magnetic fields do not require vacuum conditions for the poles and can be easily placed outside the beam pipe.

- To understand the basic equations of motion for charged particles in magnetic field configurations, which are essential for analyzing, focusing, and accelerating charged particles.
- To design magnets using Maxwell's equations.
- To perform beam optics in magnets using the COSY Infinity code.

4.1 Magnetostatic Devices

Magnets are essential components of accelerators. Different magnetic field configurations serve various functions for the beam. Dipole magnets are used for bending the beam, primarily for mass or charge state selection. Quadrupole magnets focus the beam, while solenoids also provide focusing but with the added complexity of coupling horizontal and vertical motion.

To bend an energetic beam with a constant radius, a magnetic field of 1 T is approximately equivalent to an electric field of strength $E = 3 \times 10^8$ V/m. Achieving a 1 T magnetic field strength is relatively straightforward with conventional iron-dominated magnets, while superconducting magnets can reach up to 10 T on an industrial scale. In contrast, achieving electric fields in the GV/m range is practically challenging. Therefore, magnetic fields are the preferred method for guiding beams in high-energy particle accelerators. This chapter will explore various magnetic field configurations.

Charged Particle Beam Physics: An Introduction for Physicists and Engineers, First Edition.
Sarvesh Kumar and Manish K. Kashyap.

Companion Website: https://www.wiley.com/go/Kumar_1e

4.1.1 Motion of a Charged Particle in a Magnetic Field

To describe the motion of a particle near the nominal trajectory, we introduce a Cartesian coordinate system $K = (x, y, s)$, where the origin moves along the beam's trajectory. The beam direction is along the s axis, with x and y representing the horizontal and vertical axes, respectively. We assume that the particles move primarily parallel to the s-direction, i.e. $\boldsymbol{v} = (0, 0, v_s)$, and that the magnetic field has only transverse components, such that $\boldsymbol{B} = (B_x, B_y, 0)$. For a particle moving in the horizontal plane through the magnetic field, there is a balance between the Lorentz force $F_L = qv_sB_y$ and the centrifugal force $F_r = \frac{mv_s^2}{R}$. Here, m is the particle mass, and R is the curvature radius of the trajectory. Using $p = mv_s$, this force balance leads to the following relationship:

$$\frac{1}{R(x,y,s)} = \frac{q}{p}B_y(x,y,s) \tag{4.1}$$

A similar expression can be derived for vertical deflection. Since the beam's transverse dimensions are small compared to the trajectory's curvature radius, we can expand the magnetic field around the nominal trajectory:

$$B_y(x) = B_{y0} + \frac{dB_y}{dx}x + \frac{1}{2!}\frac{d^2B_y}{dx^2}x^2 + \frac{1}{3!}\frac{d^3B_y}{dx^3}x^3 + \dots \tag{4.2}$$

Multiplying by $\frac{q}{p}$ gives:

$$\frac{q}{p}B_y(x) = \frac{e}{p}B_{y0} + \frac{e}{p}\frac{dB_y}{dx}x + \frac{1}{2!}\frac{e}{p}\frac{d^2B_y}{dx^2}x^2 + \frac{1}{3!}\frac{e}{p}\frac{d^3B_y}{dx^3}x^3 + \dots$$

which can be written as:

$$\frac{q}{p}B_y(x) = \frac{1}{R} + kx + \frac{1}{2!}mx^2 + \frac{1}{3!}ox^3 + \dots \tag{4.3}$$

The magnetic field around the beam can thus be seen as a sum of multipoles, each affecting the particle path differently. The most important multipoles and their effects are summarized in Table 4.1. Higher-order multipoles (e.g. sextupole, octupole) are typically either unwanted field errors or are introduced deliberately for compensation or field correction.

Table 4.1 Multipoles and their effects on particle motion.

Multipole	Definition	Effect
Dipole	$\frac{1}{R} = \frac{e}{p}B_{y0}$	Beam steering
Quadrupole	$k = \frac{e}{p}\frac{dB_z}{dx}$	Beam focusing
Sextupole	$m = \frac{e}{p}\frac{d^2B_y}{dx^2}$	Chromaticity compensation
Octupole	$o = \frac{e}{p}\frac{d^3B_y}{dx^3}$	Field errors or compensation

4.2 Ampere's Law for Magnet Design

The magnetic field can be described as the gradient of the magnetostatic potential:

$$B = \nabla\phi \tag{4.4}$$

In cylindrical coordinates (r, θ, z), the potential ϕ can be expressed as:

$$\phi = \sum \frac{a}{n+1}\left(\frac{r}{a}\right)^{n+1} [F_n \cos((n+1)\theta) + G_n \sin((n+1)\theta)] \tag{4.5}$$

where $G_n = \frac{B_0 b_n}{\mu_0}$ and $F_n = \frac{B_0 a_n}{\mu_0}$ are the normal and skew strengths of the $2(n+1)$th multipoles.

For a pure normal $2(n+1)$th multipole with iron approximated to infinite permeability, the pole profile can be written as:

$$\left(\frac{r}{a}\right)^{n+1} \sin[(n+1)\theta] = 1 \tag{4.6}$$

where $n = 0, 1, 2, 3, \ldots$ for dipole, quadrupole, sextupole, etc.

The pole profile for a dipole magnet with equipotential surfaces is given by:

$$r \sin(\theta) = y = a = \text{constant} \tag{4.7}$$

Applying Ampere's law to estimate the field in the gap:

For $n = 1$, the pole profile equation is:

$$r^2 \sin(2\theta) = 2xy = a^2 \tag{4.8}$$

For $n = 2$, the pole profile equation is:

$$r^3 \sin(3\theta) = 3xy = a^3 \tag{4.9}$$

4.3 Equation of Motion in a Co-moving Coordinate System

In a particle accelerator, charged particles such as protons or heavy ions travel along a well-defined, often curved path known as the reference orbit. This orbit is shaped by magnetic fields that bend and focus the beam. However, not all particles follow the exact same trajectory – some move slightly off-axis or travel at different speeds. To study this, we use a co-moving coordinate system that travels with the reference particle. In this system, the longitudinal coordinate is denoted by s, which measures distance along the reference orbit. The horizontal and vertical deviations from this path are given by $x(s)$ and $y(s)$ respectively. The curvature of the path at each point is described by the bending radius $R(s)$, while the effect of focusing magnets is captured by the focusing strength $k(s)$.

The motion of particles in this frame is governed by the following differential equations:

$$\begin{gathered} x''(s) + \left(\frac{1}{R^2(s)} - k(s)\right) x(s) = \frac{1}{R(s)} \cdot \frac{\Delta p}{p} \\ y''(s) + k(s) y(s) = 0 \end{gathered} \tag{4.10}$$

Here, $\Delta p/p$ represents the relative momentum deviation of a particle. The horizontal motion equation includes both the focusing term and the bending curvature, while the vertical motion depends only on focusing.

We simplify the horizontal equation by defining a new function $K(s) = \frac{1}{R^2(s)} - k(s)$, giving:

$$x''(s) + K(s)x(s) = \frac{1}{R(s)} \cdot \frac{\Delta p}{p} \tag{4.11}$$

The total solution is a sum of a homogeneous part $x_h(s)$ and a particular inhomogeneous part $x_i(s)$:

$$x(s) = x_h(s) + x_i(s)$$

Substituting back gives:

$$\begin{aligned} x_h''(s) + K(s)x_h(s) &= 0 \\ x_i''(s) + K(s)x_i(s) &= \frac{1}{R(s)} \cdot \frac{\Delta p}{p} \end{aligned} \tag{4.12}$$

Since $x_i(s)$ depends on $\Delta p/p_0$, we define the dispersion function:

$$D(s) = \frac{x_i(s)}{\Delta p/p_0}$$

The full expression for position and angle becomes:

$$\begin{aligned} x(s) &= C(s)x_0 + S(s)x_0' + D(s) \cdot \frac{\Delta p}{p_0} \\ x'(s) &= C'(s)x_0 + S'(s)x_0' + D'(s) \cdot \frac{\Delta p}{p_0} \end{aligned} \tag{4.13}$$

Here, $C(s)$ and $S(s)$ are solutions to the homogeneous equation and are known as the cosine-like and sine-like transfer functions:

$$\begin{aligned} C''(s) + K(s)C(s) &= 0 \\ S''(s) + K(s)S(s) &= 0 \end{aligned} \tag{4.14}$$

Their Wronskian, $W = CS' - SC'$, is constant. For normalized functions, $W = 1$.

The dispersion $D(s)$ is given by:

$$D(s) = S(s) \int_0^s \frac{1}{R(t)} C(t)\, dt - C(s) \int_0^s \frac{1}{R(t)} S(t)\, dt \tag{4.15}$$

Differentiating, we get:

$$D'(s) = S'(s) \int_0^s \frac{1}{R(t)} C(t)\, dt - C'(s) \int_0^s \frac{1}{R(t)} S(t)\, dt \tag{4.16}$$

and

$$D''(s) = -K(s)D(s) + \frac{1}{R(s)} \tag{4.17}$$

This confirms:

$$D''(s) + K(s)D(s) = \frac{1}{R(s)} \tag{4.18}$$

When K is constant, $C(s)$ and $S(s)$ take the form:
For $K > 0$:

$$\psi = s\sqrt{K}$$

$$\begin{pmatrix} C & S \\ C' & S' \end{pmatrix} = \begin{pmatrix} \cos(\psi) & \frac{1}{\sqrt{K}}\sin(\psi) \\ -\sqrt{K}\sin(\psi) & \cos(\psi) \end{pmatrix}$$

For $K < 0$:

$$\psi = s\sqrt{-K}$$

$$\begin{pmatrix} C & S \\ C' & S' \end{pmatrix} = \begin{pmatrix} \cosh(\psi) & \frac{1}{\sqrt{|K|}}\sinh(\psi) \\ \sqrt{|K|}\sinh(\psi) & \cosh(\psi) \end{pmatrix}$$

The full transport matrix becomes:

$$\begin{pmatrix} x \\ x' \\ \frac{\Delta p}{p_0} \end{pmatrix}_s = \begin{pmatrix} C & S & D \\ C' & S' & D' \\ 0 & 0 & 1 \end{pmatrix} \begin{pmatrix} x \\ x' \\ \frac{\Delta p}{p_0} \end{pmatrix}_0$$

Evaluating the integrals for $K > 0$ yields:

$$\begin{aligned} D(s) &= \frac{1}{RK}(1 - \cos\psi) \\ D'(s) &= \frac{1}{RK}\sin\psi \end{aligned} \tag{4.19}$$

The motion of charged particles in a curved accelerator beamline can be understood through linear differential equations in a co-moving frame. The particle's trajectory consists of betatron oscillations caused by magnetic focusing and a dispersion component arising from momentum deviation. The use of transfer functions $C(s)$, $S(s)$, and the dispersion function $D(s)$ enables a complete description of particle motion. These results form the foundation of linear beam optics and are widely used in designing precise beam transport systems like achromats, Wien filters, and spectrometers.

4.4 Drift

Let us consider a particle beam passing through a field-free region known as the drift region in accelerator physics. The transformation of beam coordinates (x_0, x_0') to (x_1, x_1') after passing through drift (L) is governed by the following relations:

$$x_1 = x_0 + Lx_0' \tag{4.20}$$

$$x_1' = x_0' \tag{4.21}$$

For a synchronous particle of velocity (v), the time required to cross a drift (L) is given as

$$t = \frac{L}{v} = \frac{L}{\beta c} \tag{4.22}$$

For particles arriving late or early, the above equation can be written as

$$\Delta t = -\frac{L\Delta\beta}{\beta^2 c} \tag{4.23}$$

The momentum of particle is given as

$$p_0 = m_0 c\beta\gamma = \frac{m_0 c\beta}{\sqrt{1-\beta^2}}$$

$$dp_0 = m_0 c\frac{d\beta}{(1-\beta^2)^{3/2}} = m_0 c\gamma^3 d\beta$$

Thus, we get,

$$\frac{dp_0}{p_0} = \gamma^2 \frac{d\beta}{\beta} = \gamma^2 \frac{dv}{v}$$

$$dv = \frac{v}{\gamma^2}\frac{dp_0}{p_0}$$

$$z = t\delta v = t\frac{v}{\gamma^2}\frac{dp_0}{p_0} = \frac{L}{\gamma^2}\frac{dp_0}{p_0} \tag{4.24}$$

Thus, for all three phase spaces, we can write a 6×6 order transfer matrix, representing a drift region (L) that is given as follows:

$$R = \begin{pmatrix} 1 & L & 0 & 0 & 0 & 0 \\ 0 & 1 & 0 & 0 & 0 & 0 \\ 0 & 0 & 1 & L & 0 & 0 \\ 0 & 0 & 0 & 1 & 0 & 0 \\ 0 & 0 & 0 & 0 & 1 & L/\gamma^2 \\ 0 & 0 & 0 & 0 & 0 & 1 \end{pmatrix}$$

General solution of Equation 3.84 and recalling the general transformation of Twiss parameters as follow:

$$\begin{bmatrix} u_1 \\ u_1' \end{bmatrix} = \begin{bmatrix} C & S \\ C' & S' \end{bmatrix} \begin{bmatrix} u_0 \\ u_0' \end{bmatrix} \tag{4.25}$$

$$\begin{bmatrix} \beta \\ \alpha \\ \gamma \end{bmatrix} = \begin{bmatrix} C^2 & -2SC & S^2 \\ -CC' & S'C + SC' & -SS' \\ C'^2 & -2S'C' & S'^2 \end{bmatrix} \begin{bmatrix} \beta_0 \\ \alpha_0 \\ \gamma_0 \end{bmatrix}$$

The transverse beam distribution in phase space (x, x') is characterized by its second-order moments. These are collected into the beam matrix σ, defined as:

$$\sigma = \begin{pmatrix} \langle x^2 \rangle & \langle xx' \rangle \\ \langle xx' \rangle & \langle x'^2 \rangle \end{pmatrix} \tag{4.26}$$

The beam matrix expressed in terms of the Twiss parameters and the geometric emittance ϵ:

$$\sigma_0 = \epsilon \begin{pmatrix} \beta_0 & -\alpha_0 \\ -\alpha_0 & \gamma_0 \end{pmatrix} \quad \text{where} \quad \gamma_0 = \frac{1 + \alpha_0^2}{\beta_0} \tag{4.27}$$

Consider only 2×2 Matrix form of R as follows:

$$R = \begin{pmatrix} 1 & L \\ 0 & 1 \end{pmatrix} \tag{4.28}$$

The beam matrix after the drift is given by:

$$\sigma = R\sigma_0 R^T \tag{4.29}$$

Carrying out the matrix multiplication, the updated beam matrix becomes:

$$\sigma = \epsilon \begin{pmatrix} \beta_0 - 2\alpha_0 L + \gamma_0 L^2 & -\alpha_0 + \gamma_0 L \\ -\alpha_0 + \gamma_0 L & \gamma_0 \end{pmatrix} \tag{4.30}$$

From this, the Twiss parameters after the drift are:

$$\beta = \beta_0 - 2\alpha_0 L + \gamma_0 L^2 \tag{4.31}$$

$$\alpha = \alpha_0 - \gamma_0 L \tag{4.32}$$

$$\gamma = \gamma_0 \tag{4.33}$$

These relations show how the beam envelope evolves in a drift, preserving the emittance and the identity $\beta\gamma - \alpha^2 = 1$.

The transformation can also be written in matrix form:

$$\begin{pmatrix} \beta \\ \alpha \\ \gamma \end{pmatrix} = \begin{pmatrix} 1 & -2L & L^2 \\ 0 & 1 & -L \\ 0 & 0 & 1 \end{pmatrix} \begin{pmatrix} \beta_0 \\ \alpha_0 \\ \gamma_0 \end{pmatrix} \tag{4.34}$$

This provides a compact and analytic way to propagate Twiss parameters through a drift section.

4.5 Dipole Magnets

A dipole magnet is characterized by a pure homogeneous magnetic field to analyze the charge state, energy, and mass of the charged particle beams. While producing such a field, it is obvious that fringe field effects come into the picture, so a dipole magnet is studied in radial (bending) and vertical planes for transformation of charged particle beams. It is weak, focusing on the radial plane and just drift in the vertical plane without contribution from the edges. The edges of the dipole magnets produce fringe fields that can be used for focusing or defocusing of charged particles depending upon its design with respect to beam entry. Therefore, in order to control the particle trajectory in vertical plane, usually the fringe fields or edge of the dipole magnet is used to focus on vertical plane. So, the dipole magnet can be visualized in the following manner:

1. Radial plane: object distance (drift)–entry edge (defocusing)–dipole magnet (weak focusing)–exit edge (defocusing)–image distance (drift)
2. Vertical plane: object distance (drift)–entry edge (focusing)–dipole magnet (drift)–exit edge (focusing)–image distance(drift)

Now we will try to see the evolution of particle coordinates in different parts of the dipole magnet in both radial and vertical planes.

4.5.1 Radial and Vertical Motion Inside Dipole Magnets

In many beam transport systems, sector dipole magnets are used to bend particle beams along circular arcs. These magnets not only guide the beam through the desired path but also introduce focusing effects that depend on the magnetic field gradient. This gradient is quantified using a dimensionless field index n. The purpose of this discussion is to derive the complete 6×6 first-order transfer matrix for such a dipole, including both geometric and chromatic contributions.

We consider a reference particle that follows a circular path with bending radius r_0. The bending angle, which represents the arc subtended by the particle path in the magnet, is defined as

$$\theta = \frac{z}{r_0} \tag{4.35}$$

where z is the arc length traveled through the magnet. The field index n defines the radial dependence of the magnetic field:

$$n = -r_0 \cdot \frac{1}{B} \cdot \frac{dB}{dr} \tag{4.36}$$

This field index modifies the focusing strength in both the radial and vertical directions:

$$\alpha = \sqrt{1-n}, \quad \beta = \sqrt{n} \tag{4.37}$$

We describe the state of a particle using the phase space vector $(r, r', y, y', z, \delta)$, where r and r' are the radial position and angle, y and y' are the vertical position and angle, z is the longitudinal displacement from the reference particle, and $\delta = \frac{\Delta p}{p}$ is the relative momentum deviation.

Let us begin with the radial motion. The equation of motion for a particle in the radial plane, considering both curvature and weak focusing, is given by

$$\frac{d^2r}{ds^2} + \frac{1-n}{r_0^2} r = \frac{\delta}{r_0} \tag{4.38}$$

We define $\alpha = \sqrt{1-n}$, so the equation simplifies to

$$\frac{d^2r}{ds^2} + \alpha^2 r = \frac{\delta}{r_0} \tag{4.39}$$

This is a second-order linear inhomogeneous differential equation. Its general solution is composed of a homogeneous part and a particular solution, and takes the form

$$r(s) = A\cos(\alpha s) + B\sin(\alpha s) + \frac{\delta}{\alpha^2 r_0} \tag{4.40}$$

Taking the derivative with respect to s, we obtain

$$r'(s) = -A\alpha\sin(\alpha s) + B\alpha\cos(\alpha s) \tag{4.41}$$

To determine the constants A and B, we use the initial conditions $r(0) = r_0$ and $r'(0) = r'_0$. Substituting $s = 0$, we find

$$A = r_0 - \frac{\delta}{\alpha^2 r_0}, \quad B = \frac{r'_0}{\alpha} \tag{4.42}$$

We now evaluate the position and angle at the end of the magnet, where $s = z = r_0\theta$. Let us define $\psi = \alpha\theta$ for simplicity. Substituting, the final radial position becomes

$$r = r_0\cos(\psi) + \frac{r_0}{\alpha}\sin(\psi)r'_0 + \frac{r_0}{\alpha^2}(1-\cos(\psi))\delta \tag{4.43}$$

and the final radial angle is

$$r' = -\frac{\alpha}{r_0}\sin(\psi)r_0 + \cos(\psi)r'_0 + \frac{1}{\alpha}\sin(\psi)\delta \tag{4.44}$$

Next, we consider the vertical motion. The vertical equation of motion is

$$\frac{d^2y}{ds^2} + \frac{n}{r_0^2}y = 0 \tag{4.45}$$

With $\beta = \sqrt{n}$, this simplifies to

$$\frac{d^2y}{ds^2} + \beta^2 y = 0 \tag{4.46}$$

The general solution is

$$y(s) = y_0 \cos(\beta s) + \frac{y_0'}{\beta} \sin(\beta s) \tag{4.47}$$

and the corresponding angle is

$$y'(s) = -\beta y_0 \sin(\beta s) + y_0' \cos(\beta s) \tag{4.48}$$

Evaluating at $s = r_0\theta$, define $\phi = \beta\theta$, so the vertical position becomes

$$y = y_0 \cos(\phi) + \frac{r_0}{\beta} \sin(\phi) y_0' \tag{4.49}$$

and the vertical angle is

$$y' = -\frac{\beta}{r_0} \sin(\phi) y_0 + \cos(\phi) y_0' \tag{4.50}$$

Now we turn to the longitudinal coordinate z. The geometric contribution arises from the transverse motion, approximated by

$$\Delta z = \frac{1}{2} \int_0^{r_0\theta} r'(s)^2 ds \tag{4.51}$$

From this, we find

$$M_{51} = -\frac{1}{\alpha} \sin(\psi), \quad M_{52} = -\frac{r_0}{\alpha^2}(1 - \cos(\psi)) \tag{4.52}$$

The chromatic contribution arises from the dispersion path:

$$r_\delta(s) = \frac{\delta}{\alpha^2 r_0}(1 - \cos(\alpha s)) \tag{4.53}$$

and the associated path length is

$$\Delta z = -\int_0^{r_0\theta} \frac{r_\delta(s)}{r_0} ds \tag{4.54}$$

Changing variables with $\xi = \alpha s$, this leads to

$$M_{56} = -\frac{1}{r_0^2\alpha^2} \int_0^{r_0\theta} (1 - \cos(\alpha s)) ds = -\frac{r_0}{\alpha^3}(\psi - \sin(\psi)) \tag{4.55}$$

Putting all results together, the full 6×6 transfer matrix is

$$M = \begin{pmatrix} \cos(\psi) & \frac{r_0}{\alpha}\sin(\psi) & 0 & 0 & 0 & \frac{r_0}{\alpha^2}(1-\cos(\psi)) \\ -\frac{\alpha}{r_0}\sin(\psi) & \cos(\psi) & 0 & 0 & 0 & \frac{1}{\alpha}\sin(\psi) \\ 0 & 0 & \cos(\phi) & \frac{r_0}{\beta}\sin(\phi) & 0 & 0 \\ 0 & 0 & -\frac{\beta}{r_0}\sin(\phi) & \cos(\phi) & 0 & 0 \\ -\frac{1}{\alpha}\sin(\psi) & -\frac{r_0}{\alpha^2}(1-\cos(\psi)) & 0 & 0 & 1 & -\frac{r_0}{\alpha^3}(\psi-\sin(\psi)) \\ 0 & 0 & 0 & 0 & 0 & 1 \end{pmatrix} \tag{4.56}$$

This matrix comprehensively describes the first-order beam dynamics through a sector dipole, incorporating both dispersion and geometric effects, and is widely used in designing achromats, spectrometers, and energy analyzers in beamlines.

To obtain the transfer matrix for a pure dipole magnet, we start from the general sector dipole case and apply specific substitutions. A pure dipole magnet is characterized by having no focusing effect, which corresponds to setting the field index $n = 0$. This simplifies the dynamics significantly.

First, when $n = 0$, the horizontal focusing strength becomes

$$\alpha = \sqrt{1-n} = 1, \tag{4.57}$$

and the vertical focusing strength becomes

$$\beta = \sqrt{n} = 0. \tag{4.58}$$

This means the horizontal plane still experiences bending due to the curvature of the reference orbit, while the vertical plane behaves like a drift over the same arc length.

Let r_0 be the bending radius and θ the total bending angle of the magnet. The path length through the magnet is $s = r_0\theta$. Using these substitutions in the general matrix, we obtain the first-order 6×6 transfer matrix for a pure dipole magnet as follows:

$$M_{\text{dipole}} = \begin{pmatrix} \cos\theta & r_0\sin\theta & 0 & 0 & 0 & r_0(1-\cos\theta) \\ -\frac{1}{r_0}\sin\theta & \cos\theta & 0 & 0 & 0 & \sin\theta \\ 0 & 0 & 1 & r_0\theta & 0 & 0 \\ 0 & 0 & 0 & 1 & 0 & 0 \\ -\sin\theta & -r_0(1-\cos\theta) & 0 & 0 & 1 & -r_0(\theta-\sin\theta) \\ 0 & 0 & 0 & 0 & 0 & 1 \end{pmatrix} \tag{4.59}$$

This matrix fully describes the linear beam dynamics through a pure sector dipole, including bending in the horizontal plane, drift in the vertical plane, and the path lengthening effects due to both geometry and momentum deviation.

R16 is known as lateral dispersion and R26 as angular dispersion. One can see dispersion term R16 scales with bending radius. We have maximum dispersion for 180° bending and no dispersion for 360° bending, such as cyclotron. The resolving power for a given spot size x_0 is given as follows and usually expressed in cm/%.

$$R = -\frac{R_{16}}{x_0 R_{11}} \tag{4.60}$$

General solution of Equation 3.84 and recalling the general transformation of Twiss parameters as follows:

$$\begin{bmatrix} u_1 \\ u_1' \end{bmatrix} = \begin{bmatrix} C & S \\ C' & S' \end{bmatrix} \begin{bmatrix} u_0 \\ u_0' \end{bmatrix}$$

$$\begin{bmatrix} \beta \\ \alpha \\ \gamma \end{bmatrix} = \begin{bmatrix} C^2 & -2SC & S^2 \\ -CC' & S'C + SC' & -SS' \\ C'^2 & -2S'C' & S'^2 \end{bmatrix} \begin{bmatrix} \beta_0 \\ \alpha_0 \\ \gamma_0 \end{bmatrix}$$

The transformation of Twiss parameters by 3 × 3 order matrix (T) for bending plane of dipole magnet is applied to initial Twiss parameters, resulting in new Twiss parameters as follows:

$$\begin{bmatrix} \beta \\ \alpha \\ \gamma \end{bmatrix} = \begin{bmatrix} \cos^2\theta & -2r_0 \sin\theta\cos\theta & r_0^2 \sin^2\theta \\ -\frac{1}{r_0}\sin\theta\cos\theta & \cos^2\theta - \sin^2\theta & -r_0\sin\theta\cos\theta \\ \frac{\sin^2\theta}{r_0^2} & \frac{2}{r_0}\sin\theta\cos\theta & \cos^2\theta \end{bmatrix} \begin{bmatrix} \beta_0 \\ \alpha_0 \\ \gamma_0 \end{bmatrix}$$

Figure 4.1 shows beam optics of a 90° bending magnet optimized using COSY Infinity. The entrance and exit edge angles, input and output drifts to the dipole magnet are varied in order to get double focusing in X and Y planes. The beam is 1.8 MeV/u ion beam with 1% ΔE, 2 mm and 10 mrad as beam size and divergence, charge state: +1 with varying edge angles, input and output drifts.

4.6 Design of Dipole Magnets

Electromagnets are frequently used in beam transport systems because they provide higher magnetic fields that can be easily adjusted by varying the current in the coils. For particle beams with low momentum, air-core coils might suffice. However, for higher momentum beams, iron cores are preferred for two main reasons:

1. To reduce the reluctance of the magnetic circuit, thereby decreasing the excitation needed to achieve a specific field.
2. To facilitate the shaping of the magnetic field into the desired configuration.

Let μ_o denote the permeability of free space, g the pole gap, and λ the path length within the iron core, with $\frac{\lambda}{\mu}$ representing the reluctance of the iron, where μ is the permeability of iron. Let *NI* represent the ampere-turns for the dipole magnet.

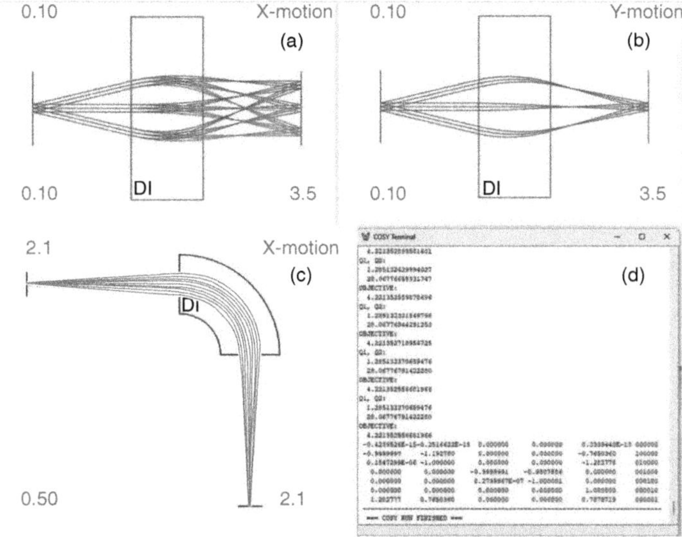

Figure 4.1 Beam optics of dipole magnet using COSY Infinity. (a) Bending X-plane with rays X,X',E:3,3,3. (b) Vertical Y-plane with rays Y,Y':3,3. and (c) Bending view in X plane of dipole with rays as X,X',E:3,3,1. (d) Transfer matrix and optimized values of drifts and edge angles as 1.285 m and 28.06°, respectively. Focusing in Y-vertical plane due to edge angles only.

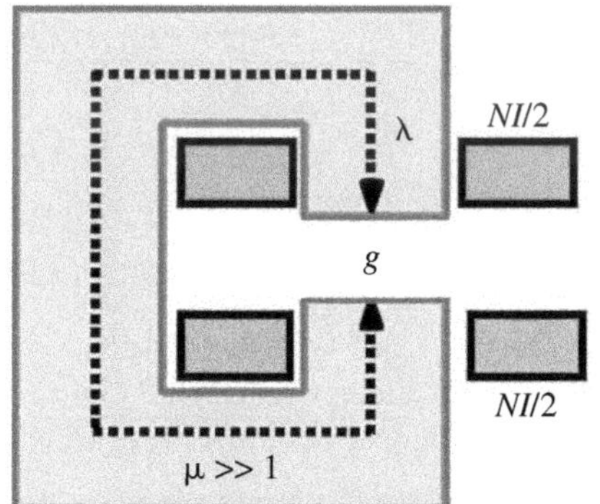

Figure 4.2 C-type magnet.

Applying Ampere's circuital law over a closed loop encompassing the entire C-magnet (as shown in Figure 4.2), we derive the following relationship:

$$B = \frac{\mu_o NI}{g + \frac{\lambda}{\mu}}$$

Given that μ is quite large, we can approximate this as:

$$NI = \frac{Bg}{\mu_o} \tag{4.61}$$

Figures 4.2 and 4.3 illustrate the C-magnet and H-magnet, Window frame magnet respectively. Each design offers unique advantages and disadvantages.

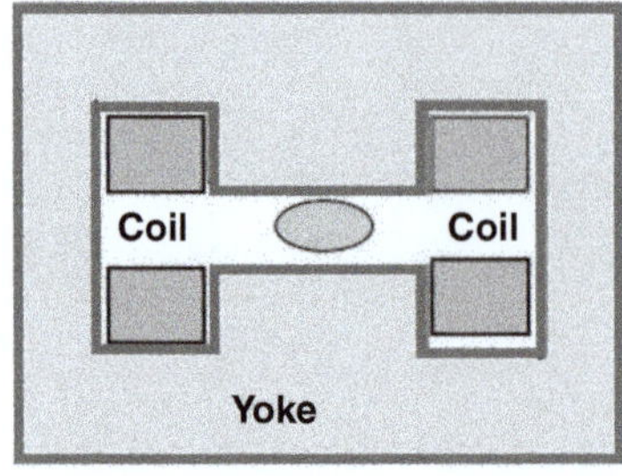

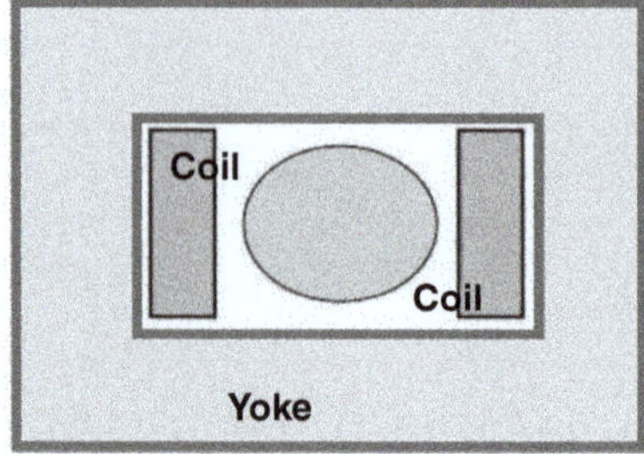

Figure 4.3 H-type magnet and window-frame magnet.

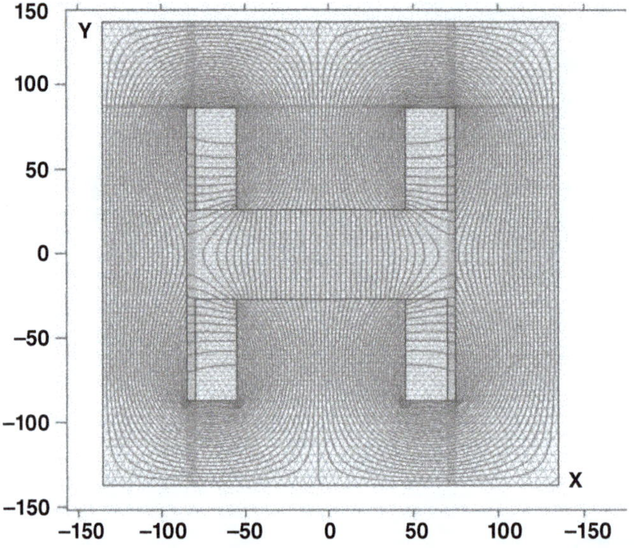

Figure 4.4 Poisson simulation [66–68] drawing of the H-type magnet (scales are in mm).

The C-type magnet provides easy access to the pole gap, but it suffers from asymmetry in the magnetic field due to flux leakage, which affects field uniformity. The window-frame or pole-less construction is highly efficient in minimizing flux leakage but has limited space for coil winding, leading to potential heat dissipation issues. Additionally, it presents challenges in accessibility. The H-type magnet, with its symmetrical design, ensures better field uniformity and is often preferred for accelerator dipole magnets. Figure 4.4 shows a Poisson simulation of the H-type magnet. The excitation of the coils and their cooling requirements depend on the power involved, which in turn is influenced by the gap field, pole gap, and available space for windings. A step-by-step approach can help in arriving at the optimal design.

4.6.1 Steps for Designing a Dipole Magnet

1. Calculate the magnet's maximum rigidity using the formula:

$$B(T)\rho(m) = 0.1437\sqrt{\frac{M(\text{amu})E(\text{MeV})}{q^2}}$$

2. Select the appropriate bending radius based on space constraints and the maximum magnetic field value, considering the available technology for iron-dominated or superconducting magnets.
3. Choose the magnet shape, typically H-type, for better field uniformity.
4. Determine the ampere-turns (NI) using the formula provided in Equation 4.61, and decide on a suitable combination of N and I (current).
5. Distribute half the total turns ($N/2$) on each pole. If A is the cross-sectional area of the coil, the current density J is defined as:

$$J = \frac{NI}{A}$$

   Distribute the number of turns in the form of layers.
6. Depending on the bending angle, draw the curved face of the poles. The ROGOWSKI pole profile [69] is often chosen, as the equation below provides the maximum rate of increase in the gap with a monotonic decrease in flux density at the surface, preventing saturation. Due to machining difficulties, this profile is sometimes approximated by two successive chamfers. If g is the pole gap, the pole profile equation can be written as:

$$y = \frac{g}{2} + \frac{g}{\pi}\exp\left(\frac{\pi x}{g} - 1\right)$$

7. Modify the pole edges to provide the edge angles for vertical focusing. The required edge angles are determined by beam optics calculations.
8. Determine the total length of the copper coil by calculating the length of one layer, multiplying it by the total number of layers, and then calculating the coil resistance using the standard formula:

$$R(\text{ohm}) = \frac{\rho_{Cu} \times L}{A}$$

   where $\rho_{Cu} = 1.68 \times 10^{-8}$ ohm-m.
9. Calculate the total power required to generate the corresponding magnetic field:

$$P(\text{W}) = V(\text{volts}) \times I(\text{Amp}) = I^2 \times R(\text{ohm})$$

10. Simulate the geometry using field computational tools and ensure the magnetic field homogeneity is maintained to 10^{-3} or better in the median plane of the dipole magnet. The field inside the core should not exceed the saturation level.

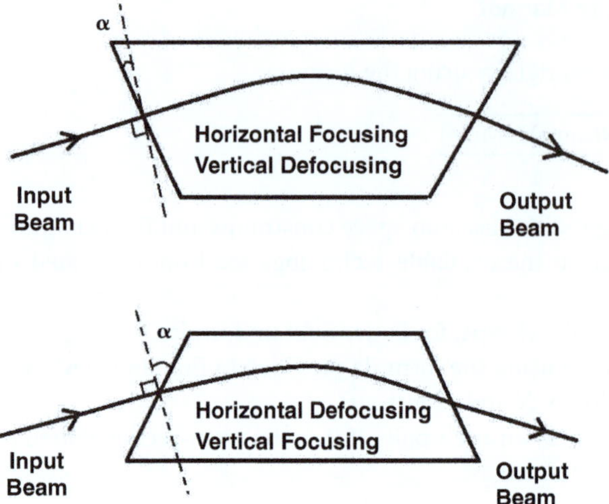

Figure 4.5 Edge focusing and defocusing in dipole magnets.

4.6.2 Edge Effects in a Dipole Magnet

While it is ideal for the beam trajectory to be perpendicular to the magnet's entrance and exit, this is often not the case, and the edge of the magnet may be tilted. The edge of a dipole magnet can either focus or defocus the beam in a given transverse plane, depending on its orientation relative to the particle direction. If a particle enters perpendicular to the magnet's edge, the entry is called normal, and the edge angle is zero. The edge may be rotated clockwise or counterclockwise from this normal entry.

A particle may experience either focusing or defocusing forces depending on whether α (the edge angle) is positive or negative.

Here, α is the edge angle, and ρ is the bending radius. If the angles are positive, the fringing field of the magnet will focus the beam in the vertical direction, as shown in Figure 4.5.

When the particle exits the edge of the dipole magnet, the radial coordinate remains unchanged, but the angular coordinate varies depending on whether the edge provides focusing or defocusing.

The change in angle of the particle at a radius r_D is given by:

$$\theta = \frac{r\tan(\alpha)}{r_D} = \frac{r\tan(\alpha)(1 - nr/r_0)}{r_0} \tag{4.62}$$

For $r \ll r_0$:

$$\theta = \frac{r\tan(\alpha)}{r_0} \tag{4.63}$$

Therefore, the equations of motion for the initial and final values (denoted by subscripts i and f) are:

$$r_f = r_i \tag{4.64}$$

$$r'_f = r'_i - r_i \frac{\tan(\alpha)}{r_0} \tag{4.65}$$

Similarly, for vertical motion, the equations are:

$$y_f = y_i \tag{4.66}$$

$$y'_f = y'_i + y_i \frac{\tan(\alpha)}{r_0} \tag{4.67}$$

Thus, the radial equations can be written as:

$$R = \begin{pmatrix} 1 & 0 & 0 & 0 & 0 & 0 \\ -\frac{1}{r_0}\tan(\alpha) & 1 & 0 & 0 & 0 & 0 \\ 0 & 0 & 1 & 0 & 0 & 0 \\ 0 & 0 & \frac{1}{r_0}\tan(\alpha) & 1 & 0 & 0 \\ 0 & 0 & 0 & 0 & 0 & 0 \\ 0 & 0 & 0 & 0 & 0 & 0 \end{pmatrix}$$

For a 90° bending magnet, the focal length is $2r_0$:

$$f = \frac{r_0}{\tan(\alpha)} = 2r_0$$

Hence, the ideal edge angle for a 90° dipole magnet to achieve edge focusing is $\alpha = 26.6°$ in drift plane.

The general solution of Equation 3.84 and the corresponding transformation of Twiss parameters can be expressed as follows:

$$\begin{bmatrix} u_1 \\ u'_1 \end{bmatrix} = \begin{bmatrix} C & S \\ C' & S' \end{bmatrix} \begin{bmatrix} u_0 \\ u'_0 \end{bmatrix}$$

The transformation of Twiss parameters by a 3×3 order matrix (T) for the entrance and exit edges of the dipole magnet, applied to the initial Twiss parameters, results in the new Twiss parameters in the transverse planes as follows:

$$\begin{bmatrix} \beta \\ \alpha \\ \gamma \end{bmatrix} = \begin{bmatrix} 1 & 0 & 0 \\ -\frac{\tan(\alpha)}{r_0} & 1 & 0 \\ \frac{\tan^2 \alpha}{r_0^2} & 2\frac{tan\, \alpha}{r_0} & 1 \end{bmatrix} \begin{bmatrix} \beta_0 \\ \alpha_0 \\ \gamma_0 \end{bmatrix}$$

$$\begin{bmatrix} \beta \\ \alpha \\ \gamma \end{bmatrix} = \begin{bmatrix} 1 & 0 & 0 \\ \dfrac{\tan(\alpha)}{r_0} & 1 & 0 \\ \dfrac{\tan^2 \alpha}{r_0^2} & -2\dfrac{tan\,\alpha}{r_0} & 1 \end{bmatrix} \begin{bmatrix} \beta_0 \\ \alpha_0 \\ \gamma_0 \end{bmatrix}$$

4.7 Quadrupole Magnets

Quadrupole magnets consist of four magnetic poles arranged in an alternating configuration as shown in Figure 4.6. They provide focusing in one plane while defocusing in the perpendicular plane relative to the beam's propagation. Depending on the orientation of the poles, they can be categorized as symmetric or skew types. In symmetric quadrupoles, the poles are aligned purely in the horizontal and vertical directions. In contrast, skew quadrupoles have poles rotated by 45° compared to the symmetric configuration, which aligns the focusing and defocusing planes with the horizontal and vertical planes, making this configuration common in most accelerators. The equations of motion for a particle (m, q, v) within the quadrupolar field in the horizontal (X-plane) and vertical (Y-plane) directions are as follows:

For a magnetic quadrupole of aperture 2a, the scalar magnetic potential has the form:

$$V_B = g_B xy$$

The components of magnetic field can be obtained as follows:

$$B_x = -\frac{\partial V_B}{\partial x} = -g_B y, \quad B_y = -\frac{\partial V_B}{\partial y} = -g_B x, \quad B_z = 0$$

Field density is maximum at the pole tip thus one can write:

$$g_B = -\frac{B_T}{a}$$

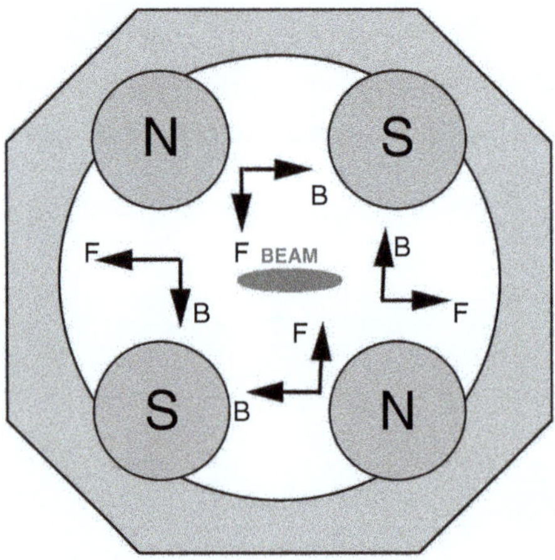

Figure 4.6 Quadrupole magnet configuration for vertical focusing and horizontal defocusing.

Equation of motion in X and Y plane can be written as follows:

$$m\ddot{x} = -qv_zB_y, \quad m\ddot{y} = qv_zB_x$$

Using $\ddot{x} = x''v_z^2$ *and* $\ddot{y} = y''v_z^2$, then one can get:

$$x'' = -k^2x, \quad y'' = k^2y$$

where $k^2 = qg_B/mv_z = qB_T/a(B\rho)$.

To solve these equations, boundary conditions are applied. At $z = 0$, the particle coordinates are (x_1, y_1), and at $z = l$, they are (x_2, y_2). The solution is obtained in matrix form as follows:

$$x = a\cos(kz) + b\sin(kz) \tag{4.68}$$

$$x' = -ak\sin(kz) + bk\cos(kz) \tag{4.69}$$

At $z = 0$, $a = x_1$, $b = x_1'/k$

Thus, the above Equations 4.68 and 4.69 can be expressed in matrix form as:

$$\begin{bmatrix} x_2 \\ x_2' \end{bmatrix} = \begin{bmatrix} \cos(kl) & \frac{1}{k}\sin(kl) \\ -k\sin(kl) & \cos(kl) \end{bmatrix} \begin{bmatrix} x_1 \\ x_1' \end{bmatrix} \tag{4.70}$$

Similarly, for vertical motion, the solution is expressed in matrix form as:

$$\begin{bmatrix} y_2 \\ y_2' \end{bmatrix} = \begin{bmatrix} \cosh(kl) & \frac{1}{k}\sinh(kl) \\ k\sinh(kl) & \cosh(kl) \end{bmatrix} \begin{bmatrix} y_1 \\ y_1' \end{bmatrix} \tag{4.71}$$

No force acts along the beam direction (z), so a magnetic quadrupole acts as a drift in that direction, and the transfer matrix is simply given as follows:

$$R_{zz} = \begin{bmatrix} 1 & L/\gamma^2 \\ 0 & 1 \end{bmatrix} \tag{4.72}$$

$$R = \begin{pmatrix} \cos(kl) & \frac{1}{k}\sin(kl) & 0 & 0 & 0 & 0 \\ -k\sin(kl) & \cos(kl) & 0 & 0 & 0 & 0 \\ 0 & 0 & \cosh(kl) & \frac{1}{k}\sinh(kl) & 0 & 0 \\ 0 & 0 & k\sinh(kl) & \cosh(kl) & 0 & 0 \\ 0 & 0 & 0 & 0 & 1 & L/\gamma^2 \\ 0 & 0 & 0 & 0 & 0 & 1 \end{pmatrix} \tag{4.73}$$

The general solution of Equation 3.84 and the transformation of Twiss parameters are given by:

$$\begin{bmatrix} u_1 \\ u_1' \end{bmatrix} = \begin{bmatrix} C & S \\ C' & S' \end{bmatrix} \begin{bmatrix} u_0 \\ u_0' \end{bmatrix}$$

$$\begin{bmatrix} \beta \\ \alpha \\ \gamma \end{bmatrix} = \begin{bmatrix} C^2 & -2SC & S^2 \\ -CC' & S'C + SC' & -SS' \\ C'^2 & -2S'C' & S'^2 \end{bmatrix} \begin{bmatrix} \beta_0 \\ \alpha_0 \\ \gamma_0 \end{bmatrix}$$

The transformation of Twiss parameters using a 3 × 3 matrix (T) for the focusing and defocusing planes of the quadrupole magnet is applied to the initial Twiss parameters, resulting in new Twiss parameters in both transverse planes as follows:

$$\begin{bmatrix} \beta \\ \alpha \\ \gamma \end{bmatrix} = \begin{bmatrix} \cos^2(kl) & -\frac{2}{k}\sin(kl)\cos(kl) & \frac{1}{k^2}\sin^2(kl) \\ k\sin(kl)\cos(kl) & \cos^2(kl) - \sin^2(kl) & -\frac{1}{k}\sin(kl)\cos(kl) \\ k^2\sin^2(kl) & k\sin(kl)\cos(kl) & \cos^2(kl) \end{bmatrix} \begin{bmatrix} \beta_0 \\ \alpha_0 \\ \gamma_0 \end{bmatrix}$$

$$\begin{bmatrix} \beta \\ \alpha \\ \gamma \end{bmatrix} = \begin{bmatrix} \cosh^2(kl) & -\frac{2}{k}\sinh(kl)\cosh(kl) & \frac{1}{k^2}\sinh^2(kl) \\ k\sinh(kl)\cosh(kl) & \cosh^2(kl) + \sinh^2(kl) & -\frac{1}{k}\sinh(kl)\cosh(kl) \\ k^2\sinh^2(kl) & -k\sinh(kl)\cosh(kl) & \cosh^2(kl) \end{bmatrix} \begin{bmatrix} \beta_0 \\ \alpha_0 \\ \gamma_0 \end{bmatrix}$$

Figure 4.7 shows beam optics of a quadrupole magnet using COSY Infinity for 1.8 MeV/u ion beam, charge state: +1. Here we vary the fields on quadrupole magnets symmetrically and finally, the focusing on both planes is achieved symmetrically.

4.7.1 Design of Quadrupole Magnets

Quadrupole magnets are essential components for focusing on charged particles in accelerators. Most quadrupole lenses are designed with pole surfaces that approximate magnetic equipotentials in the desired quadrupole field. These equipotentials are rectangular hyperbolae extending to infinity. However, poles with finite extent and deviations from the hyperbolic shape can help mitigate field distortions. Circularly approximated pole surfaces provide a reasonable uniform field and simplify the machining process, making it more cost-effective compared to other profiles. More details on this topic can be found in references [70, 71].

The equation for poles is $xy = a^2/2$. For Quadrupoles, we have from Ampere's law:

$$NI = \frac{B_y y}{\mu_0} = \frac{1}{\mu_0}\frac{gx}{a^2/2x} = \frac{ga^2}{2\mu_0}$$

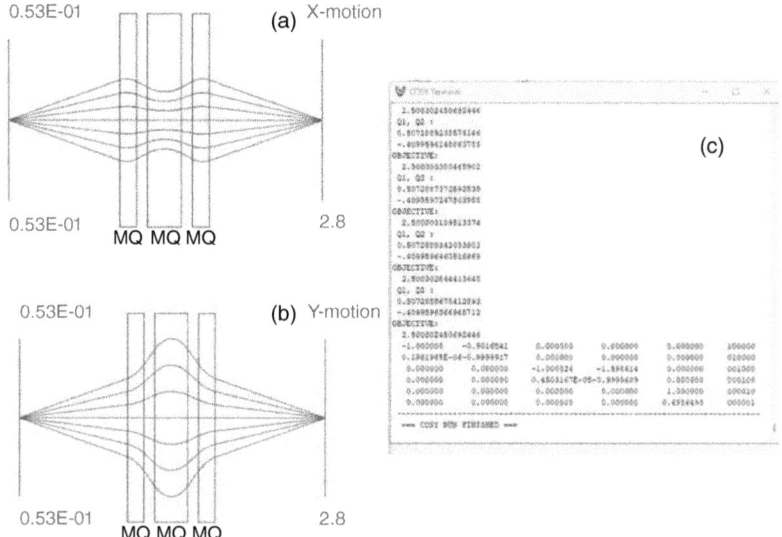

Figure 4.7 Beam optics for a quadrupole triplet magnet using COSY Infinity capable for focusing the beam on both transverse planes. (a) Beam in horizontal plane with X,X': 1,7. (b) Beam in vertical plane with Y,Y': 1,7. (c) Optimized values of quadrupole magnets are 0.5072 T and −0.4099 T, the overall transfer matrix is also shown. Detailed code is in Appendix C.

Steps to Design a Quadrupole Magnet

1. Calculate the magnetic field gradient required to focus an ion beam at a specified distance.
2. Based on the emittance of the ion beam, choose the appropriate pole gap to accommodate a 1-inch, 2-inch, or 3-inch beam pipe diameter. Typically, a 3-inch diameter beam pipe is selected for the low-energy side.
3. Calculate the required ampere-turns using the given equation and then determine a suitable combination of N and I (Amps).
4. Distribute the same number of turns NI over each pole.
5. Calculate the resistance of one coil and, consequently, the total power required.

$$R(\Omega) = \frac{\rho_{Cu} \times L}{A}, \text{ where } \rho_{Cu} = 1.68 \times 10^{-8}\,\Omega \cdot m; \;\; P(W) = VI = I^2R$$

6. Select the pole tip radius as $1.13\,a$ and design the pole to be circular up to a 90° angle. Then chamfer it at an angle to avoid field saturation. This approximation best suits a hyperbolic surface pole profile.

4.8 Beam Rotation Matrix

Let α is the rotation angle of the beam then rotation matrix for beam is as follows:

$$R = \begin{pmatrix} \cos\alpha & 0 & \sin\alpha & 0 & 0 & 0 \\ 0 & \cos\alpha & 0 & \sin\alpha & 0 & 0 \\ -\sin\alpha & 0 & \cos\alpha & 0 & 0 & 0 \\ 0 & -\sin\alpha & 0 & \cos\alpha & 0 & 0 \\ 0 & 0 & 0 & 0 & 1 & 0 \\ 0 & 0 & 0 & 0 & 0 & 1 \end{pmatrix}$$

4.9 Solenoid Magnets

The motion of a charged particle through a solenoid is commonly treated as a three-region process: an entry region, a central body region, and an exit region. Each region contributes differently to the evolution of the particle's phase-space coordinates due to the structure of the magnetic field.

In the entry and exit regions, the particle encounters fringe fields that have a radial magnetic component. This fringe field imparts angular momentum to the beam and produces coupling between the transverse coordinates. The Lorentz force equation is

$$\frac{d\vec{p}}{dt} = q\vec{v} \times \vec{B} \tag{4.74}$$

Focusing on the azimuthal component, we obtain:

$$\frac{dp_\theta}{dt} = qv_z B_r \Rightarrow \frac{dp_\theta}{dz} = qv_z\left(-\frac{r}{2}\frac{dB}{dz}\right) = -\frac{qr}{2}\frac{dB}{dz} \tag{4.75}$$

Integrating with respect to z, and assuming the change in field is through the fringe,

$$\Delta p_\theta = -\frac{qr}{2}B_z \tag{4.76}$$

This azimuthal change in momentum contributes to a transverse momentum in x:

$$\Delta p_x = -\Delta p_\theta \sin\theta = \frac{qB_z}{2}y \tag{4.77}$$

The change in x-angle is then:

$$\Delta x' = \frac{\Delta p_x}{p_z} = \frac{qB_z}{2p_z}y \tag{4.78}$$

Defining the focusing strength constant:

$$k = \frac{qB_z}{2p_z} = \frac{B_z}{2B\rho} \tag{4.79}$$

we obtain:

$$\Delta x' = ky, \quad \Delta y' = -kx \tag{4.80}$$

This implies the entry and exit matrices for the solenoid are:

$$M_{\text{entry}} = \begin{pmatrix} 1 & 0 & 0 & 0 & 0 & 0 \\ 0 & 1 & 0 & k & 0 & 0 \\ 0 & 0 & 1 & 0 & 0 & 0 \\ 0 & -k & 0 & 1 & 0 & 0 \\ 0 & 0 & 0 & 0 & 1 & 0 \\ 0 & 0 & 0 & 0 & 0 & 1 \end{pmatrix}, \quad M_{\text{exit}} = \begin{pmatrix} 1 & 0 & 0 & 0 & 0 & 0 \\ 0 & 1 & 0 & -k & 0 & 0 \\ 0 & 0 & 1 & 0 & 0 & 0 \\ 0 & k & 0 & 1 & 0 & 0 \\ 0 & 0 & 0 & 0 & 1 & 0 \\ 0 & 0 & 0 & 0 & 0 & 1 \end{pmatrix} \tag{4.81}$$

Within the body of the solenoid, the magnetic field is uniform and longitudinal. The charged particle undergoes helical motion. Its transverse velocity vector rotates at the cyclotron frequency $\omega = \frac{qB_z}{\gamma m}$. The time spent inside the solenoid is $t = \frac{L}{v_z}$, so the total rotation angle is:

$$\theta = \omega t = \frac{qB_z L}{p_z} = 2kL \tag{4.82}$$

As a result, the transverse slopes x' and y' rotate by θ:

$$x'_f = x'_i \cos\theta - y'_i \sin\theta, \quad y'_f = x'_i \sin\theta + y'_i \cos\theta \tag{4.83}$$

To calculate the position transformation, we model the motion as circular arcs in the transverse plane. The horizontal displacement is given by:

$$x_f = x_i + \rho_\theta(\cos(\phi + \theta) - \cos\phi) \tag{4.84}$$

Using trigonometric identities and geometry, we substitute:

$$\cos\phi = y'_i \cdot \frac{L}{\rho_\theta \theta}, \quad \sin\phi = -x'_i \cdot \frac{L}{\rho_\theta \theta} \tag{4.85}$$

Then:

$$x_f = x_i + x'_i \frac{L}{\theta} \sin\theta - y'_i \frac{L}{\theta}(1 - \cos\theta) \tag{4.86}$$

Similarly, for y:

$$y_f = y_i + x'_i \frac{L}{\theta}(1 - \cos\theta) + y'_i \frac{L}{\theta} \sin\theta \tag{4.87}$$

These expressions lead to the solenoid body matrix:

$$M_{\text{body}} = \begin{pmatrix} 1 & \frac{L}{\theta}\sin\theta & 0 & \frac{L}{\theta}(1-\cos\theta) \\ 0 & \cos\theta & 0 & -\sin\theta \\ 0 & \frac{L}{\theta}(1-\cos\theta) & 1 & \frac{L}{\theta}\sin\theta \\ 0 & \sin\theta & 0 & \cos\theta \end{pmatrix} \tag{4.88}$$

Now substitute $\theta = 2kL$, define $s = \sin\theta$, $c = \cos\theta$, and simplify each entry:

$$\frac{L}{\theta} = \frac{1}{2k} \tag{4.89}$$

Then the simplified matrix becomes:

$$M_{\text{body}} = \begin{pmatrix} 1 & \frac{s}{2k} & 0 & \frac{1-c}{2k} \\ 0 & c & 0 & -s \\ 0 & \frac{1-c}{2k} & 1 & \frac{s}{2k} \\ 0 & s & 0 & c \end{pmatrix} \tag{4.90}$$

The full 6×6 solenoid transfer matrix is obtained by multiplying:

$$M = M_{\text{exit}} \cdot M_{\text{body}} \cdot M_{\text{entry}} \tag{4.91}$$

This final matrix accounts for both the coupling introduced by fringe fields and the rotation from the longitudinal field. It shows that a solenoid introduces x-y coupling and preserves overall emittance, though not individually in x or y.

The full transfer matrix of a solenoid for field strength and length is thus given as the product of entry, body, and exit matrices. Using shorthand notations $s = \sin(2kL)$, $c = \cos(2kL)$, and k, the complete matrix becomes:

$$M = \begin{pmatrix} c^2 & \frac{sc}{k} & \frac{s^2}{k} & \frac{sc}{k} & 0 & 0 \\ -ksc & c^2 & -ksc & s^2 & 0 & 0 \\ -\frac{s^2}{k} & \frac{ksc}{k^2} & c^2 & \frac{sc}{k} & 0 & 0 \\ ks^2 & -sc & -ksc & c^2 & 0 & 0 \\ 0 & 0 & 0 & 0 & 1 & \frac{L}{\gamma^2} \\ 0 & 0 & 0 & 0 & 0 & 1 \end{pmatrix} \tag{4.92}$$

The determinant of the 2×2 diagonal submatrices in the x and y planes are not equal to one, which means that x- and y-emittances are not preserved individually. However, the total transverse emittance remains conserved. The off-diagonal elements such as $x' \leftrightarrow y$ and $y' \leftrightarrow x$ are nonzero, indicating transverse coupling introduced by the solenoid field.

4.9.1 Focal Length of Solenoid Magnet

Let's derive some useful identities and formulae which are helpful in particle dynamics inside solenoid of cylindrical symmetry.

In cylindrical coordinates, the position vector $\mathbf{x}$ is expressed as:

$$\mathbf{x} = r\hat{r} + z\hat{z}$$

where,

- r is the radial distance
- θ is the angular coordinate
- z is the axial coordinate
- $\hat{r}$, $\hat{\theta}$, and $\hat{z}$ are the unit vectors in the radial, angular, and axial directions, respectively.

The unit vectors in cylindrical coordinates are:

$$\hat{r} = \cos\theta\,\hat{\imath} + \sin, \quad \hat{\theta} = -\sin\theta\,\hat{\imath} + \cos\theta\,\hat{\jmath}$$

Taking the time derivative:

$$\frac{d\hat{r}}{dt} = \dot{\theta}(-\sin\theta\,\hat{\imath} + \cos\theta\,\hat{\jmath}) = \dot{\theta}\hat{\theta}$$

$$\frac{d\hat{\theta}}{dt} = \dot{\theta}(-\cos\theta\,\hat{\imath} - \sin\theta\,\hat{\jmath}) = -\dot{\theta}\hat{r}$$

The velocity vector is found by differentiating the position vector $\mathbf{x}$ with respect to time:

$$\dot{\mathbf{x}} = \frac{d}{dt}(r\hat{r} + z\hat{z})$$

Applying the product rule:

$$\dot{\mathbf{x}} = \dot{r}\hat{r} + r\frac{d\hat{r}}{dt} + \dot{z}\hat{z}$$

Substituting $\frac{d\hat{r}}{dt} = \dot{\theta}\hat{\theta}$:

$$\dot{\mathbf{x}} = \dot{r}\hat{r} + r\dot{\theta}\hat{\theta} + \dot{z}\hat{z}$$

Thus, the velocity vector in cylindrical coordinates is:

$$\dot{\mathbf{x}} = \dot{r}\hat{r} + r\dot{\theta}\hat{\theta} + \dot{z}\hat{z}$$

Next, we find the acceleration vector by differentiating the velocity vector $\dot{\mathbf{x}}$ with respect to time:

$$\ddot{\mathbf{x}} = \frac{d}{dt}(\dot{r}\hat{r} + r\dot{\theta}\hat{\theta} + \dot{z}\hat{z})$$

Applying the product rule:

$$\ddot{\mathbf{x}} = \ddot{r}\hat{r} + \dot{r}\frac{d\hat{r}}{dt} + r\frac{d(\dot{\theta}\hat{\theta})}{dt} + \ddot{z}\hat{z}$$

We already know:

$$\frac{d\hat{r}}{dt} = \dot{\theta}\hat{\theta}$$

And:

$$\frac{d(\dot{\theta}\hat{\theta})}{dt} = \ddot{\theta}\hat{\theta} + \dot{\theta}\frac{d\hat{\theta}}{dt} = \ddot{\theta}\hat{\theta} - \dot{\theta}^2\hat{r}$$

Substituting these into the expression for $\ddot{\mathbf{x}}$:

$$\ddot{\mathbf{x}} = \ddot{r}\hat{r} + \dot{r}\dot{\theta}\hat{\theta} + r(\ddot{\theta}\hat{\theta} - \dot{\theta}^2\hat{r}) + \ddot{z}\hat{z}$$

Collecting the terms involving $\hat{r}$ and $\hat{\theta}$:
So, the full expression for the acceleration vector in cylindrical coordinates is:

$$\ddot{\mathbf{x}} = (\ddot{r} - r\dot{\theta}^2)\hat{r} + (r\ddot{\theta} + 2\dot{r}\dot{\theta})\hat{\theta} + \ddot{z}\hat{z} \tag{4.93}$$

Equation of motion of a charged particle in solenoidal field:

$$\gamma m\ddot{x} = q(\dot{\vec{x}} \times \overrightarrow{B}) \tag{4.94}$$

Using the following algebra within linear beam optics:

$$\vec{B} = B_r\hat{r} + B_z\hat{z}, \quad \vec{B} = -\frac{1}{2}B_0'(z)\hat{r} + B_{0z}\hat{z}$$

which reduces to

$$m\gamma(\ddot{r} - r\dot{\theta}^2) = q(r\dot{\theta})B_0 \tag{4.95}$$

$$\ddot{r} - r\dot{\theta}^2 = \frac{q(r\dot{\theta})B_0}{\gamma m} \tag{4.96}$$

Here

$$v_\theta = -\frac{qB_0}{2m\gamma}r = r\dot{\theta}$$

$$\ddot{r} - \left(\frac{qB_0}{2m\gamma}\right)^2 r = -2\left(\frac{qB_0}{2m\gamma}\right)^2 r \tag{4.97}$$

$$\ddot{r} + \left(\frac{qB_0}{2m\gamma}\right)^2 r = 0 \tag{4.98}$$

$$v^2\frac{d^2r}{dz^2}+\left(\frac{qB_0}{2m\gamma}\right)^2 r=0 \tag{4.99}$$

$$\frac{d^2r}{dz^2}+\left(\frac{B_0(z)}{2B\rho}\right)^2 r=0 \tag{4.100}$$

It has the form of Hill's equation, so the transfer matrix is given as follows:

$$\begin{pmatrix} r_2 \\ r_2' \end{pmatrix}=\begin{pmatrix} 1 & 0 \\ -\frac{1}{f} & 1 \end{pmatrix}\begin{pmatrix} r_1 \\ r_1' \end{pmatrix}$$

If l is length of the solenoid magnet, then the focal length of solenoid magnet is given as follows:

$$\frac{1}{f}=\left(\frac{B_0(0)}{2B\rho}\right)^2 l=\left(\frac{qB_0(0)}{2mc\beta\gamma}\right)^2 \tag{4.101}$$

General solution of Equation 3.84 and recalling the general transformation of Twiss parameters as follows:

$$\begin{bmatrix} u_1 \\ u_1' \end{bmatrix}=\begin{bmatrix} C & S \\ C' & S' \end{bmatrix}\begin{bmatrix} u_0 \\ u_0' \end{bmatrix}$$

$$\begin{bmatrix} \beta \\ \alpha \\ \gamma \end{bmatrix}=\begin{bmatrix} C^2 & -2SC & S^2 \\ -CC' & S'C+SC' & -SS' \\ C'^2 & -2S'C' & S'^2 \end{bmatrix}\begin{bmatrix} \beta_0 \\ \alpha_0 \\ \gamma_0 \end{bmatrix}$$

The transformation of Twiss parameters by 3×3 order matrix (T) for the focusing and defocusing planes of quadrupole magnet is applied to the initial Twiss parameters, resulting in new Twiss parameters, respectively, as follows: Here $\theta=-kl$, where $k=B_0/B\rho$

$$\begin{bmatrix} \beta \\ \alpha \\ \gamma \end{bmatrix}=\begin{bmatrix} cos^2(\theta) & -\frac{2}{k}sin(\theta)cos(\theta) & \frac{1}{k^2}sin^2(\theta) \\ ksin(\theta)cos(\theta) & cos^2(\theta)-sin^2(\theta) & -\frac{1}{k}sin(\theta)cos(\theta) \\ k^2sin^2(\theta) & ksin(\theta)cos(\theta) & cos^2(\theta) \end{bmatrix}\begin{bmatrix} \beta_0 \\ \alpha_0 \\ \gamma_0 \end{bmatrix}$$

Here the function invokes the solenoid with the following tanh-based on-axis field model, which has been plotted in Figure 4.8:

$$B(z)=\frac{B_0}{2\cdot\tanh\left(\frac{l}{2d}\right)}\cdot\left[\tanh\left(\frac{z}{d}\right)-\tanh\left(\frac{z-l}{d}\right)\right] \tag{4.102}$$

The corresponding beam optics using COSY Infinity is shown in Figure 4.9.

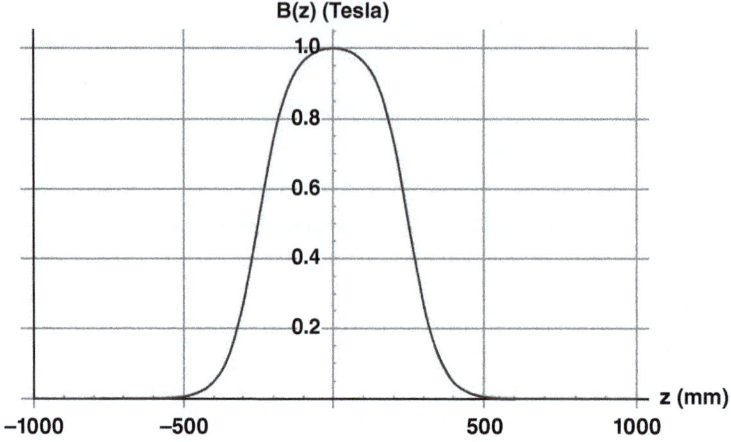

Figure 4.8 Longitudinal Field plot for Solenoid magnet as per above equation 4.101.

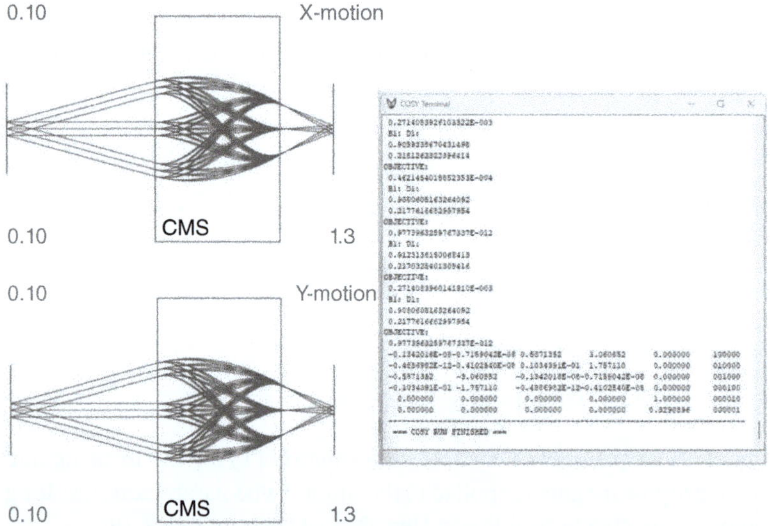

Figure 4.9 Beam optics for solenoid magnet for 1 MeV proton beam.

4.9.2 Design of Solenoid Magnets

Solenoid magnets are capable of focusing on both the X and Y planes, offering a simpler alternative to quadrupoles without the need for the precise alignment required in a FODO array. A solenoid magnet is created by winding copper coils symmetrically around a cylindrical core. To improve field homogeneity and reduce external field leakage, a return iron yoke can be employed, as illustrated in Figure 4.10. This yoke not only enhances field uniformity but also minimizes the magnetic field outside the solenoid. The magnetic field is typically simulated along the solenoid's length, as depicted in Figure 4.11, which aids in determining the effective length of the magnet and assessing field homogeneity along the longitudinal axis.

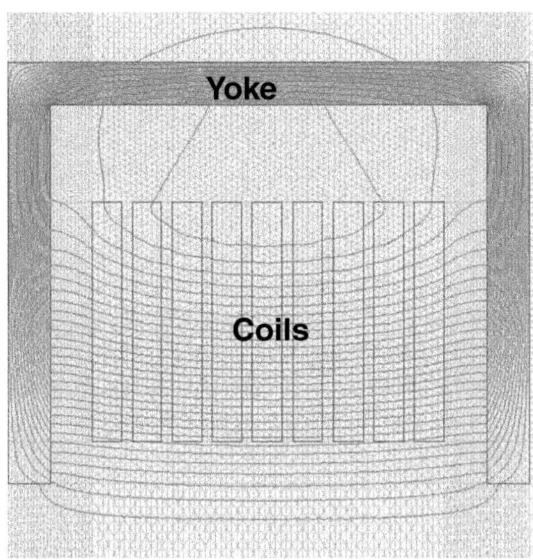

Figure 4.10 2D field configuration of solenoid magnet using Poisson/superfish code [67, 68] from LANL.

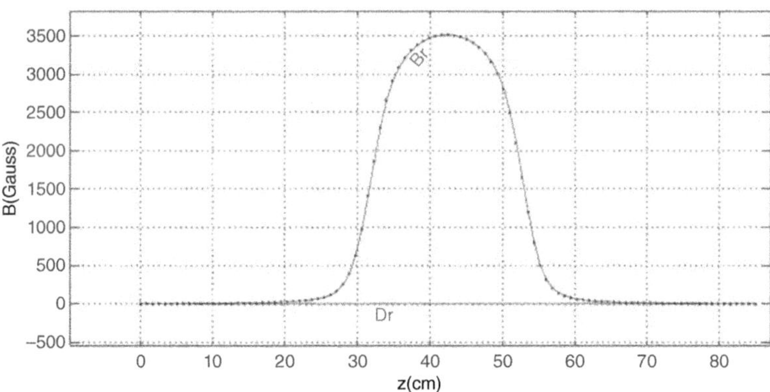

Figure 4.11 Longitudinal field mapping of solenoid magnet.

Steps to Design a Solenoid Magnet

1. Determine the number of ampere-turns required to produce the desired magnetic field within a specified aperture:

$$B = \mu_o nI$$

where n is the number of turns per unit length.

2. Calculate the focal length using the following equation:

$$\frac{1}{f} = \left(\frac{q}{2mc\beta\gamma}\right)^2 \int_{-\infty}^{+\infty} B_z^2 dz \tag{4.103}$$

3. Ensure that the length of the solenoid magnet is greater than its aperture to maintain field homogeneity within a tolerance of better than 10^{-3}.

4.10 Sextupole Magnets

The Sextupole Magnet consists of six poles with alternating north and south poles configuration, and its magnetic field intensity changes in a quadratic way as the distance from the center of the magnet increases. An F-type sextupole directs the particle beam towards the center of the ring. Its primary function is to adjust the beam at the outer regions, similarly to how an optical lens corrects chromatic aberrations. They are used in particle accelerators for the control of chromatic aberrations and for damping the head tail instability. A doublet of sextupole magnets is used in transmission electron microscopes to correct for spherical aberration. They are most widely used within the circular storage ring to correct chromatic aberrations of the electron beam. Usually, it is required to correct for the chromatic effect of a quadrupole magnet. A quadrupole system focuses a beam of charged particles at different longitudinal positions depending upon the energy spread of the beam. Thus, the beam bunch would get spread out or defocused in the longitudinal direction. To correct this chromatic effect, a sextupole magnet which has a larger focusing effect for particles that are displaced further from the central axis. This happens with the sextupole because it produces a magnetic field that varies as square of position coordinates of the beam particles as compared to linear variation of focusing power of a quadrupole system. In other words, If a sextupole is placed at a point of non-zero dispersion in the beam, then the sextupole can be used in such a way that particles within a particular energy spread window, are focused to the same point. This will counterbalance the nature of the quadrupole magnets to introduce dispersion in the beam.

We start by considering the motion of a charged particle in a sextupole magnetic field. The magnetic field components in the transverse directions are given by:

$$B_x = gxy, \quad B_y = g\left(\frac{x^2}{2} - \frac{y^2}{2}\right) \tag{4.104}$$

where g is the sextupole field gradient, and x and y are the transverse coordinates.

The equations of motion for a charged particle in these fields are obtained from the Lorentz force law:

$$m\ddot{x} = qv_zB_y, \quad m\ddot{y} = -qv_zB_x \tag{4.105}$$

Substituting the expressions for B_x and B_y, we get the equations of motion:

$$m\ddot{x} = qv_zg\left(\frac{x^2}{2} - \frac{y^2}{2}\right) \tag{4.106}$$

$$m\ddot{y} = -qv_zgxy \tag{4.107}$$

We convert the time derivatives into spatial derivatives by using the relationship:

$$\frac{d}{dt} = v\frac{d}{dz} \tag{4.108}$$

where s is the longitudinal coordinate (distance traveled by the particle) and v is the particle's velocity along the longitudinal direction (approximately v_z). Thus, the second-order time derivatives become second-order space derivatives:

$$\frac{d^2x}{dt^2} = v^2\frac{d^2x}{dz^2}, \quad \frac{d^2y}{dt^2} = v^2\frac{d^2y}{dz^2} \tag{4.109}$$

Substituting this into the equations of motion gives:

$$mv^2\frac{d^2x}{dz^2} = qv_zg\left(\frac{x^2}{2} - \frac{y^2}{2}\right) \tag{4.110}$$

$$mv^2\frac{d^2y}{dz^2} = -qv_zgxy \tag{4.111}$$

Assuming $v \approx v_z$, we simplify the equations:

$$\frac{d^2x}{dz^2} = \frac{qg}{mv^2}\left(\frac{x^2}{2} - \frac{y^2}{2}\right) \tag{4.112}$$

$$\frac{d^2y}{dz^2} = -\frac{qg}{mv^2}xy \tag{4.113}$$

These equations show that the particle's acceleration in the transverse directions depends on the squares or products of the transverse coordinates x and y. This makes the motion nonlinear. The force the particle feels becomes stronger the farther it is from the center.

Now suppose the particle is close to the axis, so that both x and y are small. In that case, the squared and product terms are even smaller. So, near the axis, the forces are very weak:

$$\frac{d^2x}{dz^2} \approx 0, \quad \frac{d^2y}{dz^2} \approx 0 \tag{4.114}$$

This means a particle near the center of the sextupole travels in a nearly straight line, unaffected by the field.

To understand what happens when the particle is farther from the center, we use the thin-lens approximation. We assume the sextupole is short enough that the particle's position doesn't change much while inside, but the direction of motion (its angle) gets a small, sudden change.

We define the integrated sextupole strength as:

$$S = \frac{qgL}{mv^2} = \frac{gL}{B\rho} \tag{4.115}$$

where L is the sextupole length and $B\rho$ is the magnetic rigidity.

Under the thin-lens model, the particle's angles change like this:

$$x'_{\text{out}} = x'_{\text{in}} + \frac{S}{2}\left(x^2 - y^2\right) \tag{4.116}$$

$$y'_{\text{out}} = y'_{\text{in}} - Sxy \tag{4.117}$$

Finally, the sextupole magnet introduces a nonlinear focusing force whose strength grows quadratically with the transverse displacement from the axis. This property enables sextupoles to correct the chromatic aberrations that arise due to energy-dependent focusing in quadrupole systems, ensuring that particles with slightly different momenta converge more accurately.

4.11 Magnetostatic Steerer/Deflector

The steerer magnet constitutes a set of four coils wounded over the four MS plates to constitute a field in the aperture region in such a way that the beam can be steered in both transverse directions independently and uniformly. It is based on the principal of solenoid magnet, where the outer lines of force do the beam steering. A magnetostatic steerer has an advantage over an electrostatic steerer as being compact. The deflection of charged particle in a plane perpendicular to magnetic field is shown in Figure 4.12.

The equation of motion of a charged particle in magnetic field is given as follows:

$$\frac{mv^2}{r} = qvB \tag{4.118}$$

$$r = \frac{mv}{qB} \tag{4.119}$$

For small angle deflection, one may write

$$\theta = \frac{L}{r} \tag{4.120}$$

Here L is the length of the steerer magnet.

Using the above equation, one may get easily:

$$\theta = \frac{LqB}{mv} \tag{4.121}$$

$$\theta(mrad) = 6950 * B(T) * L(m)\left[\frac{M(amu)E(MeV)}{q^2}\right]^{-1/2} \tag{4.122}$$

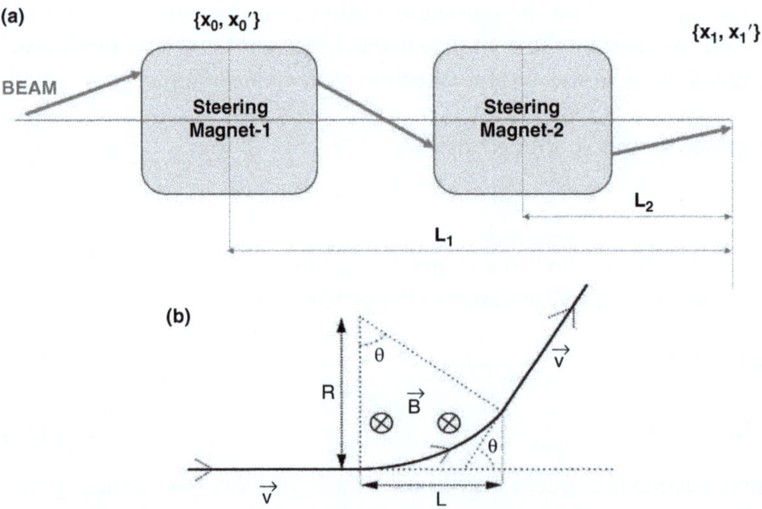

Figure 4.12 Deflection of beam by two steerer magnets and deflection of charged particle by transverse magnetic field.

A steering magnet is used to correct the trajectory of a charged particle beam by introducing small angular deflections. In beam transport systems, minor deviations from the desired path can occur due to misalignments, external disturbances, or imperfections in magnetic fields. Steering magnets provide controlled deflections to bring the beam back to its intended path.

In a general sense, let x be the transverse position of the beam and x' be the transverse angle. A steering magnet introduces a deflection θ, which affects x' directly while causing a change in x due to drift. The relationship between these parameters follows:

$$x_2 = x_1 + x_1' L,$$

where L is the drift length. When a steering magnet applies a deflection θ, the angle changes as:

$$x_2' = x_1' + \theta.$$

Thus, the position at the observation point becomes:

$$x_2 = x_1 + (x_1' + \theta)L.$$

This general form shows how a beam is influenced by a single steering magnet.

Since a single steering magnet can only correct either position or angle but not both simultaneously, two steering magnets are required. By strategically placing two steering magnets in sequence, both the position and angle deviations can be corrected, ensuring that the beam follows the desired trajectory.

The steering magnet equation [72] describe how small angular kicks correct the beam trajectory. The beam's position and angle evolve naturally due to drift, and the steering magnets introduce additional deflections.

Ignoring the steering magnets initially, the beam centroid at a distance L_1 from the first magnet follows the transport equation:

$$x_{1,0} = x_0 + x_0' L_1,$$

where x_0 and x_0' are the initial position and angle deviations. Since no other elements act, the beam angle remains unchanged:

$$x_{1,0}' = x_0'.$$

These describe the natural drift of the beam to the observation point.

Now, considering the steering magnets, the first magnet at L_1, introduces an angular deflection θ_1. This changes the angle immediately:

$$\Delta x_{1,1}' = \theta_1$$

Since the beam drifts after the deflection, the position contribution at the observation point is:

$$\Delta x_{1,1} = L_1 \theta_1.$$

The second steering magnet at L_2, similarly introduces an additional angular deflection θ_2:

$$\Delta x_{1,2}' = \theta_2,$$

affecting the beam's final position by:

$$\Delta x_{1,2} = L_2\theta_2.$$

Superimposing all contributions, the total position and angle at the end of the section become:

$$x_1 = x_0 + x_0'L_1 + L_1\theta_1 + L_2\theta_2, \quad x_1' = x_0' + \theta_1 + \theta_2.$$

Writing these equations in matrix form:

$$\begin{bmatrix} x_1 \\ x_1' \end{bmatrix} = \begin{bmatrix} L_1 & L_2 \\ 1 & 1 \end{bmatrix} \begin{bmatrix} \theta_1 \\ \theta_2 \end{bmatrix} + \begin{bmatrix} x_0 + x_0'L_1 \\ x_0' \end{bmatrix}.$$

To fully correct the beam such that $x_1 = 0$ and $x_1' = 0$, we solve:

$$\begin{bmatrix} L_1 & L_2 \\ 1 & 1 \end{bmatrix} \begin{bmatrix} \theta_1 \\ \theta_2 \end{bmatrix} = -\begin{bmatrix} x_0 + x_0'L_1 \\ x_0' \end{bmatrix}.$$

This determines the required steering magnet strengths to bring the beam to the desired trajectory.

4.12 Wien Filter

The Wien filter is a compact design with combination of electric field and magnetic field to select a particular velocity of a charged particle. Thus, it is also known as a velocity selector. If the ions in the beam have the same energy (qV) but due to different masses, they will have different velocities. It is used as a charge state analyzer, energy analyzer as well as a mass analyzer. The schematics of Wien Filter is shown in Figure 4.13. All particles of a particular velocity will pass the Wien filter irrespective of their mass and the charge. All the neutral particles will pass the Wien filter irrespective of their velocity. Let V_w is the potential applied across the plates separated by distance (d), then if E is the resulting electric field, a magnetic field (B) is also applied in a perpendicular direction and the particle is traversing perpendicular to both fields. Then the force due to electric and magnetic fields gets balanced out for a particle experiencing no force.

$$qvB = qE \tag{4.123}$$

Here $E = V_w/d$, the condition under which the Wien filter selects the velocity (v) of charged particles of mass m and charge state q is given as follows:

$$v = \frac{E}{B} = \frac{V_w}{dB} \tag{4.124}$$

In other words, the overall Lorentz force for the charged particle traveling with a specific velocity perpendicular to both fields vanishes. Neither the mass of the particle nor the charge of the particle are important for this velocity filter. All particles of particular velocity will pass the Wien filter irrespective of their mass and charge. All the neutral particles

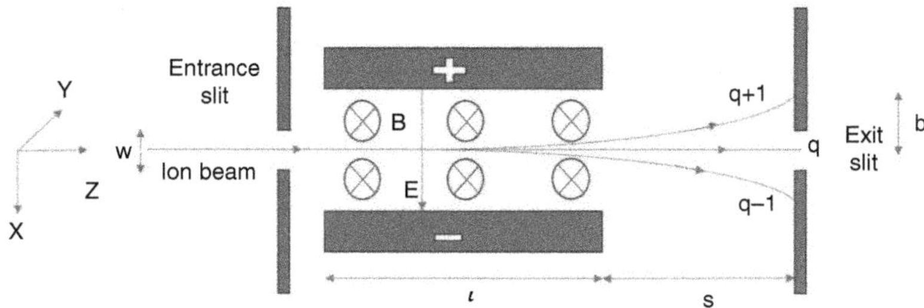

Figure 4.13 Schematics of Wien filter.

will pass the Wien filter irrespective of their velocity. In the low-energy beam accelerator, all ions are accelerated to the same energy; therefore, Wien filter can select ions based on their masses. The schematics of Wien Filter is shown in Figure 4.13 where a beam is passing through it and gets deflected as per deviation from the matching velocity criteria defined by Wien Filter.

Now the kinetic energy of particles (T) is given as:

$$T = \frac{1}{2}mv^2 \tag{4.125}$$

Using the above equations:

$$V_w = db\sqrt{\frac{2T}{m}} \tag{4.126}$$

Consider the electric field in X-direction as $\vec{E} = Ee_x$ and magnetic field in Y-direction $\vec{B} = Be_y$. To solve the equation for trajectories of charged particles within an ideal Wien filter consisting of uniform electric and magnetic fields, the basic equation needs to be solved as follows:

$$m\ddot{r} = q(\vec{E} + \vec{v} \times \vec{B}) \tag{4.127}$$

It can be expressed in terms of the following three equations:

$$m\ddot{x} = q(E - B\dot{z}) \tag{4.128}$$

$$m\ddot{y} = 0 \tag{4.129}$$

$$m\ddot{z} = qB\dot{x} \tag{4.130}$$

Consider the initial conditions as $v_{x0} = v_{y0} = 0$ and $v_{z0} = v_0$ under which the above equation can be solved to an explicit solution as follows:

$$x(t) = \omega\left(v_0 - \frac{E}{B}\right)(\cos(\omega t) - 1) \tag{4.131}$$

$$\dot{x}(t) = \left(\frac{E}{B} - v_0\right)\sin(\omega t) \tag{4.132}$$

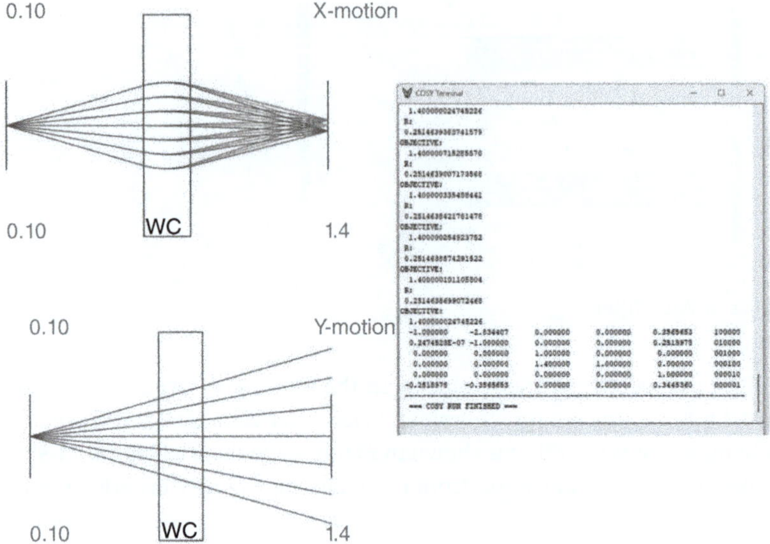

Figure 4.14 Beam optics for Wien filter for 1 MeV proton beam with energy spread 1%. It is drift in Y-plane.

$$z(t) = \omega\left(v_0 - \frac{E}{B}\right)\sin(\omega t) + \frac{E}{B}t \tag{4.133}$$

$$\dot{z}(t) = \left(v_0 - \frac{E}{B}\right)\cos(\omega t) + \frac{E}{B} \tag{4.134}$$

Here $\omega = qB/m$ is cyclotron frequency of the particle.

Also by Schmidt et al. [73], the charge state and mass resolutions are given as follows:

$$\frac{q}{\Delta q} = \frac{r_z}{2b}\left[\cos\frac{l}{r_z} - \frac{s}{r_z}\sin\frac{l}{r_z} - 1\right] \tag{4.135}$$

$$\frac{m}{\Delta m} = \frac{r_z}{2b}\left[1 - \cos\frac{l}{r_z} + \frac{s}{r_z}\sin\frac{l}{r_z}\right] \tag{4.136}$$

Here r_z is the cyclotron radius and is given as follows:

$$r_z = \frac{mv}{qB} = \frac{\sqrt{2qVm}}{qB} \tag{4.137}$$

Here V is the ion beam potential. The beam optics of Wien filter for proton beam is shown in Figure 4.14. The Wien filter is able to resolve δE of 1%.

4.13 Achromatic Magnets

An achromatic system is one in which the final transverse position of a particle is independent of its energy deviation. Such a system is particularly useful in transporting charged particle beams with momentum spread while maintaining beam focus and image quality.

4.13.1 Two Dipoles Bending in Same Directions

In this configuration, we consider two identical dipole magnets bending the beam in the same direction, separated symmetrically by a central quadrupole section and equal drifts on both sides. The dipoles introduce energy-dependent path length differences, while the quadrupole compensates for the resulting dispersion.

The total beamline transport is represented as

$$M = D \cdot L \cdot Q \cdot L \cdot D \tag{4.138}$$

with

$$D = \begin{pmatrix} 1 & \rho \tan(\theta/2) \\ 0 & 1 \end{pmatrix}, \quad L = \begin{pmatrix} 1 & d \\ 0 & 1 \end{pmatrix}, \quad Q = \begin{pmatrix} \cos\phi & \frac{1}{\sqrt{k}}\sin\phi \\ -\sqrt{k}\sin\phi & \cos\phi \end{pmatrix} \tag{4.139}$$

where θ is the total bend angle of each dipole, ρ is the bending radius, d is the drift length between dipole and quadrupole, k is the quadrupole strength, and ϕ is the total phase advance through the quadrupole section.

After multiplying all matrices, the top-right element M_{12} of the full transfer matrix is considered as 0 for point to point focusing condition. One can get:

$$\frac{1}{\sqrt{k}} \cot\left(\frac{\phi}{2}\right) = \rho \tan\left(\frac{\theta}{2}\right) + d \tag{4.140}$$

This relation defines how the quadrupole strength k, phase advance ϕ, and geometry ρ, θ, d must be related to cancel chromatic dispersion to first order in a symmetric dipole–quadrupole–dipole achromat.

4.13.2 Two Dipoles Bending in Opposite Directions

In this case, the beamline consists of two identical dipole magnets bending the beam in opposite directions by angle θ. Between these dipoles are two identical quadrupoles placed symmetrically. Each dipole is followed by a drift of length λ leading to a quadrupole, and the two quadrupoles are separated by a drift of length d. The goal is to derive the condition under which this system is achromatic, i.e. the final transverse position of a particle is independent of its energy deviation.

The full transfer matrix of the system is:

$$M = D(-\theta) \cdot L(\lambda) \cdot Q \cdot L(d) \cdot Q \cdot L(\lambda) \cdot D(\theta) \tag{4.141}$$

After multiplying all matrices, the top-right element M_{12} of the full transfer matrix is considered as 0 for point to point focusing condition. One can get:

$$\rho \tan\theta + \lambda = \frac{1}{\sqrt{k}} \cdot \frac{d\sqrt{k}\cos\phi + 2\sin\phi}{d\sqrt{k}\sin\phi - 2\cos\phi} \tag{4.142}$$

This is the achromatic condition for a system with two opposite-bending dipoles and a symmetric quadrupole arrangement. It ensures that dispersion introduced by the first dipole is fully canceled at the exit.

4.14 Septum and Kicker Magnets

Septa and kicker magnets are often utilized together for beam injection and extraction at various stages in an accelerator, respectively. To achieve high energies for charged particles in large particle accelerator complexes, multiple accelerator stages, such as circular accelerators, are cascaded, as each has a limited dynamic range. Particle beams are transferred between accelerator stages via a beam transfer line, necessitating the extraction of particles from one accelerator and their injection into the next. In circular accelerators, particles are accelerated by passing through resonant cavities. This acceleration is not continuous; the cavities operate at a specific resonant frequency, causing the particle beam to be divided into bunches. Proper timing for injecting a beam bunch into a circular accelerator or accumulator ring, as well as for extracting the bunches of particles, is crucial for the optimal performance of the accelerator. Failure in either injection or extraction can result in significant damage to the machine. Injection or extraction completed in one revolution is termed single-turn, whereas if it spans multiple revolutions, it is called multi-turn. At the injection stage for hadrons, beam density can be limited by space charge effects or the injector's capacity. If it is not possible to further increase charge density, then via multi-turn process, one can fill the horizontal phase space to increase the overall injected intensity.

Here are the main highlights of their collective operation as shown in Figure 4.15:

1. To bring the injected beam close to the circulating beam, a septum is used. A septum has a deflecting electric or magnetic field region and a field-free region, separated by a thin septum blade. The field in a septum is typically constant or slowly pulsed. Septum deflects the beam onto the closed orbit at the center of the kicker.
2. Kicker compensates for the remaining angle and thus the final adjustments of the injected beam trajectory into the circulating orbit are made using fast, pulsed kicker magnets. These magnets ensure that only the injected beam is deflected, leaving the circulating beam unaffected.

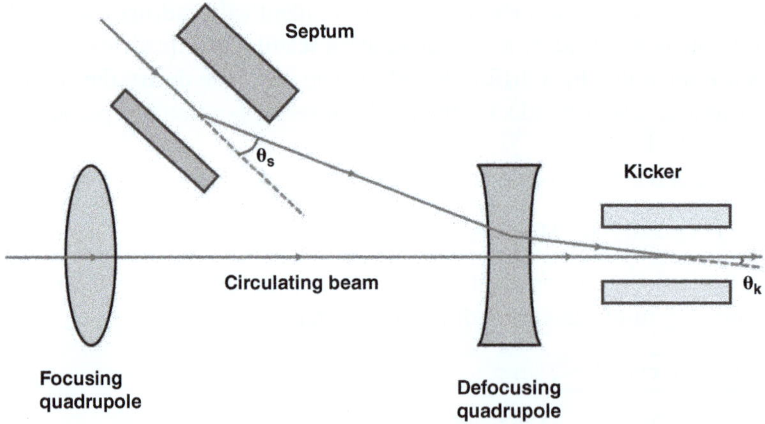

Figure 4.15 Collective action of kicker and septum over beam, θ_s and θ_k are deflection angles by septum and kicker, respectively.

3. Kicker magnets are designed to generate rectangular field pulses characterized by rapid rise and fall times, but they generally have a relatively low field strength. In order to compensate, electromagnetic septa are employed alongside kicker magnets, as they provide stronger field strength, though they operate either in DC mode or with slower pulses.
4. Both the septum and the kicker magnet can consist of multiple modules arranged in sequence, with each kicker magnet module driven by its own pulse generator. Septum and kicker are on either side of the defocusing quadrupole so as to minimize kicker strength.

Kicker magnets are essential for deflecting specific bunches of particles either from their incoming transfer line into the circulating orbit or from the circulating orbit into the extraction channel. Utilizing the Lorentz force, kicker magnets direct charged particles, which possess charge q and velocity v (close to the speed of light), through a magnetic field B. While maintaining a constant and well-defined field during the particle bunch traversal is crucial, achieving fast rise and fall times is necessary for high accelerator ring filling and minimal gaps between bunches. However, in some of the extraction kicker systems, where the entire beam is extracted, a rapid fall time is unnecessary.

Kicker magnets also play a critical role in accelerator beam dump systems. In emergencies, they swiftly extract the beam and direct it onto a specifically designed target, capable of absorbing the beam's total energy – up to hundreds of MJ per beam. Since the accelerator ring is completely emptied after beam extraction, the fall time of the dump system's kickers is not a concern. To ensure the beam dump system can be activated promptly, pulse generators for the kicker and deflection magnets are kept charged to a voltage corresponding to the beam energy, ready for immediate operation.

4.15 Glaser Lens

A solenoid is a cylindrical coil of wire that generates a uniform magnetic field along its axis when an electric current flows through it. In contrast, a Glaser magnet [74–76], is a specialized magnetic device used in particle accelerators and electron microscopes to focus and control charged particle beams. The magnetic field in a Glaser magnet is designed to vary along its axis, allowing precise focusing of particles. The magnetic field in a Glaser magnet is nearly Lorentzian. The schematics of axial symmetric Glaser magnet is shown in Figure 4.16.

Consider a charged particle with charge q and mass m moving in a magnetic field $\mathbf{B}$. The Lorentz force $\mathbf{F}$ on the particle is given by:

$$\mathbf{F} = q(\mathbf{v} \times \mathbf{B}), \tag{4.143}$$

where q is the charge, $\mathbf{v}$ is the velocity, and $\mathbf{B}$ is the magnetic field.

In cylindrical coordinates (r, θ, z), the velocity $\mathbf{v}$ and magnetic field $\mathbf{B}$ are:

$$\mathbf{v} = \dot{r}\hat{r} + r\dot{\theta}\hat{\theta} + \dot{z}\hat{z}, \quad \mathbf{B} = B_r\hat{r} + B_\theta\hat{\theta} + B_z\hat{z} \tag{4.144}$$

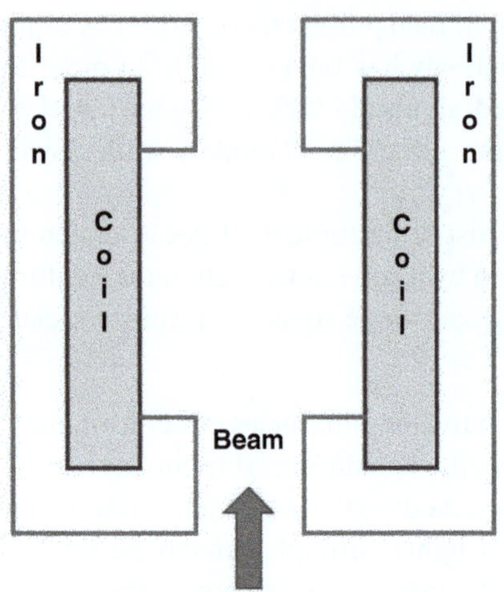

Figure 4.16 Axial symmetric Glaser magnet.

The components of the cross product $\mathbf{v} \times \mathbf{B}$ in cylindrical coordinates are obtained from the determinant:

$$\mathbf{v} \times \mathbf{B} = \begin{vmatrix} \hat{r} & \hat{\theta} & \hat{z} \\ \dot{r} & r\dot{\theta} & \dot{z} \\ B_r & B_\theta & B_z \end{vmatrix} = \left(r\dot{\theta}B_z - \dot{z}B_\theta\right)\hat{r} + (\dot{z}B_r - \dot{r}B_z)\hat{\theta} + \left(\dot{r}B_\theta - r\dot{\theta}B_r\right)\hat{z}$$

Thus, the components of the Lorentz force $\mathbf{F} = q(\mathbf{v} \times \mathbf{B})$ in the r, θ, and z directions are:

$$F_r = q\left(r\dot{\theta}B_z - \dot{z}B_\theta\right), \quad F_\theta = q\,(\dot{z}B_r - \dot{r}B_z), \quad F_z = q\left(\dot{r}B_\theta - r\dot{\theta}B_r\right) \tag{4.145}$$

The radial force is the resultant of the inward force and centrifugal force:

$$F_r = m\ddot{r} - mr(\dot{\theta})^2 \tag{4.146}$$

The time rate of change of angular momentum is equal to the moment of the force with respect to the axis of rotation:

$$\frac{d}{dt}(mr^2\dot{\theta}) = rF_\theta \tag{4.147}$$

Using Equation 4.145, we get:

$$\frac{d}{dt}(mr^2\dot{\theta}) = rq\,(\dot{z}B_r - \dot{r}B_z) \;\Rightarrow\; \frac{d}{dt}(mr^2\dot{\theta}) = -\frac{d}{dt}\left(\frac{qr^2B_z}{2}\right) \tag{4.148}$$

Thus, integrating both sides:

$$\dot{\theta} = -\frac{q}{2m}B_z \tag{4.149}$$

For the case where $B_\theta = 0$, one can get the following:

$$m\ddot{r} = qr\dot{\theta}B_z + mr\dot{\theta}^2 \;\Rightarrow\; \ddot{r} = -\frac{q^2}{4m^2} rB_z^2 \tag{4.150}$$

Converting the time coordinate into the z-coordinate with $z = v_z t$, we obtain:

$$\frac{d\theta}{dz} = -\frac{q^2}{8mE}B_z(z), \quad \frac{d^2r}{dz^2} = -\frac{q^2}{8mE}B_z(z)^2 r \tag{4.151}$$

Consider the motion of a charged particle in a magnetic field $\mathbf{B} = B_0\hat{z}$. The force acting on the particle is given by:

$$m\frac{d^2r}{dt^2} = qvB_0 \tag{4.152}$$

Using $v = \sqrt{\frac{2E}{m}}$, we rewrite the equation as:

$$m\frac{d^2r}{dt^2} = q\sqrt{\frac{2E}{m}}B_0 \tag{4.153}$$

The general solution for radial motion under the magnetic field is:

$$\frac{d^2r}{dz^2} + K^2 r = 0, \quad K^2 = \frac{q^2B_0^2}{8mE} \tag{4.154}$$

The general solution to this differential equation is:

$$r(z) = A\cos(Kz) + B\sin(Kz) \tag{4.155}$$

Using the initial conditions at $z = 0$, where $r(0) = r_0$ and $\frac{dr}{dz}(0) = v_{r0}$, we find:

$$A = r_0, \quad B = \frac{v_{r0}}{K} \tag{4.156}$$

Thus, the solution becomes:

$$r(z) = r_0\cos(Kz) + \frac{r_0\sin(Kz)}{K} \tag{4.157}$$

The magnetic field profile in the Glaser magnet is:

$$B_z(z) = \frac{B_0}{1+\left(\frac{z}{a}\right)^2} \tag{4.158}$$

Substituting into the radial motion equation:

$$\frac{d^2r}{dz^2} = -\frac{q^2B_0^2}{8mE}\frac{r}{\left[1+\left(\frac{z}{a}\right)^2\right]^2} \tag{4.159}$$

Glaser found the general solution for this as:

$$r = C\frac{\sin(\omega\phi)}{\sin\phi} + D\frac{\cos(\omega\phi)}{\sin\phi}, \quad \cot(\phi) = \frac{z}{a}, \quad \omega^2 = 1 + \frac{(qB_0a)^2}{8mE} \tag{4.160}$$

Here, C and D are two constants and related to intial values of r and r'. The solution for r and its derivative can be written in the form

$$\begin{pmatrix} r \\ r' \end{pmatrix} = M(\phi)\begin{pmatrix} C \\ D \end{pmatrix}, \tag{4.161}$$

where

$$M(\phi) = \begin{pmatrix} \dfrac{\sin(\omega\phi)}{\sin\phi} & \dfrac{\cos(\omega\phi)}{\sin\phi} \\ \omega\cos(\omega\phi)\sin\phi + \sin(\omega\phi)\cos\phi & \omega\cos(\omega\phi)\cos\phi - \sin(\omega\phi)\sin\phi \end{pmatrix}. \tag{4.162}$$

The solution at any initial point $z_0 = a\cot^{-1}(\phi_0)$ is

$$\begin{pmatrix} r_0 \\ r'_0 \end{pmatrix} = M(\phi_0)\begin{pmatrix} C \\ D \end{pmatrix}, \tag{4.163}$$

where r_0 and r'_0 are the initial values of r and r' at $\phi = \phi_0$. One can eliminate the constants C and D from equations and relate r and r' to their initial values as

$$\begin{pmatrix} r \\ r' \end{pmatrix} = M(\phi)\cdot M^{-1}(\phi_0)\begin{pmatrix} r_0 \\ r'_0 \end{pmatrix} = R\begin{pmatrix} r_0 \\ r'_0 \end{pmatrix}. \tag{4.164}$$

The transfer matrix R from the entry of the Glaser magnet (where $\phi = \phi_0$) to its exit is

$$R = \frac{1}{\omega}\begin{pmatrix} \omega\cos(\omega\phi_1) + \sin(\omega\phi_1)\cot\phi_0 & a\sin(\omega\phi_1)\cdot(\sin\phi_0)^{-2} \\ \frac{1}{a}\left[\sin(\omega\phi_1)\cdot(\omega^2\sin^2\phi_0 - \cos^2\phi_0) - \omega\cos(\omega\phi_1)\sin\phi_1\right] & \omega\cos(\omega\phi_1) + \sin(\omega\phi_1)\cot\phi_0 \end{pmatrix}, \tag{4.165}$$

where

$$\phi_1 = \phi - \phi_0. \tag{4.166}$$

If we take the entry point and the exit point of the ion to be symmetrical with respect to the centre of the Glaser magnet, then

$$\phi_1 = \pi - 2\phi_0. \tag{4.167}$$

In terms of the length L of the magnet, i.e. the distance between the entry and exit points, we have

$$\phi_0 = \cot^{-1}\left(\frac{L}{2a}\right). \tag{4.168}$$

The θ coordinate transformation is

$$\theta = -Ka\phi, \quad \theta' = -K\sin^2\phi. \tag{4.169}$$

Thus, the Glaser magnet, characterized by a Lorentzian axial magnetic field distribution, provides effective transverse focusing of charged particle beams. The field profile

$$B_z(z) = \frac{B_0}{1 + (z/a)^2}$$

leads to a radial equation of motion that admits an exact analytical solution. By expressing the solution in matrix form, the radial position and slope at any point along the magnet can be directly related to their initial values. The resulting transfer matrix formalism is useful for analyzing beam dynamics in systems where precise control over beam focusing is essential. This approach validates the Glaser magnet's lens-like behavior and supports its application in beam transport and focusing systems in accelerators and electron optics.

4.16 Undulators

An undulator is a magnetic field structure used in particle accelerators and synchrotron radiation facilities to produce highly intense, collimated, and coherent electromagnetic radiation. Charged particles such as electrons emit synchrotron radiation, typically in the X-ray or THz range, as they travel through the undulator's periodic magnetic field. A high degree of coherence means the emitted radiation from different electrons is in phase, leading to intense, narrow-bandwidth radiation. Electrons oscillate as they travel through the undulator, and the wavelength of the emitted radiation depends on their energy. By adjusting the undulator parameter K (which characterizes the strength of the magnetic field), the wavelength of the emitted radiation can be tuned.

In a free electron laser (FEL), relativistic electrons pass through an undulator, a device producing a periodic magnetic field, which causes the electrons to oscillate transversely. The parameter K defines the strength of the undulator's magnetic field and the amplitude of the electron's oscillations. The electrons travel with a longitudinal velocity $v_z = \beta c$, where β is the normalized velocity. The undulator has a wavelength λ_u, while the radiation produced by the FEL has a wavelength λ. The corresponding wave numbers are $k_u = \frac{2\pi}{\lambda_u}$ for the undulator and $k = \frac{2\pi}{\lambda}$ for the radiation. Energy exchange occurs between the electrons and the radiation field due to the interaction between the electron's transverse velocity v_x and the transverse electric field $E_x(t)$. The energy exchange rate is denoted by $\frac{dW}{dt}$, where W is the energy of the electrons.

The rate of energy exchange between the electrons and the radiation field is given by the interaction between the transverse velocity v_x of the electrons and the transverse electric field $E_x(t)$:

$$\frac{dW}{dt} = -ev_x E_x(t) \tag{4.170}$$

This equation represents the rate of energy exchange between the electron's motion and the electric field.

The transverse velocity v_x of the electrons in the undulator magnetic field can be expressed as:

$$v_x = \frac{Kc}{\gamma}\cos(k_u z) \tag{4.171}$$

Here, K is the undulator parameter, c is the speed of light, and γ is the Lorentz factor. The transverse oscillation is represented by $\cos(k_u z)$, where $k_u = \frac{2\pi}{\lambda_u}$.

The transverse electric field $E_x(t)$ of the radiation is given by:

$$E_x(t) = E_0 \cos(kz - \omega t + \phi_0) \tag{4.172}$$

This describes the time-varying electric field, where E_0 is the amplitude, $k = \frac{2\pi}{\lambda}$ is the wave number, $\omega = kc$ is the angular frequency, and ϕ_0 is the initial phase.

Substituting the expression of v_x and $E_x(t)$ into energy exchange rate equation, one can get:

$$\frac{dW}{dt} = -e\frac{Kc}{\gamma}\cos(k_u z)E_0\cos(kz - \omega t + \phi_0) \tag{4.173}$$

This equation represents the interaction between the electron's transverse velocity in the undulator and the electric field of the radiation.

Now, to simplify the product of the two cosine terms, we use the trigonometric identity:

$$\cos A \cos B = \frac{1}{2}\left[\cos(A+B) + \cos(A-B)\right]$$

Applying this identity, we simplify as follows:

$$\frac{dW}{dt} = -e\frac{Kc}{2\gamma}E_0\left[\cos((k+k_u)z - \omega t + \phi_0) + \cos((k-k_u)z - \omega t + \phi_0)\right] \tag{4.174}$$

The energy exchange rate is now written as the sum of two oscillating terms.

In above equation, the second term $\cos((k-k_u)z-\omega t+\phi_0)$ oscillates rapidly and averages out over time. The first term $\cos((k+k_u)z - \omega t + \phi_0)$ can be made constant by setting appropriate conditions, leading to continuous energy exchange. The energy exchange rate thus simplifies to:

$$\frac{dW}{dt} = -e\frac{Kc}{2\gamma}E_0\cos(\psi) \tag{4.175}$$

Here, ψ is called the ponderomotive phase and is defined as:

$$\psi = (k + k_u)z - \omega t + \phi_0 \tag{4.176}$$

This equation defines the phase relationship between the radiation field and the electron's motion.

For continuous energy exchange, the ponderomotive phase ψ must remain constant. To ensure that ψ remains constant over time, we differentiate ψ with respect to time:

$$\frac{d\psi}{dt} = (k + k_u)\beta c - \omega = 0 \tag{4.177}$$

Here, we used $\frac{dz}{dt} = \beta c$ for the longitudinal velocity of the electrons. The condition $\frac{d\psi}{dt} = 0$ ensures that the phase does not vary with time.

Since $\omega = kc$, we substitute this into above equation and solve for the resonant wave number k_{res}:

$$(k + k_u)\beta c = kc \tag{4.178}$$

Rearranging to isolate k, we get:

$$k_{\text{res}} = \frac{k_u}{1 - \beta} \tag{4.179}$$

This equation defines the resonant wave number k_{res}, which optimizes energy exchange between the electrons and the radiation.

The electron's longitudinal velocity β is slightly less than the speed of light because of the electron's transverse oscillations in the undulator. The expression for β in terms of the Lorentz factor γ and the undulator parameter K is derived from the relativistic effects:

$$\beta = 1 - \frac{1}{2\gamma^2}\left(1 + \frac{K^2}{2}\right) \tag{4.180}$$

This expression accounts for the slight reduction in the longitudinal velocity due to the oscillations in the undulator.

Using the expression for β we can now calculate the resonant wavelength λ_{res} that corresponds to the resonant wave number k_{res}. The resonant wavelength λ_{res} is given by the relationship $\lambda_{\text{res}} = \frac{2\pi}{k_{\text{res}}}$, and substituting k_{res} from Equation 4.126, we get:

$$\lambda_{\text{res}} = \frac{\lambda_u}{2\gamma^2}\left(1 + \frac{K^2}{2}\right) \tag{4.181}$$

This is the resonant wavelength at which the energy exchange between the electrons and the radiation is optimized, corresponding to the wavelength of spontaneous undulator radiation in the forward direction.

Consider an electron moving along an arbitrary trajectory $\mathbf{r}(t')$ relative to the origin O in an undulator [77] with velocity $\mathbf{v}$. The position of the observer is specified by the vector $\mathbf{x}$, and the unit vector $\mathbf{n}$ makes an angle θ with the velocity vector $\mathbf{v}$. An electromagnetic signal emitted by the electron at time t' travels in a straight line and arrives at the observer at a later time t, where

$$t = t' + \frac{|\mathbf{x} - \mathbf{r}(t')|}{c} \tag{4.182}$$

The stationary observer sees the electron's motion as a function of time t, which differs from $\mathbf{r}(t')$ due to the change in time scale represented by Equation 4.129. The scale change factor is given by

$$\frac{dt}{dt'} = 1 + \frac{d|\mathbf{x} - \mathbf{r}(t')|}{dt'}\frac{1}{c} = 1 - \mathbf{n} \cdot \beta = 1 - \beta \cos\theta \tag{4.183}$$

The Lorentz factor γ for the electron can be expressed as

$$\gamma = \frac{1}{\sqrt{1-\frac{v^2}{c^2}}} = \frac{1}{\sqrt{(1-\beta)(1+\beta)}} \tag{4.184}$$

Using the approximation $1-\beta \approx \frac{1}{2\gamma^2}$, we can rewrite Equation 4.130 as

$$\frac{dt}{dt'} = \frac{1}{2}\left[\frac{1}{\gamma^2} + \theta^2\right] \tag{4.185}$$

Consider the case of an exactly sinusoidal field and trajectory. Then, one can write:

$$x = -a\cos(k_u z) \tag{4.186}$$

$$\frac{dx}{dz} = k_u a \sin(k_u z), \quad \frac{d^2x}{dz^2} = -k_u^2 a = -\frac{1}{\rho} \tag{4.187}$$

Now, using the Lorentz force equation:

$$\frac{m_e v^2}{\rho} = e\mathbf{v}\times\mathbf{B}, \quad \rho = \frac{m_e \gamma c}{eB} \tag{4.188}$$

Using the above two equations, one can get:

$$K = \frac{eB}{k_u m_e c} = 0.934\lambda_u(\text{cm})B(\text{T}) \tag{4.189}$$

$$a = \frac{1}{\rho k_u^2} = \frac{K}{k_u \gamma}, \quad \frac{dx}{dz} = \frac{K}{\gamma}\sin(k_u z) \tag{4.190}$$

The fundamental equation of undulator action sets a relationship between the wavelength of the undulator and the wavelength of the emitted radiation, which is determined by the Doppler shift due to the motion of the radiating electron. The amount of shift or time compression factor is given by

$$\lambda_1 = \lambda_u(1-\beta_z\cos\theta) \tag{4.191}$$

The radiation emitted by the electron in the undulator is Doppler shifted due to its high longitudinal velocity. The observed wavelength in the laboratory frame is obtained by applying the relativistic Doppler shift to the fundamental wavelength as seen in the electron's instantaneous rest frame. The period of the magnetic structure in the electron's rest frame is Lorentz contracted:

$$\lambda' = \frac{\lambda_u}{\gamma} \tag{4.192}$$

The Doppler-shifted wavelength observed at an angle θ in the lab frame is:

$$\lambda = \lambda'(1+\beta_z\cos\theta) \tag{4.193}$$

The electron's longitudinal velocity β_z is reduced due to transverse oscillations in the undulator, and is approximated as:

$$\beta_z \approx 1 - \frac{1}{2\gamma^2}\left(1 + \frac{K^2}{2}\right) \tag{4.194}$$

Substituting and expanding to first order in $1/\gamma^2$ and small θ, we obtain:

$$\lambda \approx \frac{\lambda_u}{2\gamma^2}\left(1 + \frac{K^2}{2} + \gamma^2\theta^2\right) \tag{4.195}$$

Thus, one can get from the above equations:

$$\lambda_m = \frac{\lambda_u}{2m\gamma^2}\left[1 + \frac{K^2}{2} + \gamma^2\theta^2\right] \tag{4.196}$$

This is the fundamental equation describing undulator action, derived from the principle of time compression and relativistic Doppler shift.

Undulators are used in synchrotron light sources for applications like X-ray diffraction, X-ray spectroscopy, and medical imaging. They are also a key component of free electron lasers (FELs), advanced light sources that produce intense, coherent X-ray or ultraviolet radiation for ultrafastscience and medical imaging. In an FEL, relativistic electrons pass through the undulator and interact with a radiation field, exchanging energy based on their transverse oscillations induced by the undulator.

4.17 Wigglers

A wiggler [78] is an electromagnetic device used in particle accelerators and synchrotron radiation facilities to produce high-intensity and high energy electromagnetic radiation, usually X-rays. Wigglers also make electrons to oscillate as electrons travel through a periodic magnetic structure but have a different magnetic field configuration compared to undulators. As charged particles pass through the alternating magnetic field of a wiggler, they experience a transverse force. Unlike undulators, wigglers do not maintain a strict periodicity in the magnetic field, causing particles to undergo complex transverse oscillations. This motion leads to the emission of synchrotron radiation. Wigglers produce more intense radiation than undulators due to a high magnetic field and the complex motion of particles and are thus suitable for applications with high radiation flux. The radiation emitted has a relatively broad spectrum from the infrared to X-ray regions of the electromagnetic spectrum compared to the narrow-bandwidth radiation of undulators. Wigglers generate a lot of heat due to the interaction between the charged particles and the strong magnetic field, so they require efficient cooling systems. As an electron travels at nearly the speed of light through a planar wiggler, it experiences a periodic transverse magnetic field that forces it to oscillate in the horizontal plane. The magnetic field is oriented vertically and described by the sinusoidal function

$$B_y(s) = B_0 \sin(k_w s), \tag{4.197}$$

where B_0 is the peak magnetic field strength, s is the longitudinal coordinate along the beam path, and $k_w = \frac{2\pi}{\lambda_w}$ is the wiggler wavenumber corresponding to the period λ_w. Because the particle's velocity is mainly longitudinal, the Lorentz force generated by the magnetic field acts transversely, causing the electron to oscillate sideways as it moves forward.

Assuming that motion is confined to the horizontal plane, the equation of motion derived from Newton's second law in the relativistic regime becomes

$$x''(s) = -\frac{eB_0}{m_e\beta c\gamma}\cos(k_w s), \tag{4.198}$$

where e and m_e are the charge and mass of the electron, c is the speed of light, $\beta \approx 1$ is the normalized velocity, and γ is the Lorentz factor. This equation represents forced harmonic motion.

Integrating this once with respect to s gives the transverse velocity (or angular deviation):

$$x'(s) = -\frac{eB_0}{m_e\beta c\gamma k_w}\sin(k_w s). \tag{4.199}$$

A second integration gives the transverse displacement:

$$x(s) = -\frac{eB_0}{m_e\beta c\gamma k_w^2}\cos(k_w s). \tag{4.200}$$

Substituting $k_w = \frac{2\pi}{\lambda_w}$ into these expressions, we obtain:

$$x'(s) = -\frac{\lambda_w eB_0}{2\pi m_e c\gamma}\sin(k_w s), \tag{4.201}$$

$$x(s) = -\frac{\lambda_w^2 eB_0}{4\pi^2 m_e c\gamma}\cos(k_w s). \tag{4.202}$$

The maximum transverse angle of the trajectory, denoted θ_w, occurs when the sine term reaches unity:

$$\theta_w = x'_{\text{max}} = \frac{\lambda_w eB_0}{2\pi m_e c\gamma}. \tag{4.203}$$

This angle determines whether the device functions as an undulator or a wiggler. If $\theta_w \leq \frac{1}{\gamma}$, the oscillations are mild, and the radiation from each cycle adds coherently, producing a narrow radiation spectrum; such a device is an undulator. Conversely, if $\theta_w > \frac{1}{\gamma}$, the radiation adds incoherently and spans a broader spectrum; this regime defines a true wiggler.

Beyond altering the electron trajectory, the wiggler increases energy loss via synchrotron radiation. In a storage ring, the total energy loss per turn is given by

$$U_0 = \frac{C_\gamma}{2\pi}E_0^4 I_2, \tag{4.204}$$

where E_0 is the beam energy and C_γ is a universal radiation constant. The integral I_2, known as the second radiation integral, accounts for the total bending the beam undergoes:

$$I_2 = \oint \frac{1}{\rho^2}\, ds, \tag{4.205}$$

where ρ is the bending radius of the trajectory. In the case of a wiggler with constant peak field B_0, its contribution to I_2 is

$$I_{2w} = \frac{1}{(B\rho)^2} \int_0^{L_w} B^2(s)\, ds = \frac{1}{(B\rho)^2} \cdot \frac{B_0^2 L_w}{2}, \tag{4.206}$$

with L_w being the total length of the wiggler and $B\rho$ the magnetic rigidity of the particle. It is important to note that I_{2w} does not depend on the wiggler period λ_w, making the total radiation output a function only of the field strength and length. This is useful in damping rings where fast emittance reduction is desired.

Another important integral is I_3, the third synchrotron radiation integral, which influences the natural energy spread of the beam:

$$\sigma_\delta^2 = C_q \frac{\gamma^2 I_3}{J_z I_2}, \tag{4.207}$$

where C_q is a quantum radiation constant, and J_z is the longitudinal damping partition number. The integral I_3 is defined as

$$I_3 = \oint \frac{1}{|\rho|^3}\, ds. \tag{4.208}$$

Because wigglers typically have very strong magnetic fields and small bending radii, they can dominate the contribution to I_3, thereby significantly affecting σ_δ.

The final important quantity influenced by the wiggler is the natural horizontal emittance ε_0, given by

$$\varepsilon_0 = C_q \frac{\gamma^2 I_5}{J_x I_2}, \tag{4.209}$$

where J_x is the horizontal damping partition number. The integral I_5, which encapsulates dispersion and optical functions, is given by

$$I_5 = \oint \frac{H_x}{|\rho|^3}\, ds, \tag{4.210}$$

and

$$H_x = \gamma_x D_x^2 + 2\alpha_x D_x D_x' + \beta_x D_x'^2, \tag{4.211}$$

with D_x being the horizontal dispersion, D_x' its derivative, and $\beta_x, \alpha_x, \gamma_x$ the Twiss parameters. In a wiggler, due to periodic strong bending and small dispersion, the contribution to I_5 is sensitive to the local beta function.

Hence, the influence of wigglers extends beyond simply increasing radiation – they modify beam dynamics profoundly. Their effect on integrals I_2, I_3, and I_5 ties directly into energy loss, energy spread, and emittance. This is why wigglers are not just radiation sources but also indispensable beam conditioning tools in storage rings and damping rings, such as those proposed for the International Linear Collider.

4.18 Numerical Problems

1. Find out the edge angle of a dipole magnet required for double focusing on both planes with object and image distances equal to twice the bending radius for your beam?
2. Prove that the determinant of the transfer matrix of a dipole magnet and a thick quadrupole magnet is unity in both planes.
3. Calculate the ampere-turns required to design a 0.5 T dipole magnet, 0.5 T quadrupole magnet, and 0.5 T solenoid magnet. Given the full aperture as 50 mm.
4. Deflect $^{28}Si^{10+}$ beam of energy 100 MeV by 10 mrad. What is the magnetic field required for this purpose? Given the length of the magnetic steerer is 100 mm.
5. What are the field components Bx, By, and Bz at the center of dipole, quadrupole, and solenoid magnet? Consider the design field in each as 1000 G. Take the direction of beam as z and others x,y as transverse directions.
6. Consider a dipole magnet of bending radius as 500 mm and bending angle as 90°. Calculate the drift length before and after the dipole magnet for focusing in bending plane. Also calculate lateral dispersion and the resolving power of dipole magnet for a beam size of 1 mm.
7. A solenoid magnet is needed to focus a proton beam with an energy of 100 keV and a normalized emittance of 100 π mm-mrad. The required focal length of the solenoid is 500 mm. Calculate the necessary magnetic field strength, the length of the solenoid, and the number of ampere-turns required to achieve this focusing.
8. A quadrupole magnet is required to focus on a carbon ion beam with an energy of 100 keV/u and a normalized emittance of 50 π mm-mrad. The focal length of the quadrupole is 300 mm. The charge-to-mass ratio A/q of the carbon ion is 2. Calculate the necessary magnetic field gradient, the length of the quadrupole, and the number of ampere-turns required to achieve this focusing. Additionally, suggest if the quadrupole design is practical.
9. Design a Wien filter for a 100 keV proton beam with emittance of 50 π mm-mrad. Calculate the necessary magnetic field and voltage required for the proton to pass through the filter undeflected.
10. Design a magnetostatic deflector (steerer magnet) for a 100 MeV Si^{10+} ion beam with an emittance of 50 π mm-mrad. Calculate the necessary magnetic field, ampere-turns, and the length of the steerer for a deflection of 5° over a magnet length of 0.2 m.
11. A Glaser lens is required to focus a proton beam with an energy of 100 keV and a normalized emittance of 50 π mm-mrad. Calculate the magnetic field strength and focal length required for this Glaser lens.
12. A septum and kicker magnet system is used to deflect a 200 MeV proton beam into a new beamline. The septum magnet provides a deflection angle of 5° over a length of 0.5 m, while the kicker magnet provides an additional deflection of 2° over a length of 0.25 m. The proton beam has a normalized emittance of 50 π mm-mrad. Calculate the magnetic field strength required for both the septum and kicker magnets.
13. An undulator is required to produce radiation from an electron beam with energies ranging from 4 to 8 MeV. The undulator has a magnetic period of $\lambda_u = 3$ cm and an undulator parameter $K = 1$. Calculate the wavelength of the emitted radiation for both 4 and 8 MeV electron beams.

14. Design a first-order achromat using COSY Infinity code or any other beam optics software for 1.8 MeV/u ion beam with emittance as $10\,\pi$ mm-mrad with a total length of achromat as 7 m using two dipole magnets and quadrupoles as required, first try with hands on formula, and then proceed for simulations.

15. Design a first-order magnetic quadrupole triplet using COSY Infinity code or any other beam optics software for 1.8 MeV/u ion beam with emittance as $10\,\pi$ mm-mrad, first try with hands on formula, and then proceed for simulations.

5

Electrostatic Devices

"For example, the force of electricity between two charged objects looks just like the law of gravitation: the force of electricity is a constant, with a minus sign, times the product of the charges, and varies inversely as the square of the distance. It is in the opposite direction – likes repel. But is it still not very remarkable that the two laws involve the same function of distance? Perhaps gravitation and electricity are much more closely related than we think. Many attempts have been made to unify them; the so-called unified-field theory is only a very elegant attempt to combine electricity and gravitation. But, in comparing gravitation and electricity, the most interesting thing is the relative strengths of the forces. Any theory that contains them both must also deduce how strong the gravity is."

—Richard Feynman, *The Feynman Lectures on Physics*

After reading this chapter, you should be able to:

- Understand the different electric field configurations by which charged particles get affected.
- Calculate the focal length of the electrostatic devices.
- Design electrostatic components of accelerators.

5.1 Motion of a Charged Particle in an Electric Field

Consider a charged particle with charge q and mass m, initially at position $\mathbf{x}_0$ with an initial velocity $\mathbf{v}_0$. The particle moves under the influence of an electric field $\mathbf{E}$, which exerts a force on the particle. The force causes the particle to accelerate with an acceleration $\mathbf{a}$, leading to a change in its velocity and position over time t. The particle gains kinetic energy as it is accelerated, and the change in its motion is described by Newton's second law. The system's dynamics, position, velocity, and energy are functions of the particle's charge-to-mass ratio and the electric field strength. The force on a charged particle acts in the direction of the electric field, and it is independent of the velocity of the charged particle. A charged particle can gain energy from static parallel electric field. The motion of the charged particle is governed by Newton's second law, which relates the force acting on the particle to

Charged Particle Beam Physics: An Introduction for Physicists and Engineers, First Edition.
Sarvesh Kumar and Manish K. Kashyap.

Companion Website: https://www.wiley.com/go/Kumar_1e

its acceleration. The force **F** on the particle due to the electric field **E** and using Newton's second law:

$$\mathbf{F} = q\mathbf{E} \quad \text{and} \quad \mathbf{F} = m\mathbf{a} \tag{5.1}$$

Curiosity!! *Can you derive the fundamental relations above?*

$$q\mathbf{E} = m\mathbf{a} \quad \Rightarrow \quad \mathbf{a} = \frac{q}{m}\mathbf{E} \tag{5.2}$$

If the particle starts from rest, the velocity of the particle at time t can be obtained by integrating the acceleration:

$$\mathbf{v}(t) = \mathbf{v}_0 + \mathbf{a}t = \frac{q\mathbf{E}}{m}t \tag{5.3}$$

Thus, the velocity increases linearly with time as the particle is accelerated by the electric field. The position of the particle at time t can be found by integrating the velocity:

$$\mathbf{x}(t) = \mathbf{x}_0 + \mathbf{v}_0 t + \frac{1}{2}\mathbf{a}t^2 = \mathbf{x}_0 + \frac{1}{2}\frac{q\mathbf{E}}{m}t^2 \tag{5.4}$$

As the particle accelerates, it gains kinetic energy. The kinetic energy K of the particle at time t is given by:

$$K = \frac{1}{2}mv^2 = \frac{1}{2}m\left(\frac{q\mathbf{E}}{m}t\right)^2 = \frac{q^2\mathbf{E}^2t^2}{2m} \tag{5.5}$$

5.2 Electrostatic Dipole

An electrostatic dipole (ED) consists of two conducting poles separated by some gap with a potential difference between them. It can take various forms, such as parallel plates, cylindrical dipoles, and spherical dipoles, each offering unique configurations and field characteristics. It is a two-pole structure with polarity +ve and −ve with a homogeneous electric field transverse to the motion of charged particles. It can be used as a separator based on the energy-to-charge ratio of particles. In the parallel plate dipole, two flat, parallel conductive plates create a uniform electric field between them, often used for deflecting charged particles in a beamline due to its simplicity and ability to provide a well-defined, linear field. Cylindrical dipoles consist of two cylindrical electrodes, which produce a radial electric field, commonly used in beam optics for focusing and steering beams; this setup ensures rotational symmetry and allows for efficient control of circular or annular beams. Spherical dipoles, composed of spherical electrodes, generate a nonuniform electric field with radial symmetry, used less frequently but applicable in situations requiring a point-source field effect or highly nonlinear field configurations. Each type of dipole offers advantages depending on the requirements of field uniformity, beam shape, and focusing needs in electrostatic systems.

Consider a particle with charge q and mass m, moving through an electric field E. The particle's velocity is v, and its momentum is denoted by p, which is given by the product of its mass and velocity, $p = mv$. The particle travels along a curved trajectory, where ρ represents the electrostatic rigidity of the particle, which defines how much the electric field can deflect the particle. The force exerted by the electric field causes the particle to experience a centripetal acceleration as it moves in a circular path with radius r.

$$F = qE \quad \text{and} \quad F = m\frac{v^2}{r} \tag{5.6}$$

Equating these forces gives the relationship between the electric field and the particle's motion:

$$qE = m\frac{v^2}{r} \quad \Rightarrow \quad r = \frac{mv^2}{qE} \tag{5.7}$$

The electrostatic rigidity is given in terms of particle energy (K) as follows:

$$E.r = \frac{mv^2}{q} = \frac{2K}{q} \tag{5.8}$$

Curiosity!! *Can you compare it with magnetic rigidity and find out which is practical at what energy regimes of particles? And what kind of filtering do electric and magnetic fields offer independently to charged particles?*

5.2.1 Radial and Vertical Motion Inside an Electric Field

The equation of motion of a charged particle of mass m, charge q, and velocity v_0 moving in a circle of radius r_o is written as follows:

$$qE_0 = \frac{mv_0^2}{r_0} \tag{5.9}$$

The electric field can be expanded to the first order as follows:

$$E_r = E_0 + r\frac{\partial E_0}{\partial r} = E_0\left(1 - \frac{Nr}{r_0}\right) \tag{5.10}$$

where N is the electric field index here and is defined as

$$N = -\frac{r_0}{E_0}\frac{\partial E_0}{\partial r} \tag{5.11}$$

The equation of motion for radial orbit (r_0 + r) is given as follows:

$$qE_r = \frac{mv_r^2}{r_0 + r} - m\ddot{r} \tag{5.12}$$

and the velocity at the orbit (r_0 + r) is given as:

$$v_r^2 = v_0^2(1 - (2 - \beta_0^2 r/r_0) \tag{5.13}$$

Here, $\beta_0 = v_0/c$. The final equation of motion can be simplified in z derivatives:

$$r'' = -r(3 - N - \beta_0)/r_0^2 \tag{5.14}$$

It represents simple harmonic motion provided: $k_r^2 = 3 - N - \beta_0 > 0$. In the vertical direction, perpendicular to the plane of central orbit, using the Maxwell's equations, we may write ∇ .E = 0 as follows:

$$\frac{\partial(rE_r)}{\partial r} + \frac{\partial(rE_y)}{\partial y} = 0 \tag{5.15}$$

At y = 0, $E_r = 0$. Hence

$$E_y = E_r(1 - N)\frac{y}{r} \tag{5.16}$$

Force balancing equation for particle motion in vertical direction is as follows:

$$qE_y = -m\ddot{y} \tag{5.17}$$

which gives solution in terms of z-derivatives in vertical plane as follows:

$$y'' = -y\frac{N-1}{r^2} \tag{5.18}$$

It represents simple harmonic motion provided: $k_v^2 = N - 1 > 0$.

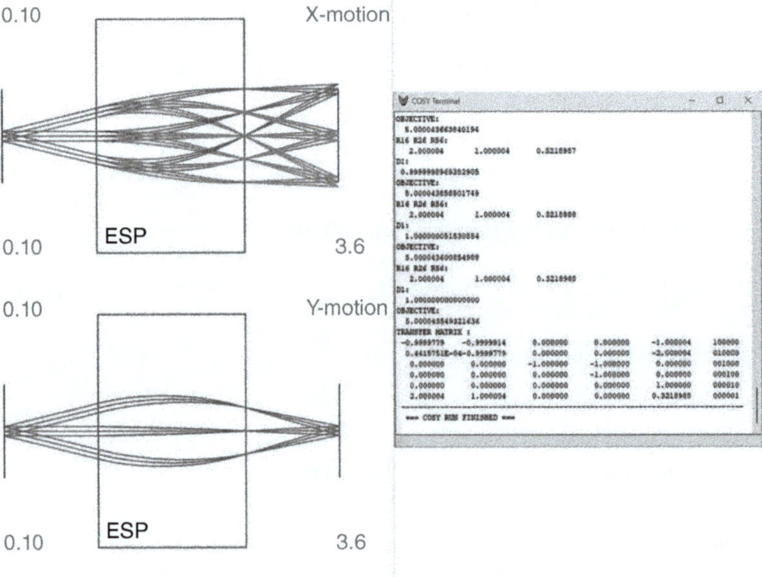

Figure 5.1 Beam optics for a 90° ESD with X profile, Y profile, and resulting transfer matrix of ESD. X plane is bending plane and hence shows dispersion of trajectories as per energy-to-charge ratios.

Hence, the combined transfer matrix is given by the solution of Equations 5.14 and 5.18 as follows:

$$R = \begin{pmatrix} \cos(k_r\theta) & r_0 k^{-1}\sin(k_r\theta) & 0 & 0 & 0 & 0 \\ -k_r r_0^{-1}\sin(k_r\theta) & \cos(k_r\theta) & 0 & 0 & 0 & 0 \\ 0 & 0 & \cos(k_v\theta) & rk_v^{-1}\sin(k_v\theta) & 0 & 0 \\ 0 & 0 & -r^{-1}k_v\sin(k_v\theta) & \cos(k_v\theta) & 0 & 0 \\ 0 & 0 & 0 & 0 & 1 & L/\gamma^2 \\ 0 & 0 & 0 & 0 & 0 & 1 \end{pmatrix}$$

Here, θ is the deflection angle for the central ray. Thus, the solution exists for double focusing in radial and vertical planes for ED only if $1 < N < 3 - \beta_0^2$.

Beam optics for a 90° electrostatic spherical deflector (ESD) is shown in Figure 5.1 for 8 keV/u ion beam with 1% energy spread using COSY Infinity code:

5.3 Electrostatic Quadrupole

Electrostatic quadrupoles (EQs) play a crucial role in various scientific and industrial applications, particularly in the manipulation and focusing of charged particles. When configured as a triplet, EQs can effectively focus particles at low energy levels (up to a few hundred keV). The electrostatic rigidity of these devices is directly proportional to the velocity of the charged particles, making them particularly effective at low energies. In ion trap devices, EQs are utilized to trap and manipulate ions. Quadrupole ion traps, such as the Paul trap, employ a combination of radiofrequency and DC voltages applied to four electrodes to confine ions within a stable region. In fusion research, EQs are used to confine and shape plasma, enabling better control over the fusion process. They are also applied in ion implantation processes in the semiconductor industry, where they precisely control ion beams on silicon wafers. Additionally, in electric propulsion systems for spacecraft, such as ion thrusters, EQ configurations are employed to control charged particle beams for propulsion, which is particularly valuable in long-duration space missions requiring high efficiency.

The electrostatic quadrupole consists of a four-pole structure arranged in a symmetric configuration, as illustrated in Figure 5.2. These devices focus on one plane and defocus on the perpendicular plane relative to the beam propagation direction. The electrodes of a quadrupole are typically approximated by cylindrical rods, with the best approximation achieved when the electrode radius is 1.147 times the aperture radius. The potential function for an electrostatic quadrupole is expressed as:

$$V(x, y) = \frac{x^2 - y^2}{a^2} V$$

The electric field derived from this potential function is given by:

$$E_x = -\frac{\partial V}{\partial x} = -\frac{2V}{a^2}x = -gx, \qquad E_y = -\frac{\partial V}{\partial y} = -\frac{2V}{a^2}y = gy$$

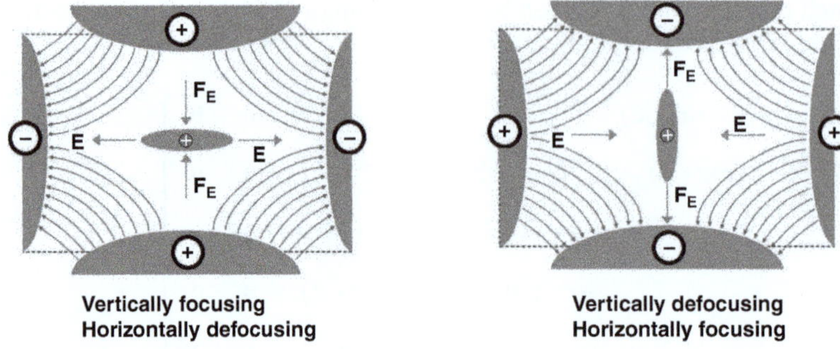

Figure 5.2 Focusing and defocusing scheme for a positively charged particle inside electrostatic quadrupoles.

The time derivative of the position coordinate can also be expressed as a space derivative:

$$\ddot{x} = \frac{d^2x}{dt^2} = \frac{v^2 d^2 x}{dz^2} = v^2 x''$$

The equation of motion for a particle with mass m, charge q, and velocity v inside a quadrupole with aperture a is written as:

$$m\ddot{x} = -qgx \qquad \text{and} \qquad m\ddot{y} = qgy$$

These equations are rewritten with respect to the longitudinal coordinate z, where $x' = dx/dz$:

$$x'' = -\frac{qg}{mv^2}x \qquad \text{and} \qquad y'' = \frac{qg}{mv^2}y$$

Here, $k^2 = \frac{2qV}{a^2mv^2} = \frac{2V}{a^2(E\rho)}$

To solve these equations, boundary conditions are applied. At $z = 0$, the particle coordinates are (x_1, y_1), and at $z = l$, they are (x_2, y_2). The solution can be represented in matrix form as follows:

$$x = a\cos(kz) + b\sin(kz) \tag{5.19}$$

$$x' = -ak\sin(kz) + bk\cos(kz) \tag{5.20}$$

At $z = 0$, $a = x_1$ and $b = x_1'/k$.

Thus, the above equations can be rewritten in matrix form as:

$$\begin{bmatrix} x_2 \\ x_2' \end{bmatrix} = \begin{bmatrix} \cos(kl) & \frac{1}{k}\sin(kl) \\ -k\sin(kl) & \cos(kl) \end{bmatrix} \begin{bmatrix} x_1 \\ x_1' \end{bmatrix} \tag{5.21}$$

Similarly, for vertical motion, the solution is expressed in matrix form as:

$$\begin{bmatrix} y_2 \\ y_2' \end{bmatrix} = \begin{bmatrix} \cosh(kl) & \frac{1}{k}\sinh(kl) \\ k\sinh(kl) & \cosh(kl) \end{bmatrix} \begin{bmatrix} y_1 \\ y_1' \end{bmatrix} \tag{5.22}$$

Since no force acts along the beam direction z, an electrostatic quadrupole functions as a drift in that direction, and the transfer matrix is simply given as:

$$R_{zz} = \begin{bmatrix} 1 & L/\gamma^2 \\ 0 & 1 \end{bmatrix} \tag{5.23}$$

The first-order transport matrix for a horizontally (radially) focusing and vertically (axially) defocusing electrostatic quadrupole is given by:

$$\begin{pmatrix} \cos(kl) & \frac{1}{k}\sin(kl) & 0 & 0 & 0 & 0 \\ -k\sin(kl) & \cos(kl) & 0 & 0 & 0 & 0 \\ 0 & 0 & \cosh(kl) & \frac{1}{k}\sinh(kl) & 0 & 0 \\ 0 & 0 & k\sinh(kl) & \cosh(kl) & 0 & 0 \\ 0 & 0 & 0 & 0 & 1 & L/\gamma^2 \\ 0 & 0 & 0 & 0 & 0 & 1 \end{pmatrix}$$

Depending on the polarity of the electric fields, the electrostatic quadrupole focuses in the x-direction and defocuses in the y-direction, or vice versa. The focal length of the quadrupole is given by:

$$f = \pm\frac{1}{kl}$$

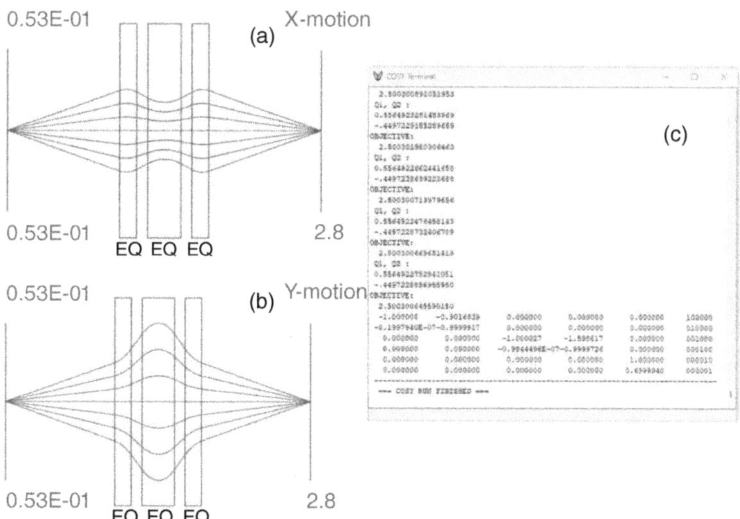

Figure 5.3 Beam optics for a electrostatic quadrupole triplet using COSY Infinity capable for focusing the beam in both transverse planes. (a) Beam in horizontal plane with X,X': 1,7. (b) Beam in vertical plane with Y,Y': 1,7. (c) Optimized values of electrostatic quadrupoles are 0.5564 and −0.4497 kV, the overall transfer matrix is also shown.

Here, a is the radial distance from the axis to the pole tip, and V represents the absolute value of the electric voltage at the pole tip. For a horizontally (radially) defocusing and vertically (axially) focusing electrostatic quadrupole, the sub-matrices for focusing and defocusing are interchanged.

Figure 5.3 shows beam optics of an electrostatic quadrupole triplet using COSY Infinity for 8 keV/u ion beam, charge state: +1. Here we vary the fields on electric fields of quadrupole symmetrically, and finally, the focusing in both planes is achieved symmetrically.

5.4 Electrostatic Thin Lens and Einzel Lens

An electrostatic thin lens is an aperture lens with some potential difference with respect to the earlier device, which is a plasma chamber mostly at ground potential. It is mainly used to extract the beam out of plasma. It is also termed as an iris aperture. Using the Taylor series expansion of potential and Laplace's equation, one can get the equation of motion as follows:

The equation of motion in a charged particle in an electric field with components E_r and E_z is given as follows:

$$\frac{d}{dt}(m\dot{r}) = -q\frac{\partial V}{\partial r}; \quad \frac{d}{dt}(m\dot{z}) = -q\frac{\partial V}{\partial z} \tag{5.24}$$

Using $\dot{r} = v_z r'$ and $\ddot{r} = v_z^2 r''$, we get the paraxial ray equation.

Let us consider the expression for the divergence in cylindrical coordinates:

$$\nabla \cdot \mathbf{E} = \frac{1}{r}\frac{\partial}{\partial r}(rE_r) + \frac{1}{r}\frac{\partial}{\partial \theta}(E_\theta) + \frac{\partial}{\partial z}(E_z)$$

Given the divergence-free condition $\nabla \cdot \mathbf{E} = 0$, we have:

$$\nabla \cdot \mathbf{E} = \frac{1}{r}\frac{\partial (rE_r)}{\partial r} + \frac{1}{r}\frac{\partial E_\theta}{\partial \theta} + \frac{\partial E_z}{\partial z} = 0$$

Given that $\mathbf{E}$ has no component in the θ direction, i.e. $E_\theta = 0$, and the divergence of $\mathbf{E}$ is zero, we have:

$$\frac{1}{r}\frac{\partial (rE_r)}{\partial r} + \frac{\partial E_z}{\partial z} = 0$$

Rewriting, we get:

$$\frac{\partial (rE_r)}{\partial r} = -r\frac{\partial E_z}{\partial z}$$

We need to integrate both sides with respect to r:

$$rE_r = -\int r\frac{\partial E_z}{\partial z}\, dr$$

Assuming $\frac{\partial E_z}{\partial z}$ is independent of r, we get:

$$rE_r = -\frac{\partial E_z}{\partial z}\int r\, dr, \qquad rE_r = -\frac{\partial E_z}{\partial z} \cdot \frac{r^2}{2}$$

Solving for E_r, we obtain:

$$E_r = -\frac{1}{2} r \frac{\partial E_z}{\partial z}$$

$$E_r = -\frac{r}{2} \frac{\partial E_z}{\partial z} = \frac{1}{2} V''(z) \tag{5.25}$$

Using equation 5.24 and the relation $\ddot{r} = v_z^2 r''$, we have:

$$r'' = \frac{q}{mv^2}\left(r' \frac{\partial V}{\partial z} - \frac{\partial V}{\partial r}\right) \tag{5.26}$$

$$r'' = \frac{q}{mv^2}\left(r' V'(z) + \frac{r}{2} V''(z)\right) \tag{5.27}$$

Also

$$r_2' - r_1' = \frac{q}{mv^2} \int_{z_1}^{z_2} E_r \, dz = -\frac{q r_1}{2mv^2}(E_2 - E_1) \tag{5.28}$$

$$r_2' - r_1' = -\frac{q r_1}{2mv^2}(E_2 - E_1) \tag{5.29}$$

This way equation arises from Newton's law applied to the radial direction, considering an electrostatic field gradient. Using the kinetic energy relation $E = \frac{1}{2} mv^2$, the expression becomes:

$$r_2' = r_1' - \frac{q}{4E}(E_2 - E_1) r_1 \tag{5.30}$$

Introducing the reference kinetic energy per unit charge as $V_0 = E/q$, we obtain:

$$r_2' = r_1' - \frac{1}{4V_0}(E_2 - E_1) r_1 \tag{5.31}$$

This change in slope can be described by a transfer matrix for a thin lens:

$$\begin{bmatrix} r_2 \\ r_2' \end{bmatrix} = \begin{bmatrix} 1 & 0 \\ -\frac{E_2 - E_1}{4V_0} & 1 \end{bmatrix} \begin{bmatrix} r_1 \\ r_1' \end{bmatrix} \tag{5.32}$$

To analyze the full Einzel lens, we now define a voltage ratio that simplifies the energy change calculations:

$$R = \sqrt{\frac{V_2}{V_1}} = \sqrt{1 + \frac{qV}{E}} = \sqrt{\frac{E + qV}{E}} \tag{5.33}$$

Let the central drift region have total length $2d$. The particle accelerates or decelerates from v_1 to v_2, and the average velocity is:

$$v_{\text{avg}} = \frac{v_1 + v_2}{2} = v_1 \cdot \frac{1 + R}{2} \tag{5.34}$$

The time taken to cross this region is:

$$\Delta t = \frac{2d}{v_{\text{avg}}} = \frac{4d}{v_1(1+R)} \tag{5.35}$$

Multiplying by the entrance velocity gives the equivalent drift length:

$$\Delta z = v_1 \cdot \Delta t = \frac{4d}{1+R} \tag{5.36}$$

Thus, the drift-in-field region has the matrix:

$$M_d = \begin{bmatrix} 1 & \dfrac{4d}{1+R} \\ 0 & \dfrac{1}{R} \end{bmatrix} \tag{5.37}$$

Next, we include the effects of the entrance and exit gaps. These cause thin-lens-like angular kicks. Define:

$$k_1 = \frac{R^2-1}{2}, \quad K_1 = \begin{bmatrix} 1 & 0 \\ -k_1 & 1 \end{bmatrix} \tag{5.38}$$

$$k_2 = \frac{1-R^{-2}}{2}, \quad K_2 = \begin{bmatrix} 1 & 0 \\ -k_2 & 1 \end{bmatrix} \tag{5.39}$$

The full Einzel lens matrix, composed of entrance kick, drift-in-field, and exit kick, is:

$$M_{db} = K_2 \cdot M_d \cdot K_1 \tag{5.40}$$

We now compute this product explicitly:
First multiply $M_d \cdot K_1$:

$$M_1 = \begin{bmatrix} 1 & L \\ 0 & 1/R \end{bmatrix} \begin{bmatrix} 1 & 0 \\ -k_1 & 1 \end{bmatrix} = \begin{bmatrix} 1-Lk_1 & L \\ -k_1/R & 1/R \end{bmatrix} \tag{5.41}$$

Now multiply $K_2 \cdot M_1$:

$$M_{db} = \begin{bmatrix} 1 & 0 \\ -k_2 & 1 \end{bmatrix} \begin{bmatrix} 1-Lk_1 & L \\ -k_1/R & 1/R \end{bmatrix} \tag{5.42}$$

$$= \begin{bmatrix} 1-Lk_1 & L \\ -k_2(1-Lk_1) - k_1/R & -k_2L + 1/R \end{bmatrix} \tag{5.43}$$

Now substitute $L = \frac{4d}{1+R}$, $k_1 = \frac{R^2-1}{2}$, $k_2 = \frac{1-R^{-2}}{2}$, and simplify. This yields:

$$M_{db} = \begin{bmatrix} \frac{1}{2}(R+1) & \frac{2d}{1+R} \\ \frac{(R^2-1)(3R+1)}{8dR^2} & \frac{3R-1}{2R^2} \end{bmatrix} \tag{5.44}$$

We now present an alternate derivation treating the Einzel lens as the product of two identical thick tube lenses:

$$M_{\text{tube}} = \begin{bmatrix} A & B \\ C & D \end{bmatrix} \tag{5.45}$$

where:

$$A = 2 - \frac{3}{4}R, \quad B = \frac{d}{R} \cdot \frac{3R-1}{1+R}, \quad C = \frac{3}{16dR}(R^2-1)(R-1)(3-R),$$
$$D = 2 - \frac{3R}{4} \tag{5.46}$$

We now square the tube matrix:

$$M_{\text{Einzel}} = M_{\text{tube}}^2 = \begin{bmatrix} A & B \\ C & D \end{bmatrix}\begin{bmatrix} A & B \\ C & D \end{bmatrix} \tag{5.47}$$

$$= \begin{bmatrix} A^2 + BC & AB + BD \\ CA + DC & CB + D^2 \end{bmatrix} \tag{5.48}$$

Now substitute expressions for A, B, C, D, expand all terms:

$$A^2 = 4 - 3R + \frac{9}{16}R^2,$$
$$BC = B \cdot C = \frac{3}{16R^2}(R^2-1)(R-1)(3-R) \cdot \frac{3R-1}{1+R} \tag{5.49}$$

Add these to find:

$$a_{11} = A^2 + BC = 4 - \frac{3R}{2} \tag{5.50}$$

Similarly,

$$a_{12} = AB + BD = B(A+D) = B\left(4 - \frac{3R}{2}\right) = \frac{2d}{R} \cdot \frac{3R-1}{1+R} \tag{5.51}$$

Also,

$$a_{21} = CA + DC = C(A+D) = C\left(4 - \frac{3R}{2}\right)$$
$$= \frac{3}{8dR}(R^2-1)(R-1)(3-R) \tag{5.52}$$

And,

$$a_{22} = CB + D^2 = C \cdot B + D^2 = 4 - \frac{3}{2R} - \frac{3R}{2} \tag{5.53}$$

Putting it all together:

$$M_{\text{Einzel}} = \begin{bmatrix} 4 - \frac{3R}{2} & \frac{2d}{R} \cdot \frac{3R-1}{1+R} \\ \frac{3}{8dR}(R^2-1)(R-1)(3-R) & 4 - \frac{3}{2R} - \frac{3R}{2} \end{bmatrix} \tag{5.54}$$

The focal strength of the Einzel lens is determined by the matrix element a_{21}:

$$\frac{1}{f} = a_{21} = \frac{3}{8dR}(R^2-1)(R-1)(3-R) \tag{5.55}$$

In the thin lens approximation, we replace d with a smaller effective length a, yielding:

$$\frac{1}{f} = \frac{3}{8aR}(R^2-1)(R-1)(3-R) \tag{5.56}$$

This derivation reveals that the Einzel lens always results in a net focusing action for $0 < R < 1$ or $1 < R < 3$, and becomes defocusing when $R > 3$. The matrix forms and focal length expressions allow for quantitative modeling of the lens behavior based on geometry and applied voltages.

For example: a (14 mm) is the longitudinal gap between the two electrodes of einzel lens, q is the charge of the beam, V_0 is the voltage on central electrode, and E is the energy of the beam. When all the geometrical parameters are considered, the voltage V [79] along the axis in an einzel lens for a particular focal length f is modified as follows:

$$\begin{aligned} V(f) = \frac{-V_0 d}{2\omega g}\left[\ln\left(\frac{\cosh(\omega(f+b/2)/d)}{\cosh(\omega(f+b/2+g)/d)}\right)\right. \\ \left. + \ln\left(\frac{\cosh(\omega(f-b/2)/d)}{\cosh(\omega(f-b/2-g)/d)}\right)\right] \end{aligned} \tag{5.57}$$

This is the expression for three-tube einzel lens as shown in Figure 5.4. Consider, for example, b (50 mm) is the length of the central electrode, g (14 mm) is the separation gap between einzel lens electrodes, d (17 mm) is the radius of all the electrodes of einzel lens, and ω (1.315) is constant. The term 8d/3f is plotted against R to find out practical parameters (Figure 5.5). A graph for the einzel lens choice configuration is shown in Figure 5.6 between einzel lens voltage and focal length. It gives us a preliminary estimate of einzel voltages to handle 45 keV H^- beams.

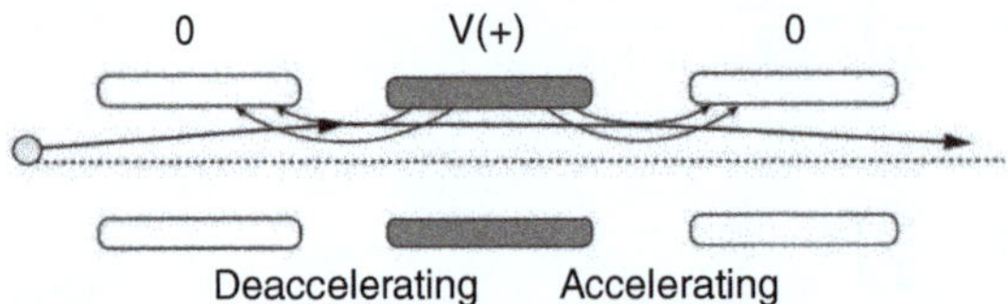

Figure 5.4 Schematics of Einzel lens in decelerating accelerating mode.

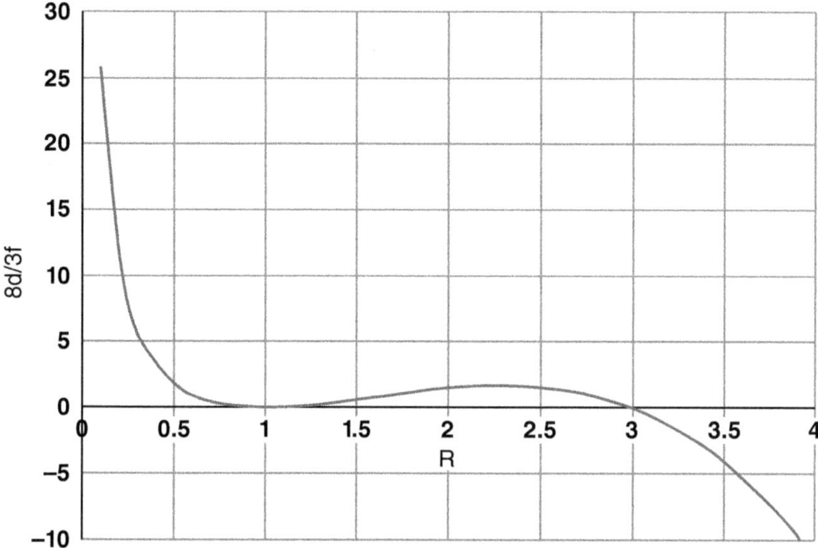

Figure 5.5 Focusing term in einzel lens versus R.

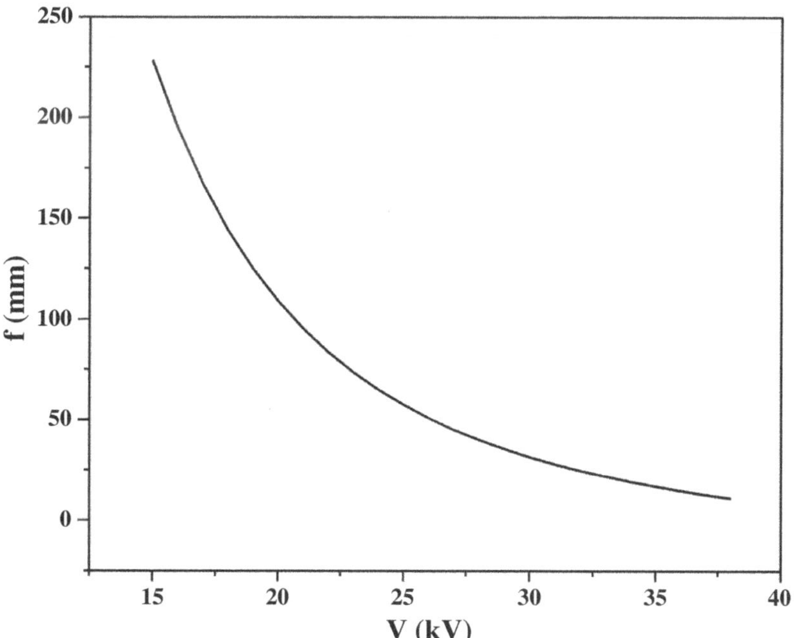

Figure 5.6 Focal length of einzel lens versus the central voltage for 45 keV H^- ion beam.

Figure 5.7 shows the two types of einzel lens configurations such as in tube and aperture form of einzel lens alongwith their beam optics using COSY infinity in Figure 5.8, respectively.

For a three-aperture einzel lens [80] as shown in Figure 5.7, voltage required on einzel lens is given as follows:

$$V(s) = \frac{V_0}{\pi c}\left[\left(s+\frac{L}{2}+c\right)\tan^{-1}\left(\frac{s+\frac{L}{2}+c}{d}\right)+\left(s-\frac{L}{2}-c\right)\tan^{-1}\left(\frac{s-\frac{L}{2}-c}{d}\right)\right.$$
$$\left.-\left(s+\frac{L}{2}\right)\tan^{-1}\left(\frac{s+\frac{L}{2}}{d}\right)-\left(s-\frac{L}{2}\right)\tan^{-1}\left(\frac{s-\frac{L}{2}}{d}\right)\right] \tag{5.58}$$

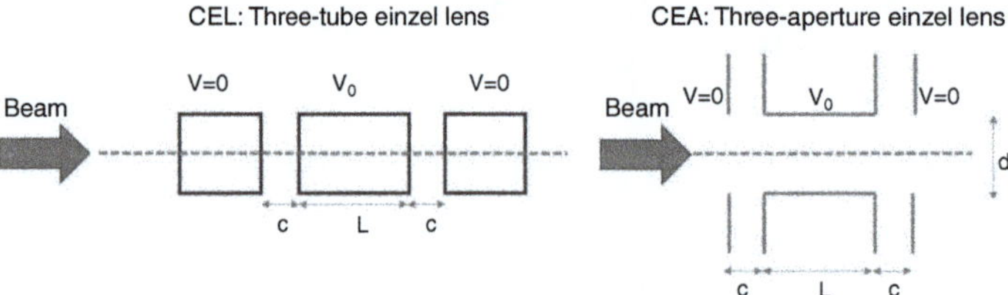

Figure 5.7 Two types of einzel configurations.

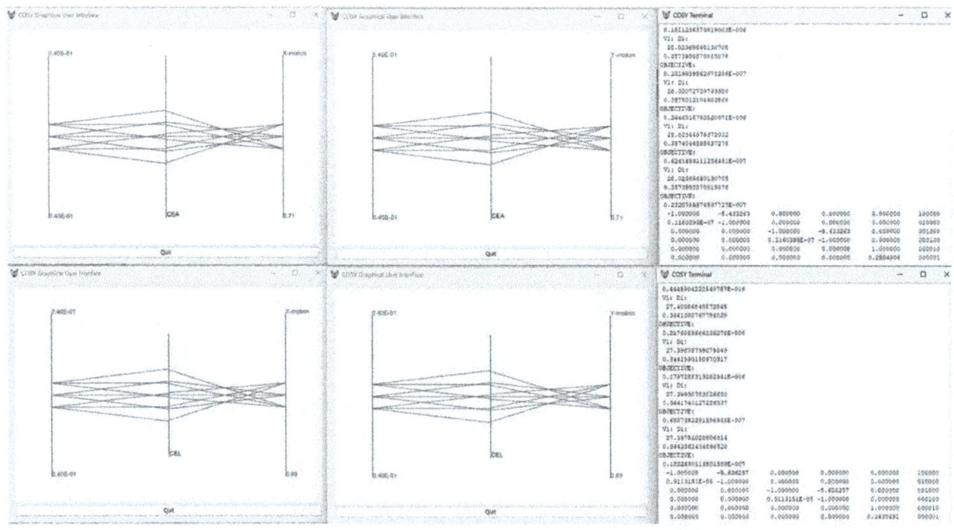

Figure 5.8 Beam optics using COSY Infinity for two types of einzel configurations as per Figure 5.7 dealing the same beam of 45 keV proton beam. Slight difference is found in terms of optimization of input, output drifts along with central einzel voltage.

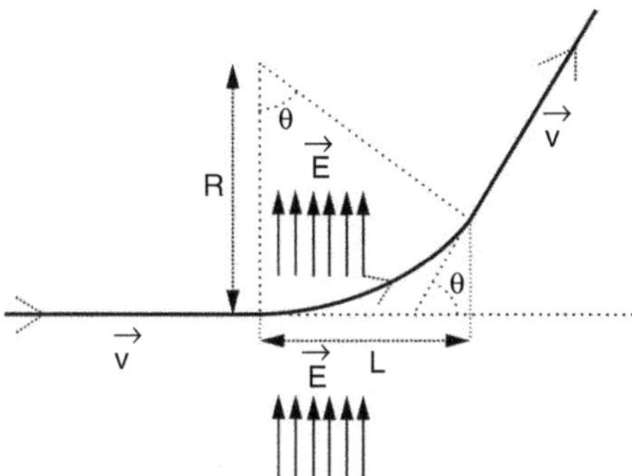

Figure 5.9 Deflection of charged particle in electric field.

5.5 Electrostatic Steerer/Deflector

Electrostatic X–Y beam steerers are used to deflect the charged particle beams in the vertical and horizontal directions by supplying high voltages to the plates. The deflection of ion beams is independent of their masses.

The equation of motion of charged particle is given as follows:

$$\frac{mv^2}{r} = qE = \frac{qV}{g} \tag{5.59}$$

$$r = \frac{\gamma m_0 \beta^2 g}{qV} \tag{5.60}$$

For small angle deflection, one may write

$$\theta = \frac{L}{r} \tag{5.61}$$

Using above equation, one may get easily:

$$\theta = \left[\frac{V.L}{\gamma E_p g \beta^2}\right] \frac{q}{A} \tag{5.62}$$

Figure 5.9 shows the schematics of ion trajectory in electrostatic deflector.

5.6 Electrostatic Accelerating Tube

Electrostatic accelerators have long been integral to research, with applications in material modification, ion implantation, ion beam analysis, etc. These applications require a high-quality beam, which depends on the ion source and the components used in the accelerators. Among these, the most crucial component is the electrostatic accelerating tube (EAT). The advancements in EAT design can lead to improved performance and cost reduction for new machines.

The total particle energy (E) for a nonrelativistic particle having charge q and potential difference U is given as follows:

$$E = qU \tag{5.63}$$

In a region of electric field $\mathbf{E}$, the applied force F for $j = x, y$, or z is

$$F_j = q\mathbf{E}_j = m\frac{dv_j}{dt} \tag{5.64}$$

where m is the mass of the particle. Take separate versions for $j = x, y$, and z, multiply it by v_j/v and then added this equation becomes:

$$q\left(\frac{v_x}{v}E_x + \frac{v_y}{v}E_y + \frac{v_z}{v}E_z\right) = m\left(\frac{v_x}{v}\frac{dv_x}{dt} + \frac{v_y}{v}\frac{dv_y}{dt} + \frac{v_z}{v}\frac{dv_z}{dt}\right) \tag{5.65}$$

For EATs E_z is the dominant term; so, left-hand side (LHS) of the above equation can be written as:

$$q\left(\frac{v_x}{v}E_x + \frac{v_y}{v}E_y + \frac{v_z}{v}E_z\right) = qE_z\frac{v_z}{v}\left(1 + \frac{v_x}{v_z}E_x + \frac{v_y}{v_z}E_y\right) \tag{5.66}$$

$$= qE_z\frac{v_z}{v}\left(1 + x'\tan\theta_x + y'\tan\theta_y\right) = qkE_z \tag{5.67}$$

where $\theta_x = arc\tan(Ex/Ez)$ and $\theta_y = arc\tan(Ey/Ez)$ represents the angles of inclination of fields with respect to the tube axis, the quantity $x' = v_x/v_z$ and $y' = v_y/v_z$ denotes the divergences in the x and y directions, respectively, and

$$k = \frac{v_z}{v}\left(1 + x'\tan\theta_x + y'\tan\theta_y\right) \tag{5.68}$$

As,

$$v^2 = v_x^2 + v_y^2 + v_z^2$$

$$\Rightarrow \frac{dv}{dt} = \frac{v_x}{v}\frac{dv_x}{dt} + \frac{v_y}{v}\frac{dv_y}{dt} + \frac{v_z}{v}\frac{dv_z}{dt} \tag{5.69}$$

Thus from above equations, one can get:

$$qkE_z = m\frac{dv}{dt} \tag{5.70}$$

$$\Rightarrow dt = \frac{m}{qkE_z}dv \tag{5.71}$$

Now, x-component of the particle velocity can be written as $v_x = dx/dt$, integrating it from initial state x_i, v_0, and t_i to the final state x_f, v, and t_f

$$\int_{x_i}^{x_f} dx = \int_{t_i}^{t_f} v_x dt \tag{5.72}$$

Put dt from Equation 5.50 in Equation 5.51

$$\Rightarrow x_f - x_i = \int_{v_i}^{v} \frac{v_x m}{qkEz} dv \tag{5.73}$$

$$\Rightarrow x_f - x_i = \int_{v_i}^{v} \frac{x_i' v_{z0} m}{qkEz} dv \tag{5.74}$$

where x_i is the divergence in x direction and v_{z0} is the z component of the velocity, respectively, at the initial state. Now, replacing $E_z = \frac{U-U_0}{L}$, $v_{z0} = \sqrt{\frac{2qU_0}{m}} = v_0$ and k = 1 ($v_z = v \Rightarrow k = \frac{v_z}{v} = 1$) where L is the length of the accelerating tube, so above equation becomes

$$x_f - x_i = \frac{x_i' L m}{q(U - U_0)} \sqrt{\frac{2qU_0}{m}} (v - v_0) \tag{5.75}$$

Now, using $v = \sqrt{\frac{2qU}{m}}$ and simplifying the above equation, this will result in

$$x_f = x_i + \frac{2L}{1 + R} x_i' \tag{5.76}$$

Here R is a figure of merit for EAT and is given as follows:

$$R = \sqrt{\frac{U}{U_0}} \tag{5.77}$$

Now, consider x_f' which is divergence in x direction at final state, which is $x_f' = \sqrt{\frac{Ex}{E}}$ and also, $x_i' = \sqrt{\frac{Ex}{E_0}}$. The quantity x_f'/x_i' is

$$\frac{x_f'}{x_i'} = \frac{1}{R} \tag{5.78}$$

For axial acceleration, linear matrix can be written by using the above equations as:

$$\begin{bmatrix} a_{x/x} & a_{x/x'} \\ a_{x'/x} & a_{x'/x'} \end{bmatrix} = \begin{bmatrix} 1 & 2L/(R+1)) \\ 0 & 1/R \end{bmatrix} == \begin{bmatrix} R11 & R12 \\ R21 & R22 \end{bmatrix} \tag{5.79}$$

This expression represents the transfer matrix for an axial EAT model applied to a non-relativistic beam. The overall transfer matrix (T) based on the schematic of the EAT, as depicted in Figure 5.10, is given by the following expression:

$$T = \begin{bmatrix} 1 & 0 \\ \frac{1}{f_2} & 1 \end{bmatrix} \begin{bmatrix} 1 & \frac{2l}{R+1} \\ 0 & \frac{1}{R} \end{bmatrix} \begin{bmatrix} 1 & 0 \\ -\frac{1}{f_1} & 1 \end{bmatrix}$$

$$T = \begin{bmatrix} -\dfrac{R-3}{2} & \dfrac{2l}{R+1} \\ -\dfrac{3(R+1)(R-1)^2}{8LR^2} & \dfrac{3R-1}{2R^2} \end{bmatrix}$$

Here

$$f_1 = \frac{4U_0L}{U-U_0} = \frac{4L}{R^2-1}$$

and

$$f_2 = \frac{4UL}{U-U_0} = \frac{4LR^2}{R^2-1}$$

Next, consider the focusing of an ion beam with a finite emittance ϵ using an accelerator tube, taking into account the effect of the random spread in ion velocities on the beam's focusing. This scenario is recognized as Elkind's problem [81] for beams with finite emittance. Assume that r_0 is the beam waist size at a distance L_1 from the entrance of the EAT, as illustrated in Figure 5.10. The distance L_2 [82], at which the beam waist occurs after passing through the EAT for a beam with finite emittance, is determined by the following expression:

$$L_2 = \frac{4NL_T\left(\sqrt{N}-1-2\xi\right)\left(r_0^4+\epsilon^2 S_1 S_2\right)}{3(N-1)\left(\xi-\sqrt{N}\right)\left(r_0^4+\epsilon^2 S_2^2\right)} \tag{5.80}$$

In this equation, L_T denotes the length of the tube, N represents the ratio of the potential at the final electrode to that at the initial electrode, and ξ is a function dependent on the aperture diameter (D) and the ratio $\phi/E_1 - E_2$ (where ϕ is the potential and E_1 and E_2 are the electric fields before and after the aperture, respectively). The terms S_1 and S_2 are defined as follows:

$$S_1 = \frac{4L_T\xi}{(\sqrt{N}+1)(\sqrt{N}-1-2\xi)} - L_1$$

$$S_2 = \frac{4L_T\xi(3\sqrt{N}-1)}{3(N-1)(\sqrt{N}-\xi)} - \frac{L_1(3\sqrt{N}-1-2\xi)}{3(\sqrt{N}-\xi)}$$

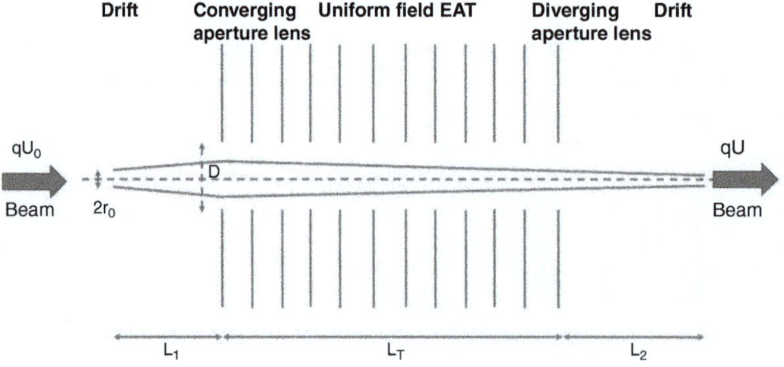

Figure 5.10 For calculations of beam focusing by an EAT.

To design an accelerating tube, one needs to have thin electrodes properly shaped in such a way that after arranging them sequentially, one should get a flat electric field profile with accuracy around 10^{-3} or better. The spacing of electrodes also needs to be done in such a way that secondary electrons generated by the impact of the beam, get suppressed. One can also use permanent magnets to deflect away such electrons. The accelerating electrodes need to be properly spaced and perfectly aligned to get a flat electric field profile for beam acceleration. Since it is an EAT, for a total voltage V_T to be generated across the total resistance of tube R_T, a current I needs to flow, and it is governed by Ohm's law as follows:

$$V_T = \frac{I}{R_T}$$

$$R_T = R1 + R2 + R3 + \ldots.. \tag{5.81}$$

where R_T is the sum of all resistances (R1, R2, R3, ...) across the individual gaps among the electrodes.

5.7 Electrostatic Septum

The electrostatic septum is an important device in particle accelerators, allowing a charged particle beam to be extracted or injected without disturbing the circulating beam. It works by placing a very thin barrier, called the septum, between two regions: one with no electric field for the circulating beam, and one with a high electric field to deflect the extracted beam. The septum is usually made of a foil or thin wires to minimize interference.

The force acting on a charged particle inside the electric field is:

$$F_x = qE_x = m\gamma a_x \quad \Longrightarrow \quad a_x = \frac{qE_x}{m\gamma} \tag{5.82}$$

where q is the particle charge, Ex is the electric field, m is the rest mass, and γ is the Lorentz factor. The time the particle spends inside the electric field of effective length l_{eff}, traveling at speed, $v = \beta c$ is:

$$t = \frac{l_{\text{eff}}}{\beta c} \tag{5.83}$$

The transverse velocity gained is:

$$\Delta v_x = a_x t = \frac{qE_x}{m\gamma} \cdot \frac{l_{\text{eff}}}{\beta c} \tag{5.84}$$

The transverse momentum becomes:

$$\Delta p_x = m\gamma \Delta v_x = qE_x \frac{l_{\text{eff}}}{\beta c} \tag{5.85}$$

The deflection angle is found by:

$$\theta_E = \tan^{-1}\left(\frac{\Delta p_x}{p}\right) \tag{5.86}$$

Substituting for Δp_x:

$$\theta_E = \tan^{-1}\left(\frac{qE_x l_{\text{eff}}}{\beta cp}\right) \tag{5.87}$$

Expressing q in units of elementary charge e and converting from eV to GeV with factor 10^9:

$$\theta_E = \tan^{-1}\left(\frac{E_x l_{\text{eff}}}{p \cdot 10^9 \cdot \beta}\right) \tag{5.88}$$

Since $E_x = \frac{V_x}{d}$, we substitute:

$$\theta_E = \tan^{-1}\left(\frac{\left(\frac{V_x}{d}\right) l_{\text{eff}}}{p \cdot 10^9 \cdot \beta}\right) = \tan^{-1}\left(\frac{V_x l_{\text{eff}}}{d \cdot p \cdot 10^9 \cdot \beta}\right) \tag{5.89}$$

5.8 Numerical Problems

1. What is the electric field required to bend a beam of 100 eV, 1 keV, 1 MeV, and 1 GeV proton beam with geometrical emittance 1 π mm-mrad by an electrostatic dipole (ED) by 90°. Discuss the practical applicability of such an electric dipole in all cases. Feel free to choose design parameters of ED to accommodate such beam.
2. Calculate the focal length of electrostatic quadrupole triplet (EQT) for 100 keV electron beam with geometrical emittance 1 π mm-mrad. Feel free to choose hardware parameters of EQT to accommodate such beam.
3. Calculate the focal length of einzel lens (EL) for 10 keV proton beam with geometrical emittance 100 π mm-mrad. Feel free to choose hardware design parameters of EL to accommodate such beam.
4. Design the various electrostatic tubes for a Tandem accelerator for 100 MeV energy gain of Si^{7+} ion beams, geometrical emittance of 1 π mm-mrad. The input energy to tandem accelerator is 250 keV for Si^{1-} ion beam.
5. What is the electric field required to steer a 100 keV O^{2+} by 5 mm at a distance of from its beam trajectory parallel to the central axis. Design a practically possible and economical electrostatic steerer for this purpose.
6. Deflect $^{28}Si^{10+}$ beam of energy 100 keV by 1 mrad. What is the voltage required by an electrostatic steerer of length 100 mm and full aperture 50 mm?
7. Design an einzel lens for 45 keV proton beams with emittance as 50 π mm-mrad as compact as possible without any electrostatic discharge problems.
8. Design an electrostatic quadrupole triplet for 2 keV/u ion beams with 100 π mm-mrad emittance to focus in both transverse planes.
9. Calculate the basic design parameters of a Tandem accelerator for its injector dipole magnet, analyzer dipole magnet as well as energy gain scheme in the accelerator. Feel free to choose bending radius for a practical design of dipole magnet. Do your calculations for obtaining 200 MeV Au and 100 MeV Ag beams. What is the terminal potential required in both cases using one charge state stripper.

6

Radio Frequency Devices

> "My little machine was a primitive precursor of this type of accelerator which today is called a 'linac' for short. However I must now emphasise one important detail. The drift tube was the first accelerating system which had earthed potential on both sides, i.e. at both the particles' entry and exit, and was still able to accelerate the particles exactly as if a strong electric field was present. This fact is not trivial. In all naivetè one may well expect that, when the voltage on the drift tube is reversed, the particles flying within would be decelerated, which is clearly not the case. After I had proven that such structures, earthed at both ends, were effectively possible, many other such systems were invented."
>
> —*Rolf Widerøe*

After reading this chapter, you should be able to:

- Understand how to accelerate charged particles using radio frequency (RF) fields.
- Define the different types of RF structures for different actions over particle beams.
- Understand longitudinal beam dynamics and the concept of synchronous particles.

Curiosity: *In particle accelerators, longitudinal beam dynamics helps answer an interesting question: how can we make sure that tiny charged particles stay in the right place and time to get accelerated? Inside the accelerator, particles move through electric fields that change very quickly. To gain energy, a particle has to arrive at each gap exactly when the electric field is pointing in the right direction. If it comes too early or too late, it won't get the boost it needs. Longitudinal beam dynamics studies how particles keep this timing while speeding up, how small differences in their energy and phase affect their motion, and how they stay together as a group. It's like making sure every runner in a race stays in rhythm with a moving walkway that speeds them up at just the right moments. This part of accelerator physics is key to making sure beams stay focused and gain energy properly, so they can be used in experiments, medical treatments, or even smashing particles to explore the universe.*

Charged Particle Beam Physics: An Introduction for Physicists and Engineers, First Edition.
Sarvesh Kumar and Manish K. Kashyap.

Companion Website: https://www.wiley.com/go/Kumar_1e

6.1 Longitudinal Beam Dynamics

Just as in transverse beam dynamics, where the beam has a finite size in transverse dimensions, the beam also forms bunches with a finite length in longitudinal phase space. A radio frequency (RF) linear accelerator (LINAC) [83, 84] is typically designed to accelerate a specific particle known as the synchronous particle, which remains in synchronization with the RF fields. The other particles in the beam bunch are distributed around this synchronous particle in terms of energy and phase. Longitudinal restoring forces are generated when the beam is accelerated by a time-varying electric field, resulting in phase and energy oscillations around the synchronous particle.

In longitudinal beam dynamics, we describe the motion of charged particles along the accelerator by comparing each one to a special particle called the synchronous particle. This reference particle travels along the ideal path with the correct energy, velocity, and radio frequency phase, interacting with the accelerating field exactly as intended.

To understand how real particles behave, we describe them in terms of small deviations from this ideal. These differences help us explain how a bunch of particles evolves, whether it spreads out, compresses, or shifts as it moves through radio frequency fields.

We start by measuring the time difference between a particle and the synchronous particle:

$$\Delta t = t - t_0 \tag{6.1}$$

This tells us whether the particle arrives early or late. A positive value means the particle is behind schedule, while a negative one means it is ahead.

This time difference leads to a longitudinal position difference:

$$\Delta z = \beta ct - \beta_0 ct_0 \tag{6.2}$$

Alternatively, if we know the positions directly, we can write:

$$\Delta z = z - z_0 \tag{6.3}$$

Here, β and β_0 are the velocities of the particle and the synchronous particle expressed as fractions of the speed of light c, and z and z_0 are their respective longitudinal positions.

The radio frequency phase a particle sees depends on when it arrives. Its deviation from the synchronous phase is expressed as:

$$\Delta\phi = \phi - \phi_s, \quad \text{with} \quad \phi = 2\pi ft \tag{6.4}$$

This phase difference determines how much energy the particle receives from the radio frequency field. If a particle is early or late compared to the ideal phase, it gains more or less energy, which causes it to oscillate around the reference phase.

Now we consider energy and momentum:

$$\Delta E = E - E_0 \tag{6.5}$$

This is the difference in total energy compared to the reference particle. It is particularly important in linear accelerators.

$$\delta = \frac{P - P_0}{P_0} \tag{6.6}$$

This is the relative momentum deviation, which is commonly used in circular accelerators such as synchrotrons. Even small differences in momentum can lead to noticeable changes in orbit length due to dispersion.

We can also express longitudinal motion in terms of fractional velocity deviation:

$$\Delta z' = \frac{v - v_0}{v_0} \tag{6.7}$$

This describes how much faster or slower a particle is compared to the synchronous particle.

Together, the quantities Δt, Δz, $\Delta\phi$, ΔE, δ, and $\Delta z'$ form the foundation of longitudinal beam dynamics. Depending on the application, we can describe the longitudinal phase space using coordinates such as $(\Delta\phi, \Delta E)$, $(\Delta z, \delta)$, or $(\Delta t, \delta)$.

These coordinates are essential for understanding how particle bunches behave in accelerators. They are used to analyze processes like bunch compression, synchrotron oscillations, and radio frequency capture, which play a central role in the design and operation of both linear and circular machines.

Longitudinal phase space is just as important as the transverse phase space, especially for bunched ion beams with significant energy spread. Beam bunching is essential for utilizing the RF accelerator to further increase energy. Important parameters like energy spread, phase spread, or time spread for a bunched beam define the longitudinal phase space, typically expressed as $\Delta E - \Delta\phi$ or $\Delta E - \Delta T$ phase spaces.

The longitudinal emittance is defined as:

$$\epsilon_n = \Delta E \times \Delta T \tag{6.8}$$

The normalized units are typically keV/u * ns. In these units, emittance remains normalized and does not change during acceleration or deceleration, provided there is no particle loss. Therefore, this quantity serves as a figure of merit for any accelerator. Normalizing emittance is crucial for making meaningful comparisons across different energy levels and accelerators. Other units of longitudinal emittance include deg*% and deg.keV. For example, a longitudinal emittance of 3 keV/u ns for a beam with an ion mass-to-charge ratio of 6 corresponds to the following at different frequencies:

$$3 \times 6 \times \frac{360}{83} = 78 \text{ deg.keV at } 12.125 \text{ MHz} \tag{6.9}$$

$$3 \times 6 \times \frac{360}{20.7} = 312 \text{ deg.keV at } 48.5 \text{ MHz} \tag{6.10}$$

$$3 \times 6 \times \frac{360}{10.3} = 624 \text{ deg.keV at } 97 \text{ MHz} \tag{6.11}$$

The longitudinal phase space is also characterized by two coordinates: $z(m)$ and $\frac{\Delta p}{p}$ (mrad). Here, z corresponds to the distance of a particle relative to the synchronous particle, directly relating to the bunch structure in terms of phase spread, time width or length of the beam bunch.

$$\Delta t = -\frac{z}{\beta c} \tag{6.12}$$

$$z = -\beta c\Delta t = \frac{\beta\lambda}{360}\Delta\phi \tag{6.13}$$

If $z \geq 0$, then $\Delta t \leq 0$, meaning the particles are ahead of the synchronous particles and therefore earlier in time. The term $\frac{\Delta p}{p}$ (mrad) corresponds to the fractional deviation of momentum relative to the synchronous particle. This is directly related to velocity deviation $\frac{\Delta\beta}{\beta}$ or energy deviation $\frac{\Delta E}{E}$.

$$\frac{\Delta p}{p} = \frac{\gamma}{1+\gamma}\frac{\Delta E}{E} \tag{6.14}$$

For a drift of length L, the transfer matrix is given by:

$$M_z = \begin{bmatrix} 1 & \frac{L}{\gamma^2} \\ 0 & 1 \end{bmatrix} \tag{6.15}$$

Consider two particles of energy E and $E + \Delta E$ traveling an equivalent distance d. Due to difference in velocities (v), a time spread will develop which can be calculated as follows:

$$v^2 = 2mE; \tag{6.16}$$

Differentiating both sides and divided by the above equation, one can get:

$$\frac{dv}{v} = \frac{1}{2}\frac{dE}{E} \tag{6.17}$$

This can also be written as follows:

$$\frac{dt}{t} = \frac{dv}{v} = \frac{1}{2}\frac{dE}{E} \tag{6.18}$$

$$dt = \frac{1}{2}\frac{d}{\beta c}\frac{dE}{E} \tag{6.19}$$

The position of buncher and its operating frequency (f) should be such that its time spread of the bunched beam is given as follows:

$$\Delta T \leq \pm\frac{\pi}{4\omega} \tag{6.20}$$

where $\omega = 2\pi f$, which gives the conditions as follows:

$$d \leq \frac{\pi}{4\omega}\frac{2E\beta c}{\Delta E}, \quad d \leq \frac{E\beta c}{4f\Delta E} \tag{6.21}$$

In summary, longitudinal beam dynamics is crucial for understanding and controlling the behavior of particle bunches in accelerators, much like transverse beam dynamics. The phase space in the longitudinal plane, defined by parameters such as energy spread, phase spread, and time spread, plays a vital role in ensuring efficient acceleration and beam quality. The longitudinal emittance, which remains constant in normalized units throughout acceleration, serves as an essential figure of merit for comparing beams across different energy regimes and accelerators. Additionally, the use of time-varying electric fields in the longitudinal plane enables particle acceleration while conserving the phase space area, ensuring beam stability. Understanding the relationship between particle position, momentum deviation, and energy spread allows for the precise control of beam bunching, which is critical for maintaining the structure and quality of high-energy particle beams during their transit through the accelerator.

6.2 Pillbox Cavity

A pillbox cavity is a fundamental resonant cavity employed in particle accelerators, distinguished by its cylindrical geometry with parallel conductive end plates and a specific radius. It acts as a basic framework for exploring electromagnetic field behavior and mode structures within accelerator systems. By applying Maxwell's equations with the relevant boundary conditions, the pillbox cavity helps elucidate essential principles of field dynamics and resonance. It accommodates various modes, such as transverse electric (TE) and transverse magnetic (TM), each exhibiting unique field patterns. This elementary structure is vital for grasping core concepts and refining more intricate accelerator designs, serving as a basis for developing advanced RF cavities and beamline configurations.

To determine the general fields of a pillbox cavity, we solve Maxwell's wave equation with boundary conditions imposed by parallel conducting plates at $z = 0$ and $z = g$, with a radius b. The pillbox geometry is crucial for study because most accelerating cells are derived from this fundamental structure, sharing the same underlying physics. Essentially, real accelerating cells are advanced versions of the pillbox design, optimized for enhanced beam acceleration efficiency. Consequently, the pillbox cavity serves as an excellent starting point due to its analytically describable features, which provide insight into the physics of more complex geometries. In cylindrical coordinates, the components E_z, E_r, and H_θ are the primary fields that influence the beam and thus play a critical role in cavity design. The pillbox cavity, with its simple geometry, serves as a foundational model to understand electromagnetic fields and resonance modes, such as TE and TM modes, which influence cavity design.

6.3 Traveling Wave Structures

Traveling wave structures function as waveguides where the injected electromagnetic wave travels with a phase velocity matching the velocity of the particle beam along the structure. Particles continuously gain energy from the longitudinal electric field of the wave, depending on their phase relative to the wave crest. As the wave travels, it experiences attenuation due to ohmic losses in the structure's walls and beam loading in high-current scenarios. After a certain distance, the remaining power is absorbed by a load matched to the waveguide's characteristic impedance. A typical traveling wave structure for electron or muon LINACs is disk-loaded. A disk-loaded traveling wave structure is an accelerating waveguide used in LINACs to boost electron speeds. It consists of a cylindrical waveguide lined with metallic disks arranged at regular intervals. These disks create periodic cavities that slow the phase velocity of the traveling electromagnetic wave, aligning it with the electrons' speed. This alignment enables electrons to continuously gain energy as they pass through the structure. This design is commonly employed in electron LINACs for applications such as medical radiotherapy, scientific research, and various industrial processes.

6.4 Standing Wave Structures

Resonant cavities support standing waves when the frequency of the injected wave matches the cavity's resonant frequency. These structures form due to the reflection of the injected wave at the ends of the cavity. A standing wave pattern can be viewed as a linear combination of two traveling waves of equal amplitude moving in opposite directions. The forward wave is synchronized with the particle beam and contributes to acceleration, while the backward wave is not synchronized and does not influence the beam.

Standing waves in an RF cavity are fundamental to the operation of many devices, including particle accelerators, microwave ovens, and various communication systems. An RF cavity is a resonant structure that confines electromagnetic fields. When RF power is supplied to the cavity, it can set up standing electromagnetic waves within the cavity.

1. When an RF signal is introduced into the cavity, it travels and reflects back and forth between the cavity walls.
2. The reflected waves interfere with the incident waves, leading to the formation of standing waves if the conditions are right.
3. Standing waves occur at specific frequencies known as resonant frequencies, which depend on the cavity's dimensions and shape.
4. The cavity must support integral multiples of half-wavelengths for standing waves to form, meaning the cavity size is a crucial factor.

When two traveling waves of the same frequency and amplitude move in opposite directions, they can interfere to form a standing wave. Consider a simple rectangular cavity of dimensions $a \times b \times d$. Let two traveling waves move along the x-axis with equal amplitude A, where the position along the wave's propagation direction is denoted by x and time by t. The wavelength of the wave is given by λ, while the wave number $k = \frac{2\pi}{\lambda}$ describes how many cycles fit into a unit length. The frequency of the wave is f, and the angular frequency $\omega = 2\pi f$ relates the time dependence of the wave to its frequency. Inside an RF cavity, the electric field $E(x, y, z, t)$ is described as a function of spatial coordinates x, y, and z and time t, where E_0 represents the maximum amplitude of the electric field. The dimensions of the cavity are given by a, b, and d, while m, n, and p are mode numbers corresponding to standing wave patterns within the cavity. The speed of light in the medium inside the cavity is represented by c, and the resonant frequency f_{mnp} depends on the cavity dimensions and mode numbers. Finally, in understanding the behavior of standing waves, we describe the stationary points as nodes, where the wave displacement is zero, and antinodes, where the wave displacement is maximum, highlighting the spatial structure of the standing wave inside the cavity.

Consider two traveling waves moving in opposite directions along the x-axis:

1. The first wave traveling to the right can be represented as:

$$y_1(x, t) = A \sin(kx - \omega t) \tag{6.22}$$

2. The second wave traveling to the left can be represented as:

$$y_2(x, t) = A \sin(kx + \omega t) \tag{6.23}$$

When these two waves overlap, the principle of superposition applies. The resulting wave is the sum of the individual waves:

$$y(x,t) = y_1(x,t) + y_2(x,t) = A\sin(kx - \omega t) + A\sin(kx + \omega t) \tag{6.24}$$

Using the trigonometric identity for the sum of sines, we get:

$$y(x,t) = A[\sin(kx - \omega t) + \sin(kx + \omega t)] = A[2\sin(kx)\cos(\omega t)] \tag{6.25}$$

This simplifies to:

$$y(x,t) = 2A\sin(kx)\cos(\omega t) \tag{6.26}$$

Here, $2A\sin(kx)$ represents the spatial part of the standing wave, and $\cos(\omega t)$ represents the temporal part.

1. **Nodes and antinodes**:
   - **Nodes**: Points where the amplitude is always zero ($y = 0$). These occur where $\sin(kx) = 0$, i.e. $kx = n\pi$ for $n = 0, 1, 2, \ldots$.
   - **Antinodes**: Points where the amplitude is maximum ($y = \pm 2A$). These occur where $\sin(kx) = \pm 1$, i.e. $kx = (n + \frac{1}{2})\pi$ for $n = 0, 1, 2, \ldots$.
2. **Stationary pattern**: The pattern of nodes and antinodes remains stationary, though the displacement at any point oscillates with time.

The electric and magnetic fields inside the cavity can be described by the wave equation, which simplifies to the following for standing wave solutions:

$$E(x,y,z,t) = E_0 \sin\left(\frac{m\pi x}{a}\right)\sin\left(\frac{n\pi y}{b}\right)\sin\left(\frac{p\pi z}{d}\right)\cos(\omega t) \tag{6.27}$$

The corresponding resonant frequency f for the mode (m, n, p) is given by:

$$f_{\mathrm{mnp}} = \frac{c}{2}\sqrt{\left(\frac{m}{a}\right)^2 + \left(\frac{n}{b}\right)^2 + \left(\frac{p}{d}\right)^2} \tag{6.28}$$

where c is the speed of light in the medium inside the cavity.

The modes in resonant cavities are categorized into **TE modes**, **TM modes**, and **transverse electromagnetic (TEM) modes**. TE modes have no electric field component in the direction of propagation and are characterized by two integers, m and n, that define the standing wave pattern in the transverse plane. TM modes, which are useful for particle acceleration, have no magnetic field component in the propagation direction and are similarly characterized by m and n. In contrast, TEM modes have both electric and magnetic fields perpendicular to the direction of propagation but cannot exist in a closed cavity; they require a waveguide with two conductors.

In particle accelerators, RF cavities are used to accelerate charged particles. The standing waves create oscillating electric fields that transfer energy to the particles. Standing waves in RF cavities are essential for many applications, from particle accelerators to microwave ovens. These cavities support resonant frequencies that depend on their dimensions and shape, leading to specific modes of resonance (TE, TM, and TEM). Understanding and controlling these standing waves allow for efficient energy transfer and manipulation of electromagnetic fields for various technological applications.

6.4.1 RF Devices

RF and microwave technology are essential in various scientific and engineering fields, especially in particle accelerators, where they accelerate charged particles like electrons and protons to high energies. RF systems are crucial for high-energy particle colliders, material research, medical treatments like cancer therapy, and isotope production. As the applications of RF accelerators continue to grow, so does the need for more powerful RF accelerator systems, despite ongoing challenges that engineers face.

To increase ion beam energy, RF voltage sources are commonly used instead of DC sources. A major challenge is ensuring particles interact with the electric fields at the correct phase for maximum energy transfer. Techniques like varying the spacing between electrodes or adjusting the RF frequency as particles accelerate are used to keep particles in sync. Some structures, like drift tubes, allow particles to pass through field-free regions when the fields would otherwise decelerate them.

At high frequencies, resonant RF cavities are often used for acceleration. Energy from a high-power source like a klystron is introduced into the cavity, where it builds up and is then transferred to the particles as they pass through. These cavities can be placed in a row to form a LINAC, where particles gain energy with each pass through a cavity. Proton LINACs, operating in the UHF range, can increase energy by 3–10 MeV/m. For electrons, which move near the speed of light at even low energies, higher frequencies like 3 GHz are used. For example, the Stanford Linear Accelerator operates at this frequency and can accelerate electrons to 50 GeV.

While the energy gradient in a LINAC increases with the RF power supplied, it's eventually limited by electrical breakdown in the RF cavities. Even though RF systems handle higher power than DC ones, space constraints limit how much energy can be transferred. Circular accelerators, like synchrotrons, bypass this limitation by guiding particles around circular paths with strong magnets, allowing them to pass through the same RF cavities multiple times. As long as particles remain synchronized with the RF fields, they continue gaining energy.

In synchrotrons, the magnetic field is gradually increased to keep particles in sync with the RF fields. This allows ion synchrotrons to achieve very high energy gradients, sometimes hundreds of MeV/m, by reusing the same cavities. However, for electrons, synchrotron radiation limits their maximum energy because they lose significant energy as they bend through the magnetic fields. This loss requires a balance between the RF power supplied and the energy lost through radiation.

In LINACs, particles are accelerated in a straight line, which makes them valuable tools for many practical applications, such as medical imaging, sterilization, and security scanning. LINACs are also essential in research for generating neutron or photon beams, which are used in material and biological sciences. Some accelerators start with electrostatic acceleration at low energies before switching to RF acceleration as the particles gain speed. This hybrid approach improves the efficiency of RF structures, making the overall system more effective for transmitting energy to the particle beam.

6.4.2 Earnshaw's Theorem

Earnshaw's theorem, formulated by Samuel Earnshaw [85] in 1842, states that a group of point charges cannot be maintained in stable stationary equilibrium solely through their electrostatic interactions. This is because it is impossible for all the second derivatives of the potential to be positive (indicating a minimum) simultaneously. In other words, the theorem asserts that a charged body cannot be kept in stable equilibrium under the influence of purely electric forces. When off-axis particles enter a gap and are accelerated by a longitudinal RF electric field, they also experience radial RF electric and magnetic forces. Although it might appear that two oppositely directed radial electric forces on the two halves of the gap would cancel out, there is generally a net radial impulse due to three mechanisms:

1. The field varies in time as the particle crosses the gap.
2. The field also depends on the radial displacement of the particle.
3. The particle's velocity increases as it crosses the gap, so it does not spend equal time on each half of the gap.

For longitudinal stability, the field should rise when the synchronous particle is injected. This ensures that the particle experiences more force in the second half of the gap than in the first half, resulting in a net defocusing force. For ion LINACs, mechanism 1 dominates and is known as the RF defocusing force. Mechanisms 2 and 3 are more significant for electron LINACs.

It has been demonstrated that the condition for longitudinal focusing, which requires RF electric fields to increase with time, is generally incompatible with local radial focusing by these fields. This is a consequence of Earnshaw's theorem. A more general principle derived from Laplace's equation states that the electrostatic potential in free space cannot have a maximum or minimum. The scheme explained above is illustrated in Figure 6.1. It concludes that RF forces are inherently transverse defocusing. From Laplace's equation:

$$\frac{\partial^2 V}{\partial^2 x} + \frac{\partial^2 V}{\partial^2 y} + \frac{\partial^2 V}{\partial^2 z} = 0 \tag{6.29}$$

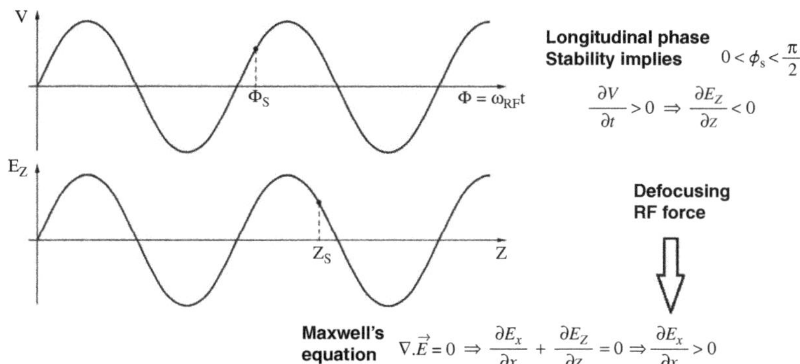

Figure 6.1 Inherent transverse defocusing effect in RF gap structure [86]/CERN.

6.4.3 Motion of a Charged Particle in an RF Field

For a very small aperture, the electric field remains nearly constant within the RF gap, and the ion gains kinetic energy approximately equal to the voltage across the gap. For wider gaps, necessary to sustain the gap voltage without sparking, the ion spends more time in the gap, where the field level is less than the peak field seen in a short gap. The field varies with time and may eventually reverse polarity. In the case of a long gap, the ion might spend more time under the reverse field polarity and could actually be decelerated, leading to $T < 0$.

An idealized calculation assumes a square field profile from $-g/2$ to $+g/2$, where g is the RF gap between two drift tubes. Real fields, however, are not ideal. The calculation also assumes no velocity change due to acceleration in the gap and that the field is at its maximum when the ion is at the center of the gap. For longitudinal focusing, the ion should enter the gap while the field is still rising. A more accurate calculation would integrate the actual field within the gap, including the fringe field in the drift tube bore along the beam axis. The relationship between the bunch centroid and the peak of the gap field determines the stable phase, with the phase being negative for longitudinal focusing.

Consider an RF gap with a length g and origin at the center of the gap. At $t = 0$, the particle is at the center of the gap ($z = 0$):

$$\Delta E = \int_{-g/2}^{g/2} qE_0 \cos\left(\frac{2\pi z}{\beta\lambda} + \phi\right) dz \tag{6.30}$$

Using trigonometric identities, we expand this expression:

$$\Delta E = \int_{-g/2}^{g/2} qE_0 \left[\cos\left(\frac{2\pi z}{\beta\lambda}\right)\cos(\phi) - \sin\left(\frac{2\pi z}{\beta\lambda}\right)\sin(\phi)\right] dz \tag{6.31}$$

Since the second term is an odd function of z, it vanishes upon integration:

$$\Delta E = qE_0 \cos(\phi) \int_{-g/2}^{g/2} \cos\left(\frac{2\pi z}{\beta\lambda}\right) dz \tag{6.32}$$

For an idealized case, we estimate the transit time factor by assuming the field is flat in the gap and zero outside, with no radial dependence. The energy gain ΔE is then given by:

$$\Delta E = qE_0 g T \cos(\phi) \tag{6.33}$$

where T is the transit time factor and ϕ is the phase of the particle relative to the RF field. The transit time factor accounts for the fact that the particle does not experience the peak field over the entire gap due to its finite transit time. The expression for the transit time factor T is:

$$T = \frac{\sin\left(\frac{\pi g}{\beta\lambda}\right)}{\frac{\pi g}{\beta\lambda}} \tag{6.34}$$

6.5 Phase Stability in LINACs

A LINAC is a device that accelerates charged particles along a straight path, allowing the particles to pass through the structure only once. Unlike circular accelerators, where particles are recycled along a curved orbit, a LINAC uses time-varying electric fields to bunch and accelerate particles to their final energy. When the LINAC operates continuously, delivering a constant particle beam, it is referred to as a continuous wave (CW) LINAC. However, in normal conducting cavities, continuous operation can lead to excessive heating, potentially causing maintenance issues and deformation. Additionally, for accelerators designed to handle only a single electron bunch or a limited number of bunches at a time, continuous operation is inefficient because much of the power is wasted. Therefore, LINACs typically operate in a pulsed mode, where the beam is generated and delivered at a specific repetition rate, usually ranging from 1 to 100 Hz.

LINACs use RF to accelerate charge particle, but particles in the beam are distributed around a special particle known as synchronous particle for which LINAC is designed. The charged particles in bunch form are accelerated across RF gap using time-dependent longitudinal electric fields. A maximum of one bunch per cycle is allowed for acceleration using RF fields. As given from above Equation 6.33, the energy gain of a synchronous particle has cosine dependence on synchronous phase of the beam ϕ, which is now designated as ϕ_{syn}. So, there are mainly four cases as follows as also shown in Figure 6.2.

1. If $\phi_{syn} = 0°$, then early and late particles around the synchronous particles get less acceleration than particles in the beam bunch center, so it leads to a simultaneous increase in the phase and energy width of the bunch and hence leads to unstable motion.
2. If $\phi_{syn} = -90°$, then early and late particles around the synchronous particles get accelerating and decelerating kick and synchronous particle gets no kick, so finally it leads to longitudinal focusing of beam after a certain distance. This is usually employed in buncher cavities.
3. If $-40° \leq \phi_{syn} \leq -30°$, then on the rising slope of the RF field, the early and late particles around the synchronous particle get less and more accelerating kick and net result is simultaneous longitudinal focusing and acceleration. The negative synchronous phase provides energy and phase compression in subsequent RF gaps. This longitudinal motion is stable and the particles around synchronous particle R will execute oscillations known as synchrotron oscillations which are also represented by the elliptical motions of each particle in the longitudinal plane of phase and energy difference with respect to synchronous particle. All the stable phase trajectories are closed curves and are confined in the separatrix as shown in Figure 6.5.
4. If $\phi_{syn} > 0°$, then the beam bunch is elongated both in energy and phase width, and thus leads to unstable motion.

Not only is the synchronous phase of beam important, but we also define transit time factor (T) for single RF gap (g), which contributes to the overall energy gain as follows:

$$T = \frac{\sin(\pi g/\beta\lambda)}{\pi g/\beta\lambda} = \frac{\sin(\theta)}{\theta} \tag{6.35}$$

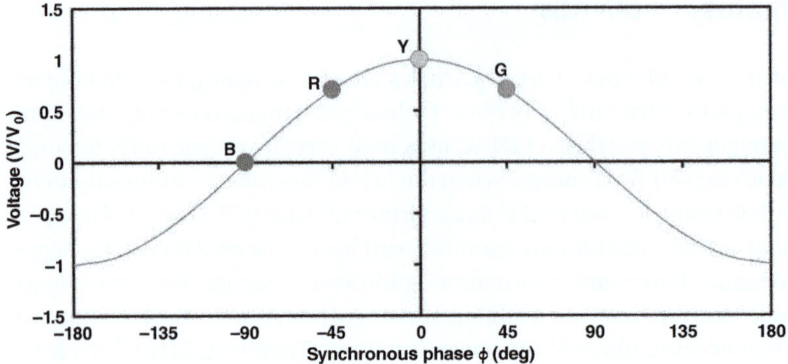

Figure 6.2 Voltage across RF gap: B dot for bunching, R dot for acceleration and longitudinal focusing, Y dot for maximum acceleration, and G dot for unstable motion.

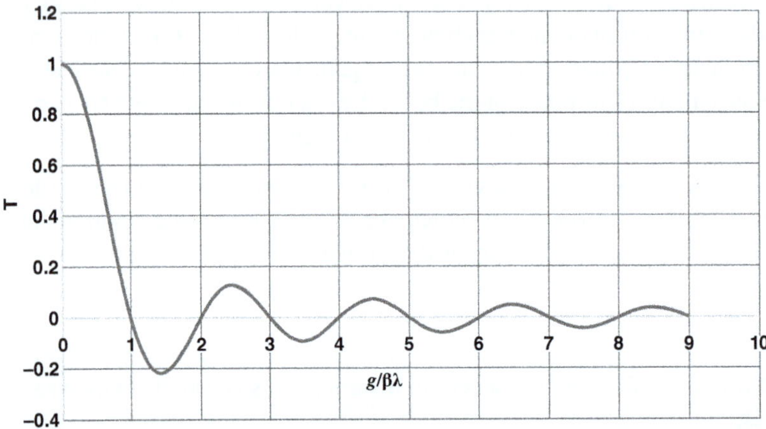

Figure 6.3 Behavior of T with geometrical parameters and incoming velocity.

Here

$$\theta = \frac{\pi g}{\beta\lambda}$$

The ideal scenario for achieving a transit time factor (T) close to 1 occurs when the argument $\theta/2$ is minimized, ideally approaching 0. This condition is met either when ω approaches 0, as in the case of DC acceleration, or when the gap g approaches 0. This implies that the RF system is most efficiently utilized when the cavity gap distance is minimized, though care must be taken to avoid breakdown fields. An optimized cavity design would thus reduce the gap to the threshold where discharges begin to occur if the resulting electric field is increased excessively. A plot of T against $g/(\beta\lambda)$ is shown in Figure 6.3.

As a particle passes through a cavity, the field fluctuates during its transit through the accelerating gap. Consequently, the particle does not always encounter the maximum field, leading to a reduction in effective acceleration by a certain factor. T quantifies the decrease in energy gain due to the particle's finite velocity in an electric field that varies sinusoidally with time. The T ranges from 0 to 1. Maximizing the T involves striking a balance between the gap width and the highest voltage across the gap, or in other words, by appropriately defining the electric field across the gap. The cosine ϕ factor mentioned above represents the energy gain factor and depends on the phase of the electric field encountered by the synchronous particles. Accordingly, the beam may be accelerated, decelerated, or bunched.

For double gap, we can write it as

$$T = \frac{\beta}{\beta_0} \frac{\sin\left(\frac{\pi\beta_0}{2x_0\beta}\right)}{\sin\left(\frac{\pi}{2x_0}\right)} \frac{\sin\left(\frac{\pi\alpha\beta_0}{2x_0\beta}\right)}{\sin\left(\frac{\pi\alpha}{2x_0}\right)} \tag{6.36}$$

Here

$$\alpha = \frac{g}{l}, \quad x_0 = \frac{\beta_0\lambda}{2l}$$

Also, for a more general case, when the electric field is not constant, the T factor is attributed to the finite time taken by the beam going through an RF gap in which the electric field changes in time. The general form of T is given as follows:

$$T = \frac{\int_{-g/2}^{g/2} E(0,z)\cos\omega t dz}{\int_{-g/2}^{g/2} E(0,z) dz} \tag{6.37}$$

where

$$\omega t = \frac{2\pi z}{\beta\lambda} = kz$$

and

$$k = \frac{2\pi}{\beta\lambda}$$

Now consider a series of drift tubes inside the LINAC with corresponding schematics as shown in Figure 6.4, the transit time between gaps (n to $n-1$) is as follows:

$$\Delta T = \frac{zl_{n-1}}{\beta_{n-1}c}$$

For an arbitrary nth particle, phase evolution is as follows:

$$\phi_n = \phi_{n-1} + \frac{\omega z l_{n-1}}{\beta_{n-1}c} \tag{6.38}$$

For synchronous particle: let phase, velocity, and energy of synchronous particle as ϕ_s, n, β_s, n, and E_s, n and similarly for other cells.

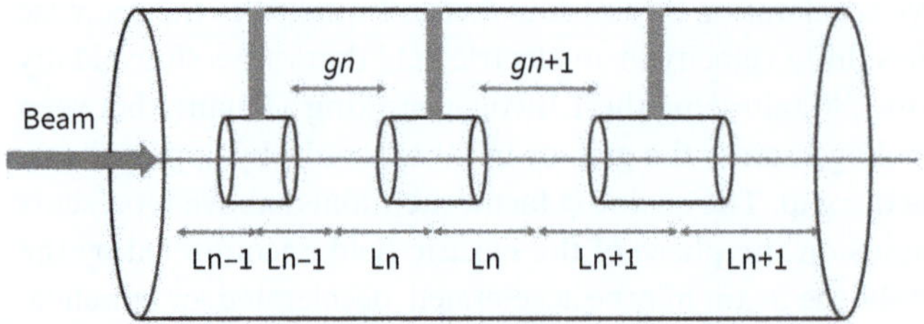

Figure 6.4 Arrangement of drift tube inside a LINAC.

The gap length and phase of a synchronous particle is changed as follows:

$$g_n = l_{n-1} + l_n = N(\beta_n + \beta_{s,n-1})\frac{\lambda}{2} \tag{6.39}$$

$$\phi_{s,n} = \phi_{s,n-1} + \omega t = \phi_{s,n-1} + \frac{2l_{s,n-1}}{\beta_{n-1}}t \tag{6.40}$$

Here $N = 1/2$ for π mode and 1 for 0 mode.
From Equation 6.40,

$$\Delta\phi_n = \phi_{s,n-1} - \phi_{s,n} = \frac{2l_{s,n-1}}{\beta_n}t$$

The change in the phase of the arbitrary particle going from n to $n-1$ gap

$$(\Delta\phi - \Delta\phi_s)_n = \Delta\phi_{n-1} - \Delta\phi_{s,n} = \frac{2l_{s,n-1}}{\beta_n}t + 2\pi N$$

$$(\Delta\phi - \Delta\phi_s)_n = \frac{2N\beta_{s,n-1}\lambda}{\beta_n}t + 2\pi N$$

$$(\Delta\phi - \Delta\phi_s)_n = 2\pi N\beta_{s,n-1}\lambda(\frac{1}{\beta} - \frac{1}{\beta_{s,n-1}}) \tag{6.41}$$

Using Taylor expansion:

$$\frac{1}{\beta} - \frac{1}{\beta_s} = \frac{1}{\beta + \delta\beta} - \frac{1}{\beta_s} = -\frac{\delta\beta}{\beta_{s,n-1}}$$

Using the above equations, one gets:

$$(\Delta\phi - \Delta\phi_s)_n = -2\pi N\frac{\delta\beta}{\beta_n - 1} \tag{6.42}$$

Now

$$E = \gamma m_0 c^2 = \frac{1}{\sqrt{1-\beta^2}}m_0 c^2$$

Differentiating the above equation:

$$\frac{\delta E}{\delta\beta} = \beta\gamma^3 m_0 c^2$$

$$\delta\beta = \frac{\delta E}{\beta\gamma^3 m_0 c^2} \tag{6.43}$$

Using the above equation, one can get the following:

$$(\Delta\phi - \Delta\phi_s)_n = \frac{2\pi N}{\beta_{s,n-1}} \frac{\delta E}{m_0 c^2 \gamma_{s,n-1}^3 \beta_{s,n-1}}$$

$$(\Delta\phi - \Delta\phi_s)_n = \frac{2\pi N}{\beta_{s,n-1}} \frac{E_n - E_{s,n-1}}{m_0 c^2 \gamma_{s,n}^3 \beta_{s,n-1}} \tag{6.44}$$

The change of energy of particles with respect to synchronous particle energy is given as follows:

$$(\Delta W - \Delta W_s)_n = qE_0 T g_n(\cos(\phi_n) - \cos(\phi_{s,n})) \tag{6.45}$$

$$\frac{d}{ds}(W - W_s)_n = qE_0 T(\cos(\phi_n) - \cos(\phi_{s,n})) \tag{6.46}$$

Total number of cells is given as follows:

$$n = \frac{s}{N\beta_s\lambda}$$

$$dn = \frac{ds}{N\beta_s\lambda}$$

Using dn in the above equations, one gets the modified equations as follows:

$$\gamma^3\beta_s^3 \frac{d(\phi - \phi_s)}{ds} = -2\pi \frac{W - W_s}{m_0 c^2 \lambda} \tag{6.47}$$

Differentiating both sides:

$$\frac{d}{ds}\left[\gamma^3\beta_s^3 \frac{d(\phi - \phi_s)}{ds}\right] = -\frac{2\pi q E_0 T}{m_0 c^2 \lambda}\Big(\cos(\phi) - \cos(\phi_s)\Big) \tag{6.48}$$

$$\begin{aligned}\left[\gamma^3\beta_s^3 \frac{d^2(\phi - \phi_s)}{ds^2})\right] &+ \left[3\gamma^2\beta_s^3 \frac{d(\gamma_s\beta_s)}{ds}\frac{d(\phi - \phi_s)}{ds}\right] \\ &+ \frac{2\pi q E_0 T}{m_0 c^2 \lambda}\Big(\cos(\phi) - \cos(\phi_s)\Big) = 0\end{aligned} \tag{6.49}$$

Let's define some constants as follows:

$$A = \frac{2\pi}{\beta^3\gamma_s^3}\lambda;\ B = \frac{qE_0 T}{m_0 c^2};\ w = \frac{W - W_s}{m_0 c^2}$$

From the above equations:

$$\phi' = -Aw \tag{6.50}$$

$$w' = B(\cos\phi - \cos\phi_s)$$

$$d\phi' = -AB(\cos\phi - \cos\phi_s)ds$$

$$\phi d\phi' = -AB(\cos\phi - \cos\phi_s)d\phi$$

$$\frac{\phi'^2}{2} = -AB(-\sin\phi - \phi\cos\phi_s) + C$$

Integrating both sides, one can get...

$$A\frac{w^2}{2} + B(\sin\phi - \phi\cos\phi_s) = H_\phi \tag{6.51}$$

At the potential maxima, $\phi = -\phi_s$, we have $\phi' = 0$ and from the above equation:

$$H_\phi = B\sin(-\phi_s) - (-\phi_s\cos\phi_s) = -B(\sin\phi_s - \phi_s\cos\phi_s) \tag{6.52}$$

Hence, we can get:

$$A\frac{w^2}{2} + B(\sin\phi - \phi\cos\phi_s) = -B(\sin\phi_s - \phi_s\cos\phi_s) \tag{6.53}$$

When $\phi = \phi_2$ and w = 0, the above equation gives:

$$\sin\phi_2 - \phi_2\cos\phi_s = \phi_s\cos\phi_s - \sin\phi_s \tag{6.54}$$

Total phase width of the sepratrix is given as:

$$\psi = -\phi_s - \phi_2 \tag{6.55}$$

Removing ϕ_2 from Equation 6.54 using Equation 6.55:

$$-\sin(\phi_s + \psi) - (\phi_s + \psi)\cos\phi_s = \phi_s\cos\phi_s - \sin\phi_s \tag{6.56}$$

Rearranging, one can get

$$\tan\phi_s = \frac{\sin\psi - \psi}{1 - \cos\psi} \tag{6.57}$$

From this, one can easily get if $\tan\phi_s \approx \phi_s$, then we have $\psi \approx 3\phi_s$ and $\phi_2 = 2\phi_s$. Thus, if $\phi_s = 90°$, then phase acceptance is maximum, 360°. One can correlate the same things from Figure 6.5.

6.6 Radial Impulse and Deflection in an RF Gap

Figure 6.6 illustrates a schematic of an RF gap, where the solid areas denote the cavity boundaries and the electric field lines are depicted. From the configuration of these field lines, it is evident that particles deviating from the axis will experience a radial force. This radial component of the electric field arises naturally when the field is concentrated in a narrow longitudinal space. According to Gauss's law, in the absence of free charges, the divergence of the electric field should be zero. If the longitudinal electric field is not uniform, it induces a radial field off-axis. The longitudinal accelerating field is zero outside the

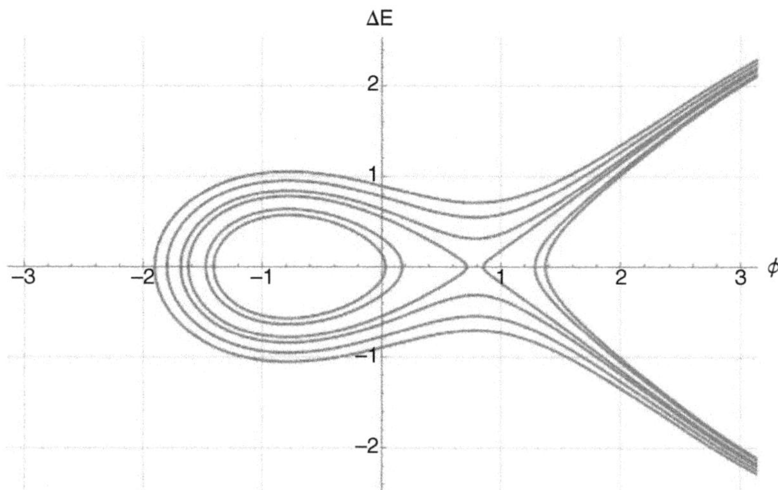

Figure 6.5 Radial and longitudinal field plot along the gap.

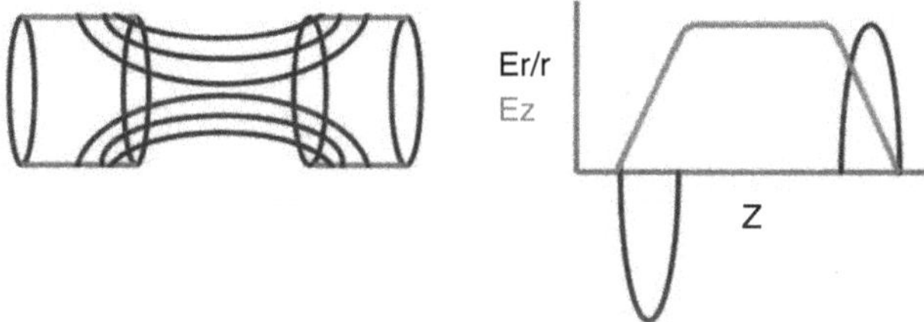

Figure 6.6 Radial and longitudinal field plot along the gap.

RF gap but peaks within it. Due to the symmetry of the gap, the radial electric fields in the two halves of the gap counteract each other. Although the fields are opposite in direction, the outward-directed field exerts a stronger force than the inward-directed field, resulting in a net defocusing effect on the beam.

Consider a cylindrical LINAC cavity that generates only the following components of electric and magnetic fields: E_z, E_r, and B_θ for the longitudinal, radial, and azimuthal components, respectively. The change in radial momentum for a particle with charge q and velocity βc within an RF accelerating gap, due to the Lorentz force impulse, is expressed by:

$$\Delta p_r = q \int_{-L/2}^{L/2} (E_r - \beta c B_\theta)\, dz \tag{6.58}$$

where $\pm L/2$ denotes the limits of the gap. Assuming E_z remains approximately constant with respect to r near the axis, Maxwell's equations yield the following relationships:

$$E_r = -\frac{r}{2}\frac{\partial E_z}{\partial z} \tag{6.59}$$

$$B_\theta = \frac{r}{2c^2}\frac{\partial E_z}{\partial t} \tag{6.60}$$

Substituting these into the previous equation, we obtain:

$$\Delta p_r = -\frac{q}{2}\int_{-L/2}^{L/2}\left(r\left(\frac{\partial E_z}{\partial z} + \beta c\frac{\partial E_z}{\partial t}\right)\right) dz \tag{6.61}$$

Noting that:

$$\frac{\partial E_z}{\partial z} + \frac{1}{\beta c}\frac{\partial E_z}{\partial t} = \frac{dE_z}{dz} \tag{6.62}$$

and using:

$$E_z = E_a(z)\cos(\omega t + \phi) \tag{6.63}$$

where E_a represents the accelerating field, ω is the RF angular frequency, and ϕ is the phase when the particle is at the gap center, we can express the change in radial momentum as:

$$\Delta p_r = -\frac{qr\omega}{2\gamma^2\beta^2c^2}\sin\phi\int_{-L/2}^{L/2} E_a(z)\cos(kz)\, dz \tag{6.64}$$

with $k = \frac{2\pi}{\lambda}$ and $\omega t = kz$.

The radial momentum of the particle is given by:

$$p_r = mc\beta\gamma r' \tag{6.65}$$

where m is the particle mass, β and γ are relativistic parameters, and $r' = \frac{dr}{dz}$.

Substituting this into the expression for Δp_r and using the effective accelerating voltage E_0TL, the radial deflection is expressed as:

$$\Delta(\beta\gamma r') = -\frac{\pi q E_0 TL r \sin\phi}{mc^2\beta^2\gamma^2\lambda} \tag{6.66}$$

Therefore, the radial deflection within the RF gap depends directly on the accelerating voltage, the radial position, and the sine of the RF phase, while it is inversely related to $(\beta\gamma)^2$ and the RF wavelength λ. To ensure stable acceleration, the phase ϕ should be negative, which leads to RF defocusing.

6.7 RF Chopper

An RF chopper is an electric field deflector that utilizes RF waves to selectively chop a particle beam. It typically consists of an RF deflector combined with slits. The RF deflector sweeps the beam in one direction, and a DC offset is applied to the RF signal to cut off one-half of the beam, effectively chopping it at the RF frequency rate. For instance, with an RF frequency of 4 MHz, the repetition rate would be 250 ns. The deflection of the ion beam in the presence of an electric field oriented vertically to the beam path is governed by:

$$\frac{mv^2}{r} = qE = \frac{qV}{g}$$

$$r = \frac{mv^2 g}{qV}$$

The deflection angle is given by:

$$\theta = \frac{LqV}{AE_P\beta^2 g} \tag{6.67}$$

where $E_0 = AE_P$, and E_P is the rest mass energy of the proton. In some cases, a high-energy sweeper has been used instead of an RF chopper, following the same principle but requiring a higher voltage. This choice is often driven by space constraints on the low-energy side.

6.8 RF Buncher

Bunching an ion beam is essential before injecting it into a LINAC. Consider a positively charged ion beam. The RF is injected at the zero crossover so that the ion beam experiences an electric field that is half accelerating and half decelerating. The synchronous particle (green) lies exactly at the zero crossover, as shown in Figure 6.7, and thus neither accelerates nor decelerates. Particles that arrive earlier than the synchronous particle are decelerated by the positive electric field, causing them to slow down and approach the synchronous particle after a certain drift distance. Conversely, particles (red) that arrive later are accelerated by the negative electric field and also approach the synchronous particle after some drift. As a result, a beam bunch forms, confined to a very small time width, typically in the range of a few nanoseconds to picoseconds.

For a beam of particles with synchronous velocity v passing through an RF gap of length L, the time taken by the synchronous particle is given by:

$$t_0 = \frac{L}{v} = \frac{L}{\sqrt{2qV_0/m)}} \tag{6.68}$$

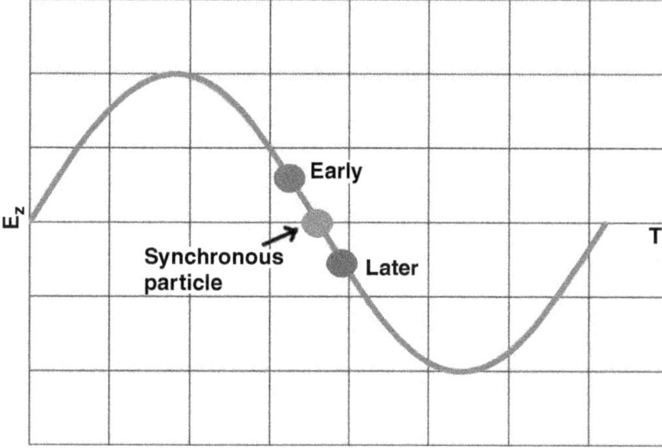

Figure 6.7 Particle riding over RF field.

For particles receiving an RF kick of voltage $\Delta V/2$, the time taken is:

$$t - \frac{\Delta t}{2} = \frac{L}{\sqrt{2q(V_0 + \Delta V/2)/m}} = \sqrt{\frac{m}{2qV_0}} L \left(1 + \frac{\Delta V}{2V_0}\right)^{-1/2}$$
$$= \sqrt{\frac{m}{2qV_0}} L \left(1 - \frac{\Delta V}{4V_0}\right)$$

Solving this yields:

$$\frac{dV}{dt} = \frac{2.8V_0}{L}\sqrt{\frac{qV_0}{m}} \tag{6.69}$$

In practical units:

$$V(\text{volts}) = \frac{140E(\text{keV})^{1.5}}{qf(\text{MHz})L(\text{m})\sqrt{m(\text{amu})}} \tag{6.70}$$

6.9 RF Acceleration

To overcome the limitations of electrostatic fields, RF fields are employed for acceleration within RF cavities. Consider a series of drift tubes with gaps between them, as shown in Figure 6.8. An RF oscillator feeds the tubes alternately. Inside the tubes, the particle is shielded from the external field. If the field's polarity reverses while the particle is inside the tube, the particle gets accelerated at each gap.

This leads to the synchronism condition, where the distance L between the gaps must satisfy:

$$L = \frac{1}{2}vT$$

where $v = \beta c$ is the particle velocity and T is the period of the RF oscillator. This arrangement cannot accelerate a continuous beam – only particles within a specific RF phase range will be accelerated, requiring the beam to be bunched. As the particle's velocity increases, the drift spaces must lengthen, reducing efficiency. While increasing the RF could counteract this effect, it would result in significant power loss due to radiation at higher frequencies. To address this, the accelerating gap is enclosed within a resonant cavity that holds electromagnetic energy as a magnetic field. Multiple such cavities can be arranged with a specific phase relationship.

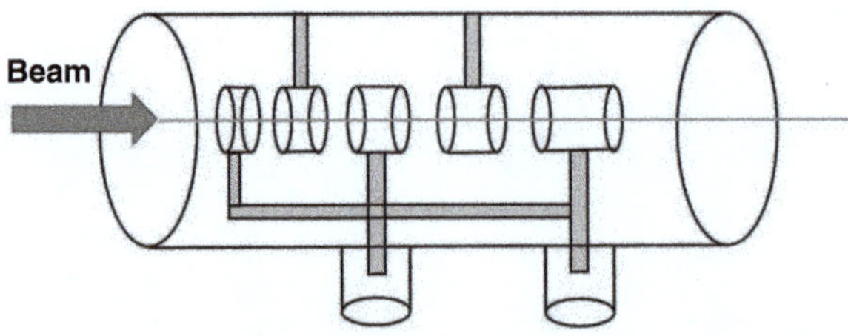

Figure 6.8 Wideroe structure.

6.9.1 Wideroe Structure

In the Wideroe structure, particles are accelerated by the electric field in the gap between electrodes connected alternately to an AC generator. This idea, initially proposed by Ising in 1924 and implemented by Wideroe in 1927, uses sine-wave voltage across a sequence of drift tubes. The particles experience no force inside the tubes (equipotential regions) and are accelerated across the gaps, creating a drift tube LINAC (DTL). In the Wideroe accelerator, the drift tubes are arranged such that they alternate between being connected to an excited coaxial line and the grounded outer wall. The coaxial line, which energizes the drift tubes, is supported by quarter-wave stubs that also serve as conduits for the cooling lines. These stubs provide structural support for the excited drift tubes. Typically, focusing devices are installed only in the drift tubes that are grounded. A schematic for this structure is shown in Figure 6.8.

6.9.2 Sloan–Lawrence Structure

The Sloan–Lawrence structure, first used in the 1920s [87], is particularly effective for low-energy particles. The radio frequency quadrupole (RFQ) is an example of this structure. Drift tubes are connected to an RF source to provide voltage V across the gap, acting as Faraday cages. Ions are first accelerated by the RF gaps and then enter the drift tubes when the polarity reverses. When they exit the tubes, the accelerating voltage is correctly phased to further accelerate them. To maintain synchronism, the length of the drift tubes must increase, requiring:

$$\frac{L}{v} = \frac{T}{2} = \frac{1}{2f} = \frac{\lambda}{2c}$$

$$L = \frac{v\lambda}{2c}; \quad L = \frac{\beta\lambda}{2} \tag{6.71}$$

If V_m is the maximum voltage in the gap, the energy gain per unit length or average accelerating gradient is:

$$\frac{E}{L} = \frac{qV_m}{L} = \frac{2qV_m}{\beta\lambda} \tag{6.72}$$

Higher accelerating gradients require smaller RF wavelength structures, which correspond to higher RF frequencies. However, at small RF wavelengths, the concept of equipotential drift tubes becomes impractical due to increased radiation losses, as the drift tubes start to behave like antennas.

6.9.3 Alvarez Structure

The Alvarez structure is a specialized DTL where the electrodes form part of a resonant macrostructure. Alvarez proposed enclosing the gaps between metallic tubes within an outer resonant cavity, which has a high quality factor (Q). In this setup, the drift tubes are placed within a single resonant tank, ensuring that the field has the same phase in all gaps. This resonant accelerating structure is used for proton acceleration in the energy range of 50–200 MeV, overcoming radiation issues by housing the tubes within a resonant cavity.

The standing wave setup's time-varying electric field is used for particle acceleration across the gaps. To maintain synchronism, the length of drift tubes must satisfy:

$$\frac{L}{v} = T = \frac{1}{f} = \frac{\lambda}{c}$$

$$L = \frac{v\lambda}{c}$$

$$L = \beta\lambda \tag{6.73}$$

If V_m is the maximum voltage in the gap, the energy gain per unit length or average accelerating gradient is:

$$\frac{E}{L} = \frac{qV_m}{L} = \frac{2qV_m}{\beta\lambda} \tag{6.74}$$

The resonant modes as shown in Figure 6.9, namely standing wave (SW) modes, are excited by RF generators coupled to the cavities via waveguides and coaxial cables. The synchronism condition depends on the mode:

1. π mode: $L = \frac{1}{2}vT$
2. 2π mode: $L = vT$

In the 2π mode, the resulting wall current between the cavities is zero, making the common cavity walls unnecessary.

For efficient acceleration by RF fields, it is crucial to match the gap length L to the distance that the particle travels n one RF wavelength. The particle passing through the gap experiences a varying field, which is not always at its peak value. The transit time factor (T) accounts for this variation, considering the time-dependent nature of the field during particle transit through the gap. Unlike linear transverse restoring fields, longitudinal fields

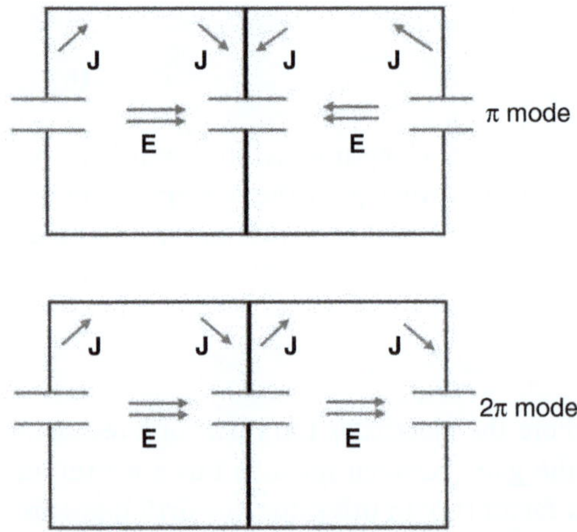

Figure 6.9 Different modes of the cavity.

are nonlinear and are used for both acceleration and focusing. Longitudinal focusing is achieved by a rising field as the bunch enters the accelerating gap. As shown in Figure 6.7, late-arriving particles receive an additional kick to rejoin the bunch, while early-arriving particles receive less acceleration and fall back into the bunch. Because the field variation is sinusoidal rather than linear, the restoring force on the bunch is also nonlinear.

6.10 Quarter-wave and Half-wave Resonators

In the design of low-velocity linear accelerators, radiofrequency (RF) cavities play a critical role in imparting energy to particle beams. Among the most effective structures for low-velocity ions are the quarter-wave resonator (QWR) and half-wave resonator (HWR), both operating in the TEM (transverse electromagnetic) mode. These resonators derive their names from the relationship between the cavity length and the RF wavelength [88]. Their compact geometries, high shunt impedance, and efficient acceleration characteristics make them essential components in heavy-ion and rare-isotope facilities.

The quarter-wave resonator achieves resonance when the cavity length satisfies $L = \lambda/4$, while the half-wave resonator resonates at $L = \lambda/2$. The "length" L refers to the axial distance between the shorted end and the open end of the inner conductor physically, the height of the inner conductor (the center conductor inside the outer cylindrical cavity). The axial electric field along the beam axis can be expressed as:

$$E_z(z) = E_0 \sin\left(\frac{\pi z}{L}\right) \tag{6.75}$$

leading to an accelerating voltage:

$$V_{acc} = \int_0^L E_z(z) e^{i\phi(z)} dz \tag{6.76}$$

attenuated by the transit time factor:

$$T = \frac{1}{V_0} \int_0^L E_z(z) \cos\left(\omega \frac{z}{v}\right) dz \tag{6.77}$$

The accelerating field E_a is defined as effective accelerating voltage per unit length along the beam path. Assuming the effective accelerating length is approximated by the inner diameter $2b$, and incorporating the transit time factor $T(\beta)$, we obtain:

$$E_a = \frac{V_{acc}}{b} = \frac{V_0 T(\beta)}{b} \tag{6.78}$$

where V_0 is the peak RF voltage at the open end of the resonator.

A critical challenge in QWRs and HWRs is multipacting, a resonant electron amplification process between cavity surfaces. Multipacting occurs when electrons emitted from a surface are accelerated by the RF electric field and strike the opposing wall in synchrony with the RF cycle, releasing secondary electrons. If conditions allow $\delta > 1$ (secondary emission coefficient), an electron avalanche builds up, leading to increased power loss, heating, and degraded cavity performance [89].

We derive the voltage condition for multipacting from Newton's law for an electron under an oscillating RF field:

$$m\frac{dv}{dt} = -eE_0 \sin(\omega t) \tag{6.79}$$

Integrating for an electron starting at rest:

$$v(t) = -\frac{eE_0}{m\omega}(1 - \cos \omega t) \tag{6.80}$$

and integrating position:

$$x(t) = -\frac{eE_0}{m\omega}\left(t - \frac{\sin \omega t}{\omega}\right) \tag{6.81}$$

Multipacting occurs when the electron traverses a distance d in time $t_f = \frac{(2n-1)\pi}{\omega}$, corresponding to odd multiples of half an RF period. Substituting t_f and noting $\sin((2n-1)\pi) = 0$:

$$d = -\frac{eE_0}{m\omega}t_f \quad \Rightarrow \quad d = \frac{eE_0}{m\omega}t_f \tag{6.82}$$

Solving for E_0:

$$E_0 = \frac{m\omega d}{et_f} \tag{6.83}$$

Substitute $t_f = \frac{(2n-1)\pi}{\omega}$:

$$E_0 = \frac{m\omega^2 d}{e(2n-1)\pi} \tag{6.84}$$

The accelerating voltage is $V_n = E_0 d$:

$$V_n = E_0 d = \frac{m\omega^2 d^2}{e(2n-1)\pi} \tag{6.85}$$

representing the threshold voltage for nth order two-point multipacting.

We derive the *final velocity* v_f at impact by substituting t_f into $v(t)$:

$$v_f = -\frac{eE_0}{m\omega}(1 - \cos \omega t_f) \tag{6.86}$$

Since $\cos((2n-1)\pi) = -1$:

$$v_f = -\frac{eE_0}{m\omega}(1 - (-1)) = -\frac{2eE_0}{m\omega} \tag{6.87}$$

We derive the impact energy:

$$K_n = \frac{1}{2}mv_f^2 = \frac{1}{2}m\left(-\frac{2eE_0}{m\omega}\right)^2 = \frac{2e^2E_0^2}{m\omega^2} \tag{6.88}$$

Substitute $E_0 = \frac{V_n}{d}$:

$$K_n = \frac{2e^2}{m\omega^2}\left(\frac{V_n}{d}\right)^2 = \frac{2e^2V_n^2}{m\omega^2 d^2} \tag{6.89}$$

showing the impact energy depends quadratically on V_n and inversely on $\omega^2 d^2$.

Physically, multipacting sustains when K_n aligns with the material's energy range for $\delta > 1$, making secondary electron production self-reinforcing. Avoiding or conditioning these voltage bands is critical, especially in superconducting cavities where multipacting can quench the superconducting state. Multipacting mitigation involves geometric optimization, surface treatment, detuning, or RF processing.

6.11 Radio Frequency Quadrupole

The operational principle of RFQ was first presented by Kapchinskiy and Tepliakov (K–T) in 1969. An RFQ is a LINAC) used in particle physics and nuclear physics to accelerate, focus, and bunch the charged particles. It is mostly a front-end accelerator to provide suitable injection energy of beam into further LINACS. It uses alternating electric fields termed as RF fields. It consists of four cylindrical electrodes in a quadrupole fashion, which are typically hyperbolic or parabolic in shape. Quadrupolar focusing forces keep the particle beam focused and directed toward the central axis of the RFQ, avoiding the beam dispersion. Before entering the RFQ, the charged particles need to be injected at suitable energy and within the RFQ's acceptance using an ion source and an electrostatic accelerator. It is suited for acceleration of beams with low velocities (1 to 6%) velocity of light. It does not work for electrons. It may employ four rods or four vane type structures with advantages over each other.

A two-term potential function in cylindrical coordinates in the form of Bessel's functions is written as follows:

$$V(r,\theta,z) = A_0 r^2 \cos(2\theta) + A_{10} I_0(kr)\cos(kz) \tag{6.90}$$

A_0, A_{10} are constants that depend on the electrode geometry.

$k = 2\pi/\text{L}$, $L = \beta_s\lambda$, Here, β_s is the velocity of the synchronous particle.

For $\theta = 0$ and $\pi/2$, the potential is solved for the horizontal and vertical vane tip and solving those equations, we get the following:

$$\begin{aligned} A_0 &= \frac{V_0}{2a^2}\frac{I_0(ka)+I_0(kma)}{m^2 I_0(ka)+I_0(kma)} = \frac{XV_0}{2a^2} \\ A_{10} &= \frac{V_0}{2}\frac{m^2-1}{m^2 I_0(ka)+I_0(kma)} = \frac{AV_0}{2} \end{aligned} \tag{6.91}$$

Here, X and A are defined as follows:

$$\begin{aligned} X &= \frac{I_0(ka)+I_0(kma)}{m^2 I_0(ka)+I_0(kma)} \\ A &= \frac{m^2-1}{m^2 I_0(ka)+I_0(kma)} \end{aligned} \tag{6.92}$$

Using above equations the Equation 6.90 becomes with time dependent potential

$$U(r,\theta,z,t) = \frac{V_0}{2}\left[X\left(\frac{r^2}{a^2}\right)\cos(2\theta) + AI_0(kr)\cos(kz)\right]\sin(\omega t + \phi) \tag{6.93}$$

It is expressed in Cartesian coordinates as follows:

$$U(x, y, z, t) = \frac{V_0}{2}\left[X\left(\frac{x^2 - y^2}{a^2}\right) + AI_0(kr)\cos(kz)\right]\sin(\omega t + \phi) \tag{6.94}$$

The electric field can be obtained by the gradient of this potential function as $E = -\nabla V$. From Equation 6.94, we get the electric field components as follows:

$$\begin{aligned} E_x &= \frac{-XV_0}{a^2}x - \frac{kAV_0}{2}I_1(kr)\frac{x}{r}\cos(kz) \\ E_y &= \frac{XV_0}{a^2}y - \frac{kAV_0}{2}I_1(kr)\frac{y}{r}\cos(kz) \\ E_z &= \frac{kV_0A}{2}I_0(kr)\sin(kz) \end{aligned} \tag{6.95}$$

The energy gain for the particle with normalized velocity (β_1) can be calculated as follows:

$$\Delta W = \int_0^l E_z \sin(kz + \phi)\, dz = \frac{qkAV_0I_0(kr)}{2}\int_0^l \sin(kz)\sin(k_1 z + \phi)\, dz \tag{6.96}$$

Here, $k = 2\pi/\beta_s\lambda$, $k_1 = 2\pi/\beta_1\lambda$

For synchronous particle, $\beta_1 = \beta_s$, the energy gain for a synchronous particle is as follows:

$$\Delta W = \frac{\pi qAV_0I_0(kr)\cos(\phi)}{4} \tag{6.97}$$

Average peak longitudinal accelerating field is as follows:

$$E_0 = \frac{1}{l}\int_0^l E_z\, dz = \frac{2AV_0}{\beta\lambda} \tag{6.98}$$

The transit time factor for the synchronous particle is given as follows:

$$T = \frac{\int_0^l E_z \sin(kz)\, dz}{\int_0^l E_z\, dz} = \frac{\pi}{4} \tag{6.99}$$

Energy gain of a synchronous particle per cell of the length l if given finally:

$$\Delta W = qE_0TI_0(kr)l\cos\phi_s \tag{6.100}$$

The derived potential and electric field expressions demonstrate how the RFQ simultaneously provides transverse focusing and longitudinal acceleration. Analytical integration shows how the energy gain depends on the vane voltage, modulation, Bessel function, and synchronous phase. The average accelerating field and transit time factor quantify the effective energy transfer per cell. The final expression confirms that energy gain per cell is directly proportional to the field strength, cell length, and phase factor. This theoretical framework forms the basis for practical RFQ design and optimization for efficient beam capture and acceleration.

6.12 Drift Tube LINACS

A LINAC consists of many drift tubes. Particles may enter from one side and exit from the other side. Inside the drift tube, particles are like in a Faraday cage. The charged particles experience a regularly oscillating electrical field through a series of tubes and get accelerated by the electric fields in the gap between the tubes. As the RF field changes in polarity, particles move into the next tube and come out with right phase and amplitude of RF field in the next gap. The drift tube potential changes, but the velocity of the particles in the drift tube remains unchanged if one neglects the fringe fields. The length of a drift tube depends on the velocity of the particles traveling through it and the operating RF. The tubes become longer as particles keep on accelerating. As explained above in RF acceleration, the Alvarez type of structure is most commonly used for this purpose. They are more efficient in power and beam delivery and come out of subsequent evolution of Wideroe and Sloan Lawrence type structures. The combined zero degree structure KONUS (German "Kombinierte Null-Grad Struktur") [90–94] is used to get high acceleration efficiency from H-mode cavities. Using two transverse focusing lenses and a cavity in between, RF defocusing effects get reduced by passing through many RF gaps of the cavity. Moreover, the beam is accelerated at a synchronous phase of 0°, so maximum acceleration is achieved. To avoid debunching of the beam, the shorter rebuncher sections with negative synchronous phase at the cavity entrance are used. There are two types of phase that characterize KONUS beam dynamics as follows:

1. Synchronous phase $-40^\circ \leq \phi_{syn} \leq -30^\circ$
2. Average synchronous phase $-20^\circ \leq \overline{\phi_{syn}} \leq -10^\circ$

Over the past 25 years, H-mode cavities have been effectively developed to cater to a wide range of applications in ion acceleration. The IH-DTL is particularly efficient for energies ranging from 100 AkeV to 15 AMeV, especially when paired with KONUS beam dynamics. Even at room temperature, IH-DTLs can be compared with existing superconducting LINACs in terms of plug power consumption for beam energies up to approximately 2 AMeV. In other words, KONUS beam dynamics represents an advanced method for accelerating beams in LINACs, especially within DTLs. The essence of KONUS beam dynamics is its innovative integration of longitudinal acceleration and transverse focusing. This approach is designed to effectively minimize emittance growth and enhance beam transport efficiency. Here are the main points about KONUS beam dynamics:

1. Beam is injected into multigap cavity like DTLs at 0° synchronous phase such that there is no longitudinal focusing. In this way, synchronous particles lie at that point. As the bunch passes through the first gap of the 0° section, it is near the 0° RF phase and undergoes longitudinal defocusing. This effect is partially mitigated in the final gaps of the 0° section, as the bunch approaches negative RF phases. By the time it exits, the bunch may even achieve slight longitudinal focusing.
2. The KONUS approach is designed to achieve both transverse and longitudinal beam focusing in extended H-mode LINAC sections. This method compensates for the defocusing effects caused by transverse RF fields and space charge, eliminating the need for quadrupole magnet focusing within the drift tubes. As a result, the diameter of the

drift tubes can be reduced, leading to a lower capacitance and significantly enhancing the efficiency of the accelerating structure.

3. The KONUS approach includes many multigap cavities back to back such that their length is approximately one quarter of longitudinal focusing period.
4. Synchronous phase for each multigap cavity is 0°, which means a point of zero transverse defocusing and zero longitudinal focusing.
5. After each multigap cavity, a transverse focusing element, typically a quadrupole triplet, is used to counter the RF-induced defocusing effects at the gaps and refocus the beam. In terms of longitudinal motion, the lens serves as a drift space and needs to be as short as possible to limit the phase spread of the bunch.
6. However, the beam bunch experiences longitudinal defocusing at the exit of the quadrupole triplet lens, making it necessary to immediately follow with a rebunching section. This rebunching section typically consists of multiple gaps set at a specific synchronous phase at the entrance. By redefining the synchronous phase and energy for the new synchronous particle, the bunch centroid and the synchronous particle are aligned, similar to conventional designs. This alignment is consistent across all KONUS rebunching sections. After passing through the rebunching section, the beam is longitudinally focused and prepared for injection into the next 0° section, for which suitable initial parameters must be defined.

In nutshell, The KONUS method significantly reduces the overall RF defocusing effect, allowing for the construction of long multicell structures with narrow drift tubes that do not require internal focusing elements. This design takes full advantage of the high efficiency of H-mode cavities with minimal capacitive load.

H-mode cavities function based on H_{n10}-modes within cylindrical resonators. The "interdigital H-type structure (IH)," which operates in the H_{110}-mode, is particularly efficient at low to medium energies ($\beta = 0.01 - 0.2$) and works within a resonance frequency range of 30–250 MHz.

However, as resonance frequencies increase, the IH structure's cavity diameter becomes too small, making the "crossbar H-type structure (CH)," which operates in the H_{210}-mode, a better fit for KONUS designs at higher velocities ($\beta = 0.1 - 0.5$) and resonance frequencies typically between 150 and 700 MHz. The CH structure also provides excellent mechanical stability due to its crossed stems and can be efficiently cooled, making it well-suited for high-duty cycle or superconducting multicell cavity applications.

The high efficiency (shunt impedance) of H-mode resonators for $\beta \leq 0.3$ is attributed to the low RF wall losses due to the cross-sectional RF current flow and the low capacitive load per unit length, resulting from the slim drift tube design – an excellent advantage of being KONUS. Additionally, these slim drift tubes are highly resistant to voltage breakdown, due to the uneven field distribution and concentrated high-field zones at the edges of the tubes.

In a Drift Tube Linac (DTL), ions or particles are accelerated by using an oscillating radiofrequency (RF) electric field. To ensure that the particle arrives at the correct accelerating phase at every gap between the drift tubes, the length of each drift tube must be carefully calculated. The formula for the drift tube length is:

$$d_n = \beta_n \frac{c}{2f} \tag{6.101}$$

where d_n is the length of the n-th drift tube, β_n is the velocity of the particle divided by the speed of light at that position, c is the speed of light, and f is the RF frequency. This ensures that the particle stays inside the drift tube for half of an RF period, avoiding the decelerating phase.

Each time the particle crosses a gap between drift tubes, it gains energy from the electric field. The energy gained at the n-th gap is given by:

$$\Delta W_n = qV_n T \cos\phi_s \tag{6.102}$$

Here, q is the charge of the particle, V_n is the peak voltage at the gap, T is the transit time factor, and ϕ_s is the synchronous phase angle. The transit time factor T corrects for the fact that the particle experiences a changing electric field as it crosses the gap, and it is calculated as:

$$T = \frac{\sin\left(\frac{\pi d}{\beta\lambda}\right)}{\frac{\pi d}{\beta\lambda}} \tag{6.103}$$

where d is the gap length and $\lambda = \frac{c}{f}$ is the RF wavelength.

The resonant frequency of the accelerating cavity, usually operating in the TM_{010} mode, depends on the geometry of the cavity and is given by:

$$f = \frac{c}{2\pi}\sqrt{\left(\frac{\pi}{b}\right)^2 + \left(\frac{X_{01}}{a}\right)^2} \tag{6.104}$$

where a is the cavity radius, b is the bore radius, and X_{01} is the first zero of the Bessel function J_0.

Finally, the total energy after passing through N accelerating gaps is:

$$W_N = W_0 + \sum_{n=1}^{N} \Delta W_n \tag{6.105}$$

This shows that each gap contributes a step of energy to the particle as it accelerates along the linac.

In this way, the Drift Tube Linac combines precise timing, electric fields, and magnetic focusing to accelerate and guide particles effectively.

6.13 Numerical Problems

1. Calculate the time, energy gain, and quadrupole focusing strength for a proton in a radio frequency quadrupole (RFQ) operating at a frequency of 97 MHz with a peak electric field of 20 MV/m. The quadrupole electrodes have a length of 1.5 m, and the inter-electrode spacing is 5 mm. The injection energy of proton is 8 keV.
2. Calculate the time, energy gain, and average accelerating gradient for a proton in a drift tube LINAC (DTL) operating at a frequency of 97 MHz with a peak electric field of 15 MV/m. The total length of the DTL is 3 m, and the drift tubes have an inner diameter of 3 cm. The proton is injected with an energy of 100 keV.

3. Calculate the velocity modulation and voltage required for a 8 keV/u ion beam in a radio frequency (RF) buncher operating at a frequency of 12.125 MHz across an RF gap of 2 mm longitudinal length to bunch it at 3 m distance.
4. Calculate the synchronous phase ϕ_s, synchronous energy E_s, energy gain ΔE, and phase slip $\Delta\psi$ for an accelerator operating at a peak electric field $E_{\text{peak}} = 5\,\text{MV/m}$, with a total length of 1 m. The RF frequency is 100 MHz, and the phase deviation ψ is considered for the values 30°, 60°, and 90°.
5. An RF chopper operates at a frequency of 100 MHz and generates a transverse electric field of $E_{\text{peak}} = 1.5\,\text{MV/m}$. The length of the chopper region is 0.25 m, and protons enter the chopper with an energy of 100 keV. Calculate the time it takes for the proton to traverse the chopper region, the deflection angle of the proton due to the transverse electric field, and the final transverse position of the proton at the exit of the chopper.
6. Calculate the speed of an electron and the distance it travels during half an RF period if it is accelerated in an RF cavity operating at 97 MHz and is in resonance with the RF field, gaining a kinetic energy of 480 eV. Also, calculate the kinetic energy in electronvolts if the electron travels a distance of 6.7 cm during half an RF period at the same frequency.
7. Calculate the multipacting voltage and the corresponding impact energy for first-order two-point multipacting in a quarter-wave resonator operating at 97 MHz, where the gap distance between the resonator surfaces is 3.35 cm. Assume that the electron moves in resonance with the RF field and crosses the gap in half an RF period. Express the multipacting voltage in volts and the impact energy in electronvolts.

7

Beam Diagnostics

> "If it disagrees with experiment, it's wrong. In that simple statement is the key to science. It doesn't make a difference how beautiful your guess is, it doesn't make a difference how smart you are, who made the guess, or what his name is. If it disagrees with experiment, it's wrong."
>
> —*Richard P. Feynman*

After reading this chapter you should be able to:

- Diagnose the particle beams.
- Define the main beam parameters to be diagnosed.
- Define destructive and nondestructive beam diagnostics.

To ensure the optimal functioning of accelerators, beam diagnostic tools play a crucial role by providing detailed insights into beam characteristics and particle behavior within the system. This field is highly diverse, utilizing various physical phenomena and offering abundant opportunities for innovation. Every diagnostic measurement affects the beam to some extent – some disturbances are minimal, while others can be significantly disruptive. It's often stated that an accelerator's performance is only as strong as its diagnostic capabilities, with precise instrumentation serving as the foundation for success. Beam diagnostics are vital as they directly or indirectly reflect the accelerator's overall performance. Diagnostic tools can be categorized as either destructive or nondestructive, depending on their impact on the beam. The primary parameters that need monitoring or measurement include beam current, transverse and longitudinal emittance, beam brightness, bunch width, energy spread, transverse profile, stability in position and current, and polarization.

The transverse profile of a beam is shaped by several key factors that control the beam's spread and distribution across its cross-sectional area. Beam pulse width in the transverse direction determines how focused or dispersed the particles are, with narrower pulses creating a tighter beam and wider ones causing greater spread. The shape of the beam pulse plays a significant role in particle distribution, where profiles like circular, elliptical, or Gaussian affect the beam's focus and uniformity. Maintaining stability and repeatability is critical for ensuring consistent transverse characteristics, as variations can lead to changes

Charged Particle Beam Physics: An Introduction for Physicists and Engineers, First Edition.
Sarvesh Kumar and Manish K. Kashyap.

Companion Website: https://www.wiley.com/go/Kumar_1e

in beam size, position, and particle density, impacting precision. Focusing and alignment systems, such as magnets or lenses, adjust the transverse profile by guiding and focusing the beam to achieve the desired size and shape. These elements are essential for achieving an accurate and controlled transverse beam profile, which is important in applications like particle accelerators, ion beam systems, and precision medical treatments.

The longitudinal profile of a beam depends on several critical factors that influence its control and behavior over time. Beam pulse duration refers to the time the beam is active, affecting the particle spread along its path, with longer beam pulses leading to a broader temporal spread and shorter ones resulting in more compact particle bunches. Beam pulse shape plays a role in particle distribution during emission, where different shapes, such as Gaussian or rectangular, create varying effects on particle spread and uniformity. Consistent stability and repeatability are essential for ensuring each beam pulse retains the same characteristics; fluctuations in pulse parameters can cause variations in energy and particle distribution, reducing precision. Triggers are responsible for synchronizing beam emission with external systems, ensuring beam pulses occur at the correct time and conditions, aligning with equipment to maintain beam quality and reduce energy spread. These factors collectively contribute to an accurate and controlled longitudinal beam profile, which is crucial in precise applications like particle accelerators and ion sources.

In a bunched beam, each group of particles, called a *bunch*, carries a certain amount of charge, denoted as q. This is the *bunch charge*. These bunches do not follow one after the other continuously; instead, they are separated by a time interval known as the *bunch period*, which is represented by T. The bunch period T is the time between two consecutive bunches.

Each bunch has a certain length in space, called the *root-mean-square (RMS) bunch length*, and we denote it as σ_z. Since these bunches travel at a certain fraction of the speed of light, we also use β_0, the *velocity factor*. This factor β_0 indicates how fast the bunches are moving compared to the speed of light c. From the bunch length σ_z, we can calculate the *RMS bunch duration*, represented by σ_τ, using the following formula:

$$\sigma_\tau = \frac{\sigma_z}{\beta_0 c} \tag{7.1}$$

This gives the duration of the bunch as it moves.

Now, when we measure the current in the beam, there are different ways to define it, depending on the time window of the measurement.

Peak current: First, if we measure over a very short time at the center of a bunch, we get the *peak current*, denoted as I_p. The peak current is the maximum current within the bunch, and we calculate it using the formula:

$$I_p = \frac{q}{\sqrt{2\pi}\sigma_\tau} \tag{7.2}$$

The peak current depends on both the bunch charge q and the bunch duration σ_τ. A shorter bunch duration results in a higher peak current for the same charge.

Pulse current: If we look at the entire bunch duration, we calculate the *pulse current*, denoted by I_{pulse}. This is the current averaged over the entire bunch. It is given by:

$$I_{\text{pulse}} = \frac{q}{T} \tag{7.3}$$

where T is the bunch period, which is the time interval between successive bunches.

Average current: Finally, if we zoom out further and consider many bunches over a longer time, we calculate the *average current*, denoted by $\langle I \rangle$. This is the average current over a sequence of pulses. If there are N_b bunches in a single pulse and the *pulse repetition period* is T_r (the time between two consecutive pulses), then the average current is given by:

$$\langle I \rangle = \frac{N_b q}{T_r} \tag{7.4}$$

7.1 Faraday Cups

Faraday cups (FC) are destructive devices that physically intercept the beam and measure the corresponding electronically equivalent current. Basically, beam current is the number of particles hitting the cup per unit time. The equivalent particle current of charge state q for an ion beam can be expressed as follows:

$$I_{pA} = \frac{I_{eA}}{q} \tag{7.5}$$

Beam current measurement frequently utilizes these devices. A particle beam is directed into a metallic cup, which acts as a stopper, and the corresponding electronic current within the cup is measured. When the beam impacts the cup, secondary electrons are emitted, and these need to be suppressed. This suppression is typically achieved using a negatively biased ring placed in front of the metallic stopper for positive ion current measurements. These parasitic currents can arise not only from the beam's impact but also from leakage currents from cooling water and suppression electrodes, electronic thermal emission, and charged particles emitted from residual gases. The beamline at ultra-high vacuum ensures the prevention of residual gaseous ions and thus its interference with the main beam current. A magnetic field may also be applied for the suppression of secondary electrons. To minimize the effect of the leakage current, any direct connection between the suppression electrode and the measuring cup must be avoided. The FC is removed from the beamline using electric motors so as to perform the action of cup in and out of position.

More the area of the beam stopper (FC) in the transverse direction to the beam, more its acceptance, and it can be used for bigger sized beams. They normally use logarithmic amplifiers in their outer circuit so as to process a wide range of beam currents from pA to mA. Finally, we summarize the specifications of the FC as follows:

1. They are constructed usually from high Z material to minimize secondary electron emission.
2. They may be capable of measuring currents from 1 nA to 1 mA with a resolution of 0.1 nA.
3. Their response time must be fast, like less than 1ms for transient measurements.
4. They must be compatible with standard flanges for easy integration into existing setups.
5. They should be operable within a temperature range of −20 to 60°C and under ultra-high vacuum environments (down to 10^{-9} Torr).

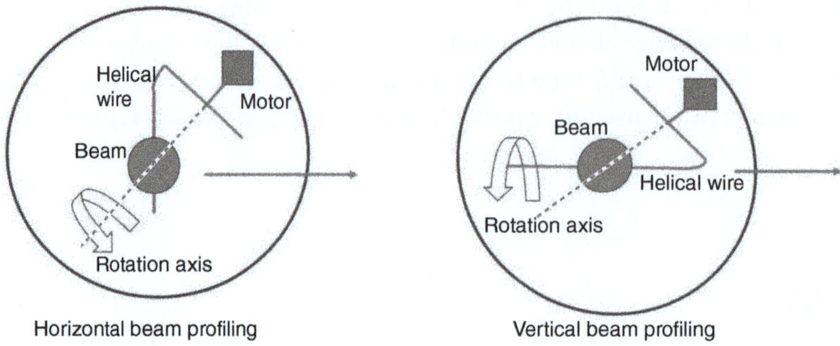

Figure 7.1 Beam profiling in transverse directions by wire scanner.

7.2 Beam Profile Monitor

Beam profile monitors are mainly to measure beam profile in transverse direction along with centroid position. These are also known as wire scanner-type monitors and nondestructive. A wire scanner measures the beam profile by traversing a thin wire [95] through the particle beam and detecting the scattered or absorbed particles or radiation. Typically, these wires are composed of materials like tungsten, chosen for their high melting points and low vapor pressures. It can measure the position and intensity distribution inside an ion beam. A motor rotates a helical grounded wire with its rotation axis oriented at a 45° angle to both horizontal and vertical directions. This setup allows the wire to sweep across the beam twice per revolution, capturing the vertical profile in one-half and the horizontal profile in the other. Secondary electrons generated on the wire are collected by a cylindrical electrode, providing information about the beam intensity at the wire's position. The resulting signals are displayed in both directions on an oscilloscope. The schematics of BPM is shown in Figure 7.1.

7.3 Beam Position Monitors

Beam position monitors (BPMs) [96] are essential tools used in nearly all linear accelerators (LINACs), cyclotrons, and synchrotrons. These devices provide information about the beam's center of mass and serve as monitors for the longitudinal bunch shape. It is basically an electromagnetic pickup signal from the beam. The signal generated by the beam's electromagnetic field is explained using the concept of transfer impedance. To achieve this, four pickup plates are installed crosswise on the beam pipe wall, and the difference in signals from opposite plates provides the beam's center of mass in both transverse planes. This setup is known as a BPM or position pickup. They operate by measuring the charges induced by the electric field of beam particles on an insulated metal plate. Since the electric field of a bunched beam varies with time, an alternating current (AC) signal is generated on the plate. In simple words: a BPM uses sensors called pickups, which are attached to the vacuum chamber, to detect the electrical current created by the beam as it passes through. The closer the beam is to a pickup, the stronger the signal from that pickup will be. These pickups are placed symmetrically around the center of the chamber. The pickups used in

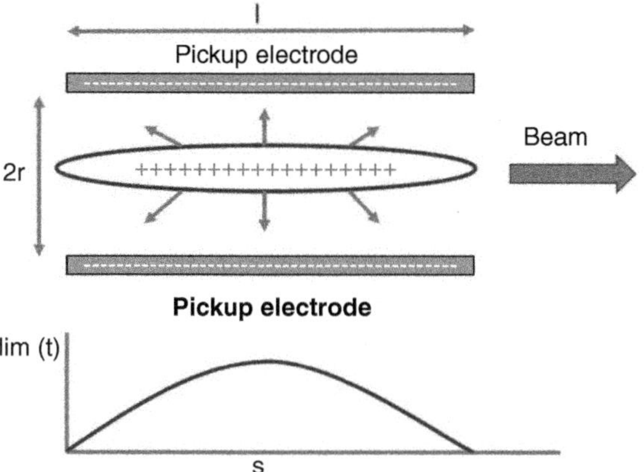

Figure 7.2 Beam pickup signal by pickup electrodes.

BPMs can either be button-shaped or stripline types. If the beam moves away from the center, the signal strength from the pickups will change. By comparing the differences in these signals and normalizing them, the exact position of the beam can be determined. Button BPMs are commonly used in electron accelerators.

Here are their merits:

1. These devices are primarily used to determine the beam position (center of mass).
2. These devices are widely used as nondestructive diagnostic tools in accelerators with bunched beams.

Consider a pickup plate of area A and length l with distance from the central axis of the beam pipe as r as shown in Figure 7.2, then the image current is given as follows:

$$I_{im}(t) = \frac{A}{2\pi rl}\frac{dQ_b}{dt} = \frac{A}{2\pi rl}\frac{l}{v}I_b(t) \tag{7.6}$$

Hence, the developed voltage across resistor R due to the above image current by beam of velocity (v) is given as follows:

$$V = R.I_{im}(t)$$

In the frequency domain

$$I = I_b e^{-\iota\omega t}$$

Then the voltage becomes:

$$V = Z.I_{im}(\omega) = \frac{A}{2\pi rvC}\frac{\iota\omega RC}{1+\iota\omega RC}I_b \tag{7.7}$$

Consider the case as a first-order high-pass filter with a cutoff frequency

$$\omega_c = 1/RC$$

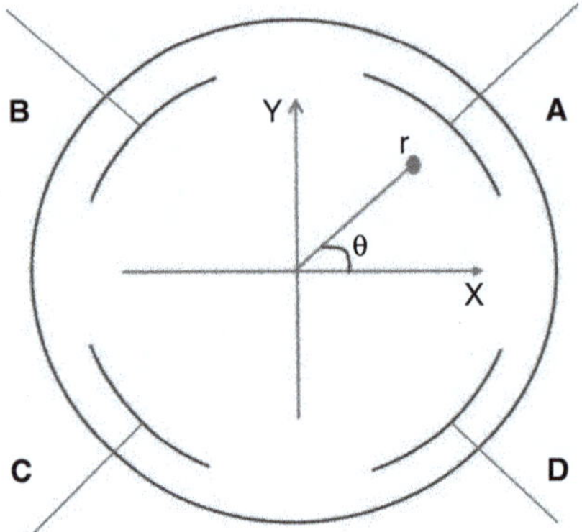

Figure 7.3 Schematics of button BPM where A, B, C, and D are pickup electrodes.

Then one can write the above expression as follows:

$$V = \frac{A}{2\pi rvC}\frac{\omega/\omega_c}{\sqrt{(1+(\omega/\omega_c)^2)}}I_b \tag{7.8}$$

for $\omega >> \omega_c$ then from equation 7.8, one can get:

$$V = \frac{A}{2\pi rvC}(\omega/\omega_c)I_b \tag{7.9}$$

for $\omega << \omega_c$ then from equation 7.8, one can get:

$$V = \frac{A}{2\pi rvC}\frac{\iota\omega}{\omega_c}I_b = \frac{RA}{2\pi rvC}\frac{dI_b}{dt} \tag{7.10}$$

The schematics of button BPM is shown in Figure 7.3. In a button BPM [97], there are usually four electrodes positioned diagonally on the walls of the vacuum chamber. The signal from each button in a BPM can be calculated by integrating the surface charge density over the area of the button. To protect the buttons from damage caused by synchrotron radiation, they are often rotated by 45° from the horizontal plane in electron storage rings. The image charge density on the walls of a circular vacuum chamber is given by the equation:

$$\sigma(\phi) = -\frac{\lambda}{2\pi R}\frac{R^2 - r^2}{R^2 + r^2 - 2rR\cos(\phi - \theta)} \tag{7.11}$$

where R is the radius of the vacuum chamber, λ is the line charge density of the beam current, and (r, θ) represents the beam's center in polar coordinates. The total signal, $S = A+B+C+D$, is proportional to the beam intensity and the sum of the angles made by the buttons. The horizontal and vertical difference signals are defined as $S_x = A+D-B-C$ and $S_y = A+B-C-D$, respectively, and these depend on both the beam's position and intensity. The ratios $\frac{S_x}{S}$ and $\frac{S_y}{S}$ are related to the beam's horizontal and vertical positions by the equations

$$\frac{S_x}{S} = \frac{\sqrt{2}\sin(\theta)}{R\theta}x + O(r^3), \quad \frac{S_y}{S} = \frac{\sqrt{2}\sin(\theta)}{R\theta}y + O(r^3), \tag{7.12}$$

where θ is half the angle made by each button, and $O(r^3)$ represents higher-order terms. These relations provide a linear approximation of the beam position near the center of the chamber. However, when the beam is far from the center, the higher-order terms become significant, leading to deviations from the actual beam position.

7.4 Beam Current Transformers

When an electric current flows through a conductor, it creates a magnetic field around it. Similarly, moving a conducting loop through a magnetic field induces a current in the loop. These basic principles of induction are utilized in the beam current transformer (BCT). In this context, the particle beam acts as the primary current I_b, which can be expressed as:

$$I_b = \frac{qeN}{t} \tag{7.13}$$

where N is the number of particles with charge q passing per unit time t, and e is the charge of an electron. As the beam moves through the accelerator, it generates a magnetic field. This magnetic field can be measured by placing a toroid with high permeability around the beam. Windings are placed around the toroid and connected to an electrical circuit to read the signal. However, measuring very small beam currents with these devices can be quite challenging.

To derive the formula for the minimum current detectable by a BCT, we start by considering the signal-to-noise ratio (SNR) of the system. The output voltage signal generated by the BCT in response to a beam current can be expressed as:

$$V_{\text{signal}} = G \cdot n \cdot I_{\text{beam}} \tag{7.14}$$

where V_{signal} is the voltage signal output from the BCT, G is the gain of the measurement system, n is the number of turns in the transformer winding, and I_{beam} is the beam current.

The noise voltage V_{noise} in the system is typically considered as the RMS noise voltage over a bandwidth B. Assuming the noise is white noise, the noise power P_{noise} is proportional to the bandwidth B and is given by:

$$P_{\text{noise}} = V_{\text{noise}}^2 = N_0 \cdot B \tag{7.15}$$

where N_0 is the noise power spectral density (noise power per unit bandwidth) and B is the bandwidth of the system.

The SNR can then be expressed as:

$$\text{SNR} = \frac{V_{\text{signal}}^2}{V_{\text{noise}}^2} \tag{7.16}$$

For the minimum detectable current, the SNR is set to 1, meaning that the signal is just equal to the noise:

$$\frac{(G \cdot n \cdot I_{\min})^2}{V_{\text{noise}}^2} = 1 \tag{7.17}$$

Solving for the minimum current I_{min}, we obtain:

$$I_{\text{min}} = \frac{V_{\text{noise}}}{G \cdot n} \tag{7.18}$$

Considering that the noise voltage V_{noise} is distributed across a bandwidth B, and the measurement is taken over an integration time T, the effective noise is reduced by a factor of $\sqrt{B \cdot T}$, yielding:

$$I_{\text{min}} = \frac{V_{\text{noise}}}{G \cdot n \cdot \sqrt{B \cdot T}} \tag{7.19}$$

This formula indicates that the minimum current detectable by the BCT is influenced by the noise voltage, the system's gain, the number of turns in the transformer winding, the bandwidth, and the integration time.

7.5 Capacitive Pick-up Probes

A capacitive pick-up probe consists of an electrode placed inside or alongside the beam pipe, separated by an insulating layer from the vacuum chamber wall. As the charged particle bunch passes through the beam pipe, it induces an image charge on the electrode due to its moving electric field. The electrode acts as a capacitor plate, with the beam serving as the other plate, coupled through the vacuum gap. The geometry is typically cylindrical, matching the shape of the beam pipe, with the electrode either forming a button, stripline, or an annular ring.

The probe detects the time-varying electric field by capacitive coupling. The induced voltage $V(t)$ relates to the beam current $I_b(t)$ as:

$$V(t) = \frac{1}{C} \int I_b(t)\, dt \tag{7.20}$$

The output signal is a bipolar pulse, corresponding to the time derivative of the beam current. Integrating $V(t)$ recovers the bunch current profile, and the pulse width reflects the bunch length.

The probe's length L must be chosen short enough so that the bunch does not overlap the electrode for too long, avoiding signal smearing. To preserve time resolution, L should be smaller than the bunch's spatial length σ_z. Since spatial length relates to time duration Δt and particle velocity v by:

$$\sigma_z = v \cdot \Delta t \tag{7.21}$$

the length is chosen as:

$$L \lesssim v \cdot \Delta t \tag{7.22}$$

This shows that higher-energy (faster) beams require longer electrodes to match their spatial bunch length, while lower-energy beams need shorter electrodes for the same time duration. However, L is usually kept smaller to maintain bandwidth, balancing between resolution and signal strength.

The frequency response must cover:

$$f_{max} \approx \frac{1}{2\pi \Delta t} \tag{7.23}$$

The measured pulse width Δt_{meas} includes contributions from the bunch width, pick-up response, and detector resolution:

$$\Delta t_{meas}^2 = \Delta t_{bunch}^2 + \Delta t_{pickup}^2 + \Delta t_{detector}^2 \tag{7.24}$$

The diameter of the electrode affects capacitive coupling and frequency response. A larger diameter increases coupling (stronger signal) but lowers bandwidth; a smaller diameter reduces signal strength but improves high-frequency response. The diameter is chosen as a compromise between amplitude and bandwidth, usually similar to the beam pipe size.

By using two pick-ups separated by distance d, the time-of-flight $\Delta t = t_B - t_A$ provides the beam velocity:

$$v = \frac{d}{\Delta t} \tag{7.25}$$

and thus energy, either relativistic:

$$E = \gamma mc^2, \quad \gamma = \frac{1}{\sqrt{1 - (v/c)^2}} \tag{7.26}$$

or non-relativistic:

$$E = \frac{1}{2}mv^2 \tag{7.27}$$

Capacitive pick-ups measure both bunch width and energy through the timing of induced signals, with electrode size optimized to balance sensitivity, frequency response, and the beam's velocity.

7.6 Fast Faraday Cups

The fast Faraday cups (FFC) are capable of measuring beam currents as well as the time width of bunches. They work with 50 ohm input and output impedance to ensure proper beam signal transmission without reflection. The device can be made as compact as possible depending on the particle beam's range, and it consists of two conical cylinders with a proper diameter ratio such that the characteristic impedance turns out to be 50 ohms.

The characteristic impedance (Z_0) of a coaxial line [98] is derived based on the physical dimensions of the conductor and the properties of the materials used in its construction. Figure 7.4 summarizes the schematics and signal processing of beam bunch measurements using an FFC.

The electric field E between the conductors can be derived using Gauss's law. For a coaxial conductor, the electric field is radial and given by:

$$E(r) = \frac{V}{r \ln(b/a)}, \quad C' = \frac{2\pi\epsilon}{\ln(b/a)}, \quad L' = \frac{\mu}{2\pi} \ln(b/a) \tag{7.28}$$

where V is the potential difference between the inner and outer conductors, r is the radial distance from the center of the inner conductor, μ is the permeability of the dielectric

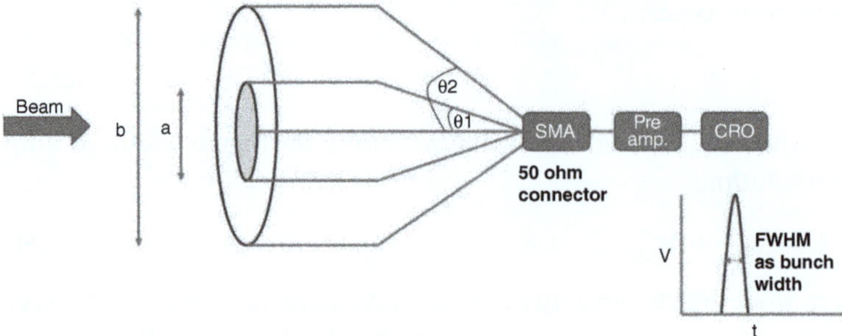

Figure 7.4 Beam bunching using a 50 ohm FFC.

material, ϵ is the permittivity of the dielectric material, a is the radius of the inner conductor, and b is the inner radius of the outer conductor. C' is the capacitance per unit length, and L' is the inductance per unit length.

The characteristic impedance Z of the coaxial conductors is given by:

$$Z = \sqrt{\frac{L'}{C'}} = \frac{1}{2\pi}\sqrt{\frac{\mu}{\epsilon}}\ln(b/a) \tag{7.29}$$

Finally, the characteristic impedance Z of a coaxial conductor for air can be expressed as:

$$Z = \frac{Z_0}{2\pi}\ln\left(\frac{b}{a}\right) = 60\ln\left(\frac{b}{a}\right) \tag{7.30}$$

where $Z_0 = \sqrt{\frac{\mu_0}{\epsilon_0}} = 377\,\Omega$ is the impedance of free space.

For a conical coaxial conductor with half-angles θ_1 and θ_2, the radii a and b are related to the cone angles as:

$$a = L\tan(\theta_1), \quad b = L\tan(\theta_2), \quad \frac{b}{a} = \frac{\tan(\theta_2)}{\tan(\theta_1)} \tag{7.31}$$

where L is the length of the cones, and θ_1 and θ_2 are the half-angles of the inner and outer cones, respectively.

Substituting this into the impedance formula:

$$Z_0 = 60\ln\left(\frac{\tan(\theta_2)}{\tan(\theta_1)}\right) = 60\ln\left(\frac{\cot(\theta_1)}{\cot(\theta_2)}\right) = 60\ln\frac{\cot\left(\frac{\theta_1}{2}\right)}{\cot\left(\frac{\theta_2}{2}\right)} \tag{7.32}$$

For an FFC, the coaxial and conical conductor combination must terminate to a 50 ohm SMA connector for further signal processing. Thus:

$$\frac{\cot\left(\frac{\theta_1}{2}\right)}{\cot\left(\frac{\theta_2}{2}\right)} = \frac{b}{a} = 2.3 \tag{7.33}$$

where the ratio $b/a = 2.3$ ensures a 50 ohm impedance matching for the SMA connector.

Normally, we consider the beam signal as a Gaussian distribution. A Gaussian function in the time domain can be written as:

$$g(t) = e^{-\frac{t^2}{2\sigma_t^2}}, \quad G(f) = e^{-\frac{f^2}{2\sigma_f^2}}, \quad \sigma_t \cdot \sigma_f = \frac{1}{2\pi} \tag{7.34}$$

where $g(t)$ represents the time-domain signal with standard deviation σ_t, and $G(f)$ is its Fourier transform in the frequency domain, with σ_f as the frequency-domain standard deviation.

The full width at half maximum (FWHM) of a Gaussian function in the time domain is:

$$\text{FWHM}_t = 2\sqrt{2\ln 2} \cdot \sigma_t \approx 2.355 \cdot \sigma_t, \quad \text{FWHM}_f = 2.355 \cdot \sigma_f \tag{7.35}$$

Using $\sigma_t \cdot \sigma_f = \frac{1}{2\pi}$, we get:

$$\text{FWHM}_f = 2.355 \cdot \frac{1}{2\pi\sigma_t} \approx 0.375 \cdot \frac{1}{\sigma_t} \tag{7.36}$$

Thus, the bandwidth B is approximately equal to FWHM_f, and the pulse width T is approximately equal to FWHM_t. Therefore:

$$B \cdot T \approx 0.35, \quad B(GHz) = \frac{0.35}{t_r(ns)} \tag{7.37}$$

where t_r is the rise time of the signal.

Let V_i be the incident voltage signal due to beam impact on the cup and V_r be the reflected voltage signal. At the load impedance Z_L, the total voltage V and total current I are given by:

$$V = V_i + V_r, \quad I = \frac{V_i}{Z_0} - \frac{V_r}{Z_0}, \quad V = Z_L \cdot I \tag{7.38}$$

where Z_0 is the characteristic impedance of the transmission line.

Substitute the expressions for V and I:

$$V_i + V_r = Z_L\left(\frac{V_i}{Z_0} - \frac{V_r}{Z_0}\right) \quad \Rightarrow \quad V_i\left(1 - \frac{Z_L}{Z_0}\right) = V_r\left(\frac{Z_L}{Z_0} + 1\right) \tag{7.39}$$

The voltage reflection coefficient (Γ) is defined as the ratio of the reflected voltage to the incident voltage:

$$\Gamma = \frac{V_r}{V_i} \quad \Rightarrow \quad 1 - \frac{Z_L}{Z_0} = \Gamma\left(1 + \frac{Z_L}{Z_0}\right) \quad \Rightarrow \quad \Gamma = \frac{Z_0 - Z_L}{Z_0 + Z_L} \tag{7.40}$$

The maximum voltage ($V_{\max}$) occurs when the incident and reflected waves are in phase, while the minimum voltage ($V_{\min}$) occurs when they are out of phase:

$$V_{\max} = V_i(1 + |\Gamma|), \quad V_{\min} = V_i(1 - |\Gamma|), \quad \text{VSWR} = \frac{V_{\max}}{V_{\min}} = \frac{1 + |\Gamma|}{1 - |\Gamma|} \tag{7.41}$$

where VSWR stands for voltage standing wave ratio, which describes the impedance matching of the system.

7.7 Phase Detector Cavity

In radiofrequency (RF) accelerators, precise timing is critical. The particle beam must arrive at the accelerating gaps at exactly the right moment to gain energy efficiently from the oscillating RF fields. Even slight deviations in arrival time can lead to energy loss, phase slippage, or beam degradation. To monitor and control the phase of the beam with respect to the RF system, a diagnostic device known as the phase detector cavity [99] is employed. This cavity, designed to resonate at the RF frequency, measures the phase difference by detecting the voltage induced as the beam traverses the structure. As the charged particle bunch passes through, it excites electromagnetic fields within the cavity, generating an output voltage that carries information about the beam's arrival time relative to the RF reference. The sensitivity of the phase detector cavity – defined as the output voltage per unit beam current – depends on both the electrical characteristics and the geometric configuration of the cavity. Understanding the relationship between sensitivity and these parameters requires a derivation based on fundamental electromagnetic and beam-cavity interaction principles.

Consider a beam current I_b passing through cavity field V_{eff}. The power transferred from the beam to the cavity fields is given by:

$$W_b = I_b V_{eff} \tag{7.42}$$

This power is lost partly as heat in the cavity walls (W) and partly extracted to an external load (W_l):

$$W_b = W + W_l \tag{7.43}$$

Under critical coupling:

$$W_l = W \quad \Rightarrow \quad W_b = 2W \tag{7.44}$$

Substitute this into the earlier equation:

$$I_b V_{eff} = 2W \tag{7.45}$$

We relate power loss W to cavity voltage through the shunt impedance:

$$R_0 = \frac{V^2}{2W} \quad \Rightarrow \quad W = \frac{V^2}{2R_0} \tag{7.46}$$

Substitute W into the power balance:

$$I_b V_{eff} = 2\left(\frac{V^2}{2R_0}\right) = \frac{V^2}{R_0} \tag{7.47}$$

Solve for voltage:

$$V^2 = I_b V_{eff} R_0 \tag{7.48}$$

and take the square root:

$$V = \sqrt{I_b V_{eff} R_0} \tag{7.49}$$

Only a fraction *r* of the cavity voltage is coupled out:

$$V_{out} = \sqrt{r}V \tag{7.50}$$

Substitute *V*:

$$V_{out} = \sqrt{r}\sqrt{I_b V_{eff} R_0} \tag{7.51}$$

Sensitivity is defined as output voltage per unit beam current:

$$S = \frac{V_{out}}{I_b} \tag{7.52}$$

Substitute V_{out}:

$$S = \frac{\sqrt{r}\sqrt{I_b V_{eff} R_0}}{I_b} = \sqrt{r}\sqrt{\frac{V_{eff} R_0}{I_b}} \tag{7.53}$$

From the power balance $I_b V_{eff} = 2W$, solve for V_{eff}:

$$V_{eff} = \frac{2W}{I_b} \tag{7.54}$$

Substitute into *S*:

$$S = \sqrt{r}\sqrt{\frac{(2W/I_b)R_0}{I_b}} = \sqrt{r}\sqrt{\frac{2WR_0}{I_b^2}} \tag{7.55}$$

This simplifies to:

$$S = \frac{\sqrt{2WR_0 r}}{I_b} \tag{7.56}$$

Recognizing that *W* scales with I_b^2, the I_b terms cancel:

$$S = \sqrt{2Rr} \tag{7.57}$$

Finally, we include the transit time factor *TTF* to account for the reduction in effective voltage due to finite particle velocity:

$$S = 2\sqrt{R_0 r}TTF \tag{7.58}$$

This final equation shows that sensitivity improves with the square root of the shunt impedance and coupling, and scales linearly with the transit time factor.

In this way, the phase detector cavity transforms the beam's passage into a measurable signal. Its design depends on optimizing the shunt impedance, coupling, and transit time factor to maximize sensitivity. A well-designed cavity ensures even a low beam current produces a detectable voltage, allowing precise measurement of the beam's phase relative to the RF. Without it, an accelerator would have no way to "listen" to the beam's timing, making this device a quiet and critical ear in the accelerator's operation.

7.8 Transverse Beam Emittance by Quadrupole Magnet Scan

Transverse emittance, an important parameter in accelerator physics, describes the distribution of particle beams in position and momentum space, providing insights into beam quality and dynamics. One effective method to measure transverse emittance is the quadrupole scan technique, which involves varying the strength of a quadrupole magnet and measuring the resulting beam size at a downstream location. By analyzing how the beam size changes with different quadrupole settings, the transverse emittance and Twiss parameters can be determined. In this method, a quadrupole magnet is positioned on the beamline, followed by a beam profile monitor at a particular distance. The focusing strength (k) of the quadrupole magnet is varied, altering its focusing properties and thus the beam size observed at the downstream monitor. The square of the measured beam size σ^2 is plotted against the quadrupole strength, typically resulting in a parabolic relationship that can be analyzed to extract the desired beam parameters.

The schematic of emittance measurements using the quadrupole scan method is shown in Figure 7.5. When a beam passes through a quadrupole magnet, the focusing effect of the quadrupole changes the beam size downstream. The transfer matrix M for a quadrupole of field gradient k and length l is given as follows:

$$M_q = \begin{pmatrix} 1 & 0 \\ kl & 1 \end{pmatrix} \tag{7.59}$$

In general, beam matrix is written as follows:

$$\sigma_q = \begin{pmatrix} \sigma_{11} & \sigma_{12} \\ \sigma_{21} & \sigma_{22} \end{pmatrix} \tag{7.60}$$

where

$$\sigma_{11} = \langle x_i^2 \rangle = \epsilon\beta; \qquad \sigma_{22} = \langle x_i'^2 \rangle = \epsilon\gamma; \qquad \sigma_{12} = \sigma_{21} = \langle x_i x_i' \rangle = -\epsilon\alpha \tag{7.61}$$

To calculate the beam size σ_x at a location downstream of the quadrupole magnet, consider a drift space of length d after the quadrupole. The beam matrix after the quadrupole and drift space is given by:

$$\sigma^{\text{diag}} = M_d \cdot M_q \cdot \sigma_q \cdot M_q^T \cdot M_d^T \tag{7.62}$$

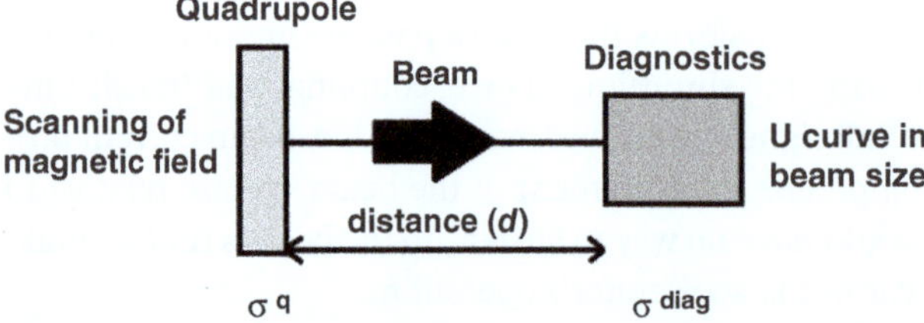

Figure 7.5 Schematic of emittance measurements using quadrupole scan method.

where:

- M_d is the transfer matrix for the drift space.
- M_q is the transfer matrix for the quadrupole.
- Σ_0 is the initial beam matrix.

Using the drift and quadrupole matrix product, one can get

$$M_d \cdot M_q = \begin{pmatrix} 1 + dkl & dkl \\ kl & 1 \end{pmatrix} \tag{7.63}$$

Multiplying transfer matrices the equation for $\sigma_{11}^{\text{diag}}$ can be expressed as:

$$\sigma_{11}^{\text{diag}} = (1 + dlk)^2 \sigma_{11}^{\text{q}} + 2(1 + dlk)d\sigma_{12}^{\text{q}} + d^2\sigma_{22}^{\text{q}} \tag{7.64}$$

$$\sigma_{11}^{\text{diag}} = \sigma_{11}^{\text{q}} d^2 l^2 k^2 + (2d\sigma_{11}^{\text{q}} + 2d^2 l\sigma_{12}^{\text{q}})k + (\sigma_{11}^{\text{q}} + 2d\sigma_{12}^{\text{q}} + d^2\sigma_{22}^{\text{q}}) \tag{7.65}$$

$$\sigma_{11}^{\text{diag}} = Ak^2 + Bk + C \tag{7.66}$$

where

$$A = \sigma_{11}^{\text{q}} d^2 l^2; \qquad B = 2d\sigma_{11}^{\text{q}} + 2d^2 l\sigma_{12}^{\text{q}}; \qquad C = (\sigma_{11}^{\text{q}} + 2d\sigma_{12}^{\text{q}} + d^2\sigma_{22}^{\text{q}}) \tag{7.67}$$

$$\sigma_{11}^{\text{q}} = \frac{A}{d^2 l^2}; \qquad \sigma_{12}^{\text{q}} = \frac{B - 2dl\sigma_{11}^{\text{q}}}{2d^2 l}; \qquad \sigma_{22}^{\text{q}} = \frac{C - \sigma_{11}^{\text{q}} - 2d\sigma_{12}^{\text{q}}}{d^2} \tag{7.68}$$

And finally calculate the normalized emittance:

$$\epsilon_n = \beta\gamma\sqrt{\sigma_{11}^{\text{q}}\sigma_{22}^{\text{q}} - (\sigma_{12}^{\text{q}})^2} \tag{7.69}$$

and the Twiss parameters at the quadrupole location are as follows:

$$\alpha_{\text{q}} = -\frac{\sigma_{12}^{\text{q}}}{\epsilon}; \qquad \beta_{\text{q}} = \frac{\sigma_{11}^{\text{q}}}{\epsilon}; \qquad \gamma_{\text{q}} = \frac{\sigma_{22}^{\text{q}}}{\epsilon} \tag{7.70}$$

The quadrupole scan technique provides a practical method to measure the transverse emittance and initial beam parameters by analyzing the beam size variation as a function of the quadrupole strength. The method is widely used in accelerator systems for beam diagnostics.

7.9 Longitudinal Beam Emittance by Buncher Scan

Longitudinal emittance is a crucial parameter in accelerator physics that characterizes the temporal and energetic spread of particle bunches, which is vital for understanding their dynamics within accelerators. The buncher scan method, a common approach for measuring longitudinal emittance, involves varying the phase or amplitude of a buncher cavity and observing how these changes affect the bunch length. This technique uses a buncher cavity, often a radio frequency (RF) cavity, placed in the beamline along with a longitudinal profile monitor (such as a streak camera or time-of-flight system) located downstream. By adjusting the buncher cavity settings and measuring the resulting bunch length, one can plot the squared bunch length against the buncher phase or amplitude. The resulting graph typically shows a parabolic relationship, allowing for the determination of longitudinal emittance and other related parameters.

In the thin lens approximation, the transfer matrix for an RF gap is written as follows:

$$\begin{pmatrix} 1 & 0 \\ -\dfrac{2\pi qV}{mc^2\lambda\beta^3\gamma} & 1 \end{pmatrix} \tag{7.71}$$

where q is the charge state of the ion beam, V is the effective voltage at the RF buncher, mc^2 is the rest mass of the ion, and λ is the free-space wavelength. Gamma (γ) is defined by $1/\sqrt{1-\beta^2}$, where $\beta = v/c$, and v is the velocity of the ion beam and c is the speed of light. The square of the bunch length measured after the bunching cavity σ_z^2 can be written as a quadratic function of the effective voltage and the Twiss parameters (α_0, β_0, and γ_0) immediately before the cavity.

Finally, the longitudinal emittance can be calculated as follows:

$$\sigma_z^2 = \epsilon_0 \left[\beta_0 R_{11}^2 - \frac{2L}{\gamma^2}\alpha_0 R_{11} + \frac{L^2}{\gamma^4}\gamma_0\right] \tag{7.72}$$

$$\epsilon_z = \frac{\gamma^2}{L}\sqrt{AC - \frac{B^2}{4}} \tag{7.73}$$

where A, B, and C are the coefficients of the quadratic equation for σ_z^2, and R_{11} is given as follows:

$$R_{11} = 1 - \frac{2\pi qVL}{mc^2\lambda\beta^3\gamma^3} \tag{7.74}$$

7.9.1 Bunch Length Measurements

Let's assume a thin buncher approximation. The squared bunch length (Δt^2), measured at a distance L downstream from the buncher, is modeled as a quadratic function of the effective voltage. This relationship is expressed as:

$$\Delta t^2 = a_2 \sin^2\phi_s V_{\text{eff}}^2 + a_1 \sin\phi_s V_{\text{eff}} + a_0 \tag{7.75}$$

where $\phi_s = \pm 90°$. The coefficients a_0, a_1, and a_2 can be linked to the Twiss parameters before the buncher and the beam's velocity β as follows:

$$a_2 = \left(\frac{q}{A}\right)^2 \left(\frac{\pi L}{\beta\lambda W_0 A}\right)^2 \beta_0 \tag{7.76}$$

$$a_1 = \left(\frac{q}{A}\right) \frac{2\pi L}{\beta\lambda W_0 A}\left(\frac{\pi L}{\beta\lambda W_0 A}\alpha_0 + \beta_0\right) \tag{7.77}$$

$$a_0 = \beta_0 + 2\frac{\pi L}{\beta\lambda W_0 A}\alpha_0 + \left(\frac{\pi L}{\beta\lambda W_0 A}\right)^2 \gamma_0 \tag{7.78}$$

The bunch length formula is more intricate compared to the energy spread since it depends on the downstream distance L. The beam's emittance is related to the fit parameters by:

$$\epsilon_0 = \left(\frac{A}{q}\right)\left(\frac{\beta\lambda W_0 A}{\pi L}\right)^2 \sqrt{a_2 a_0 - \frac{a_1^2}{4}} \tag{7.79}$$

7.10 Energy Spread Measurements

Let's assume a thin buncher approximation. The square of the energy spread measured after the buncher, ΔW^2, can be expressed as a quadratic function of the effective voltage V_{eff} and the Twiss parameters (α_0, β_0, γ_0, and ϵ_0) immediately before the buncher:

$$\frac{\Delta W^2}{A^2} = \epsilon_0 \left(\left(\frac{q}{A}\right)^2 \beta_0 \sin^2\phi_s V_{\text{eff}}^2 + 2\left(\frac{q}{A}\right)\alpha_0 \sin\phi_s V_{\text{eff}} + \gamma_0\right) \tag{7.80}$$

where $\phi_s = \pm 90°$ is the synchronous phase. The energy spread is independent of the drift distance L after the buncher and can be measured at any point along the transfer line. Experimentally, the emittance can be determined from a quadratic fit of the form $y = a_2 x^2 + a_1 x + a_0$, where $x = V_{\text{eff}} \sin\phi_s$ and $y = \frac{\Delta W^2}{A^2}$. Therefore,

$$\epsilon_0 = \left(\frac{A}{q}\right)\sqrt{a_0 a_2 - \frac{a_1^2}{4}} \tag{7.81}$$

From the same fit parameters a_0, a_1, and a_2, the other Twiss parameters can be written as:

$$\beta_0 = \left(\frac{A}{q}\right)^2 \frac{a_2}{\epsilon_0} \tag{7.82}$$

$$\alpha_0 = \frac{1}{2}\left(\frac{A}{q}\right)\frac{a_1}{\epsilon_0} \tag{7.83}$$

$$\gamma_0 = \frac{a_0}{\epsilon_0} \tag{7.84}$$

7.11 Ion Bunch Width Using Detectors

An ion beam's bunch width can be measured by detecting fast signals generated from its interaction with a target. Two methods are employed: detecting prompt gamma rays emitted from an aluminum target, and detecting scattered ions at a known angle using a solid-state barrier detector (SSBD) in a Rutherford-like scattering setup.

In the first method, an ion beam striking aluminum excites nuclei, which promptly emit gamma rays as they return to the ground state. This emission occurs within femtoseconds, preserving the bunch's time structure. The total number of gamma rays produced per second is:

$$N_\gamma = N_i \cdot n_t \cdot \sigma_\gamma \cdot \Delta x \tag{7.85}$$

where N_i is the number of incident ions per second, $n_t = \frac{N_A \cdot \rho}{A}$ is the number density of aluminum atoms calculated from Avogadro's number N_A, density ρ, and molar mass A; σ_γ is the gamma-ray production cross-section, and Δx is the target thickness.

Not all gamma rays escape the target; attenuation follows:

$$T = e^{-\mu \Delta x} \tag{7.86}$$

where μ is the linear attenuation coefficient. The detected gamma-ray count is:

$$N_{\gamma,detected} = N_\gamma \cdot \epsilon(E_\gamma) \cdot \frac{\Omega}{4\pi} \cdot T \tag{7.87}$$

with $\epsilon(E_\gamma)$ as the detection efficiency at energy E_γ, and Ω as the detector's solid angle. Gamma emission is fast because nuclear de-excitation occurs nearly instantaneously. A barium fluoride detector, with a fast decay of 0.6 nanoseconds, achieves timing resolution in hundreds of picoseconds.

In the second method, a thin foil scatters a small fraction of ions, detected by an SSBD at a known angle, similar to Rutherford's scattering experiment. The detected scattered ion count per second is:

$$N_s = N_i \cdot n_t \cdot \sigma(\theta) \cdot \Delta x \cdot \epsilon \cdot \frac{\Omega}{4\pi} \tag{7.88}$$

where $\sigma(\theta)$ is the differential scattering cross-section at angle θ, and ϵ is the detector efficiency. Elastic scattering happens within femtoseconds, preserving the bunch's time structure, while the SSBD responds within few nanoseconds, sufficient for nanosecond-scale measurements.

In both methods, the measured signal's time width, Δt_{meas}, is influenced by the bunch width Δt_{bunch} and the detector's time resolution $\Delta t_{detector}$, related by:

$$\Delta t^2_{meas} = \Delta t^2_{bunch} + \Delta t^2_{detector} \tag{7.89}$$

From this, the true bunch width is obtained:

$$\Delta t_{bunch} = \sqrt{\Delta t^2_{meas} - \Delta t^2_{detector}} \tag{7.90}$$

Both approaches rely on fast physical processes – prompt gamma emission and elastic scattering on femtosecond scales – and fast detector technologies, ensuring that the time structure of the ion bunch is faithfully captured. Many detectors besides BaF_2 and SSBD are used to measure ion beam bunch width, depending on required time resolution and particle type. Fast plastic scintillators, Cherenkov detectors, diamond detectors, microchannel plate (MCP) detectors, and streak cameras are common alternatives. Cherenkov and MCP systems achieve picosecond resolution, while plastic scintillators and SSBD work at nanosecond scales. Detectors are chosen based on time resolution, radiation hardness, sensitivity, and signal type (ion, gamma, optical). Often, systems combine detectors (e.g. scintillator + PMT) to balance speed, efficiency, and detection geometry.

7.12 Numerical Problems

1. A BPM in a circular accelerator measures the horizontal displacement of the beam from the center. If the BPM records a signal corresponding to a displacement of 1.5 mm, calculate the actual displacement if the calibration factor of the BPM is 1.2 mm per unit signal.
2. The BCT measures a beam current of 100 mA in a proton accelerator. Given the charge of a single proton, determine the number of protons passing through the BCT per second.
3. A synchrotron radiation monitor measures the horizontal beam size in a storage ring. If the beam profile fits a Gaussian with a standard deviation of 0.3 mm, evaluate the FWHM of the beam's horizontal distribution.
4. An optical transition radiation (OTR) monitor is used to measure the beam size in a LINAC. The OTR screen gives a beam image with a spot size of 0.4 mm. Calculate the beam emittance if the beta function at the monitor location is 5 m and the beam divergence is 0.05 mrad.
5. A fast beam loss monitor detects a sudden drop in beam intensity during an operation. If the beam power decreases by 20% over 5 ms in a LINAC operating at 50 MW, find the amount of energy lost by the beam.
6. The longitudinal phase space of a particle bunch is measured using a streak camera, and the bunch length is found to be 10 ps. If the energy spread of the bunch is 0.01% and the accelerator is operating at an energy of 5 GeV, show the relationship between the bunch length and energy spread using the longitudinal emittance formula.
7. A beam profile monitor measures a non-Gaussian beam distribution in a particle accelerator. The beam has a core size of 0.2 mm and tails extending up to 1 mm. Calculate the RMS beam size by integrating the beam distribution over the measured range.
8. An electrostatic beam deflector is used to measure the energy spread of an electron beam in a diagnostic section. If the deflector causes a horizontal displacement of 2 mm at a screen 1 m downstream, and the deflection angle is 0.5 mrad, determine the energy spread of the electron beam.
9. A wire scanner is used to measure the transverse profile of the beam. The scanner moves a wire across the beam, and the beam intensity drops by 30% when the wire is positioned 0.5 mm from the center of the beam. Estimate the transverse beam size assuming a Gaussian distribution.
10. A beam charge monitor detects a charge of 2 nC per bunch in a LINAC. If the accelerator operates at a repetition rate of 1 kHz, it determines the average beam current delivered by the accelerator.

8

Vacuum Devices

> "One never notices what has been done; one can only see what remains to be done... in the vacuum left by the absence of others."
>
> —*Marie Curie*

Upon completing this chapter, you should be able to:

- Understand how to create and maintain a vacuum in particle accelerators.
- Identify and locate leaks in the beamline or accelerator components.
- Diagnose and monitor the vacuum system.

In the fields of engineering and physics, a vacuum is defined as a space where the pressure is below atmospheric pressure and is quantified by its absolute pressure. Vacuum systems play a crucial role in particle accelerators, providing the necessary environment for particles to travel unobstructed through the beam pipe. Achieving and maintaining a vacuum on the order of 10^{-9} Torr is essential for the efficient operation of any accelerator. Various types of vacuum pumps are used to create and sustain these vacuum levels, while gauges are employed to measure and monitor the vacuum. Proper placement of vacuum pumps along the length of the accelerator ensures the maintenance of adequate conductance in the beam pipe. Before a stage of the particle accelerator is considered suitable for operation, vacuum leak test using leak detectors is essential. Key terms in vacuum technology include:

1. Vacuum pumps
2. Vacuum gauges
3. Outgassing
4. Conductance
5. Vacuum breakdowns
6. Vacuum valves and gaskets
7. Helium leak detectors

Charged Particle Beam Physics: An Introduction for Physicists and Engineers, First Edition.
Sarvesh Kumar and Manish K. Kashyap.

Companion Website: https://www.wiley.com/go/Kumar_1e

8.1 Basics of Vacuum Technology

8.1.1 Units in Vacuum

One atmosphere (atm) is a unit of pressure that is defined as the pressure exerted by the Earth's atmosphere at sea level. It is a standard reference pressure used in various scientific fields, including physics, chemistry, and engineering. Here, it is a listing of comparisons of 1 atm pressure with various pressure units.

1. 1 atm pressure is equal to 101325 pascals (Pa).
2. 1 atm pressure is equal to 760 millimeters of mercury (mmHg).
3. 1 atm pressure is equal to 760 Torr.
4. 1 atm pressure is equal to 1.01325 bar.
5. 1 atm pressure is equal to 14.696 pounds per square inch (psi).
6. 1 atm pressure is equal to 29.9213 inches of mercury (inHg).
7. 1 atm pressure is equal to 1.03323 kilogram-force per square centimeter (kgf/cm^2).
8. 1 atm pressure is equal to 1.01325×10^6 dynes per square centimeter (dyn/cm^2).
9. 1 atm pressure is equal to 406.78 inches of water (inH_2O).
10. 1 atm pressure is equal to 33.90 feet of water (ftH_2O).

Different units of pressure, which are very common in accelerator physics, are compared here so as to get the feeling of vacuum in numbers.

1. 1 Pa is equal to 1×10^{-5} bar, 9.8692×10^{-6} atm, 1.4504×10^{-4} psig, and 7.5006×10^{-3} Torr.
2. 1 bar is equal to 1×10^5 Pa, 0.98692 atm, 14.5038 psig, and 750.06 Torr.
3. 1 atm is equal to 1.01325×10^5 Pa, 1.01325 bar, 14.696 psig, and 760 Torr.
4. 1 psig is equal to 6.89476×10^3 Pa, 6.89476×10^{-2} bar, 6.8046×10^{-2} atm, and 51.715 Torr.
5. 1 Torr is equal to 133.322 Pa, 1.33322×10^{-3} bar, 1.3158×10^{-3} atm, and 1.9349×10^{-2} psig.

Percentage by volume (% by volume): Percentage by volume refers to the proportion of a specific gas or substance in a mixture relative to the total volume of the mixture, expressed as a percentage. It indicates how much of the total volume of the mixture is made up of the particular gas or substance. Mathematically, it is expressed as:

$$\%\text{ By volume} = \left(\frac{\text{Volume of gas}}{\text{Total volume of mixture}}\right) \times 100 \tag{8.1}$$

Partial pressure: Partial pressure is the pressure exerted by a single gas in a mixture of gases. It represents the pressure that the gas would exert if it occupied the entire volume of the mixture by itself, at the same temperature. According to Dalton's law of partial pressures, the total pressure of a gas mixture is the sum of the partial pressures of all the individual gases present in the mixture:

$$P_{\text{total}} = P_1 + P_2 + P_3 + \cdots + P_n \tag{8.2}$$

where P_{total} is the total pressure of the gas mixture, and $P_1, P_2, P_3, \ldots, P_n$ are the partial pressures of the individual gases in the mixture.

The following data corresponds to the typical composition of Earth's atmosphere at sea level.

1. Nitrogen (N_2) constitutes 78.09% by volume, with a partial pressure of 791.96 mbar.
2. Oxygen (O_2) constitutes 20.95% by volume, with a partial pressure of 212.35 mbar.
3. Argon (Ar) constitutes 0.93% by volume, with a partial pressure of 9.42 mbar.
4. Carbon dioxide (CO_2) constitutes 0.04% by volume, with a partial pressure of 0.41 mbar.
5. Neon (Ne) constitutes 0.0018% by volume, with a partial pressure of 0.018 mbar.
6. Helium (He) constitutes 0.0005% by volume, with a partial pressure of 0.005 mbar.
7. Methane (CH_4) constitutes 0.00017% by volume, with a partial pressure of 0.0017 mbar.
8. Krypton (Kr) constitutes 0.0001% by volume, with a partial pressure of 0.001 mbar.
9. Hydrogen (H_2) constitutes 0.00005% by volume, with a partial pressure of 0.0005 mbar.
10. Xenon (Xe) constitutes 0.000009% by volume, with a partial pressure of 0.00009 mbar.

Atmospheric air primarily consists of a mixture of gases, with nitrogen and oxygen making up over 99% of its composition. In contrast, the residual gas in high vacuum and UHV environments is predominantly hydrogen. This predominance of hydrogen occurs because it can easily diffuse through walls, tends to adsorb onto all surrounding surfaces, and is more challenging to pump out compared to gases with larger molecular sizes.

Gas molecules are in constant random motion, with their velocity determined solely by the mass of a gas molecule m and the temperature T. For a given temperature, molecules with higher molar mass move more slowly. Regarding molecular velocities, three key types are typically considered: the most probable velocity, the arithmetic mean velocity, and the root-mean-square (RMS) velocity, defined as follows:

Most probable velocity: The most probable velocity (v_{mp}) is the speed at which the maximum number of gas molecules is moving. It is given by:

$$v_{\mathrm{mp}} = \sqrt{\frac{2k_B T}{m}} \tag{8.3}$$

where k_B is the Boltzmann constant, T is the temperature in Kelvin, and m is the mass of a gas molecule.

Arithmetic average velocity: The arithmetic average velocity (v_{avg}) is the mean speed of the gas molecules. It is given by:

$$v_{\mathrm{avg}} = \sqrt{\frac{8k_B T}{\pi m}} \tag{8.4}$$

RMS velocity: The RMS velocity (v_{rms}) is the square root of the average of the squares of the velocities of the gas molecules. It is given by:

$$v_{\mathrm{rms}} = \sqrt{\frac{3k_B T}{m}} \tag{8.5}$$

RMS velocity for various gases

The following table provides the RMS velocity for different atmospheric gases at room temperature (298 K) and at 0°C (273 K).

1. The speed of the nitrogen molecule (N_2) is 517 m/s at 298 K and 493 m/s at 273 K.
2. The speed of the oxygen molecule (O_2) is 482 m/s at 298 K and 458 m/s at 273 K.
3. The speed of the argon molecule (Ar) is 431 m/s at 298 K and 408 m/s at 273 K.
4. The speed of the carbon dioxide molecule (CO_2) is 412 m/s at 298 K and 394 m/s at 273 K.
5. The speed of the hydrogen molecule (H_2) is 1936 m/s at 298 K and 1845 m/s at 273 K.
6. The speed of the helium molecule (He) is 1250 m/s at 298 K and 1186 m/s at 273 K.

8.1.2 Ideal Gas Law

Let's focus on a single gas molecule of mass m moving inside a cubic container with side length L. The molecule has velocity components v_x, v_y, and v_z along the x-, y-, and z-axes, respectively. When the gaseous molecule collides with a wall perpendicular to the x-axis, it transfers momentum to that wall. The change in momentum due to the collision with the wall is given as follows:

$$\Delta p_x = -mv_x - (mv_x) = -2mv_x \tag{8.6}$$

The time t between successive collisions with the same wall is given as follows:

$$t = \frac{2L}{v_x} \tag{8.7}$$

The force F_x exerted by the molecule on the wall is given as follows:

$$F_x = \frac{\Delta p_x}{t} = \frac{2mv_x}{\frac{2L}{v_x}} = \frac{mv_x^2}{L} \tag{8.8}$$

The pressure due to a single molecule on the wall is given by:

$$P = \frac{F_x}{A} = \frac{mv_x^2}{L}.\frac{1}{L^2} = \frac{mv_x^2}{L^3} = \frac{mv_x^2}{V} \tag{8.9}$$

where $V = L^3$ is the volume of the container.

Let the gas be isotropic (having no preferred direction), such that the mean square velocity $\overline{v^2}$ as:

$$\overline{v^2} = \overline{v_x^2} + \overline{v_y^2} + \overline{v_z^2} = 3\overline{v_x^2} \tag{8.10}$$

$$\overline{v_x^2} = \frac{\overline{v^2}}{3}$$

Considering the contributions of all N molecules in the gas, the total pressure (P) is given as follows:

$$P = \frac{1}{V}\sum_{i=1}^{N} m\overline{v_{x_i}^2} = \frac{1}{V}Nm\overline{v_x^2} = \frac{1}{V}Nm\frac{\overline{v^2}}{3} \tag{8.11}$$

This simplifies to the final expression for the pressure exerted by an ideal gas:

$$P = \frac{1}{3}\frac{N m \overline{v^2}}{V} \tag{8.12}$$

Here, the number of gas molecules N can also be expressed in terms of the number of moles n and Avogadro's number N_A as follows:

$$N = nN_A \tag{8.13}$$

Substituting this into the pressure equation gives:

$$P = \frac{nN_A k_B T}{V} \tag{8.14}$$

The product $N_A k_B$ is recognized as the universal gas constant R, so the equation simplifies to:

$$P = \frac{nRT}{V} \tag{8.15}$$

Finally, we derive the ideal gas law:

$$PV = nRT \tag{8.16}$$

This equation describes the relationship between pressure, volume, and temperature for an ideal gas, assuming that the gas particles have negligible volume and no intermolecular forces.

The Van der Waals equation modifies the ideal gas law to account for the finite size of gas molecules and the intermolecular forces between them. It is given by:

$$\left(P + \frac{a}{V_m^2}\right)(V_m - b) = \mathrm{RT} \tag{8.17}$$

where P is the pressure of the gas, V_m is the molar volume, T is the temperature, and R is the universal gas constant. The constant a corrects for the attractive forces between gas molecules. The term $\frac{a}{V_m^2}$ reduces the effective pressure because these forces pull molecules closer together. The constant b accounts for the finite volume occupied by the gas molecules. The term $V_m - b$ subtracts the volume occupied by the molecules from the total volume of the gas, which is ignored in the ideal gas law. This equation provides a more accurate description of real gas behavior, especially at high pressures and low temperatures where deviations from ideal gas behavior are significant.

In summary, while the ideal gas law provides a simplified description of gas behavior, the real gas law is more accurate and relevant in vacuum conditions. Understanding both laws is essential for designing and optimizing vacuum systems. Accurate modeling of gas behavior is essential for designing efficient vacuum pumps. Knowledge of gas behavior is vital for predicting gas flow and transport in vacuum systems.

8.1.3 Mean Free Path, Knudsen Number, Reynolds Number, Coefficient of Viscosity

Consider a gas enclosed in a volume. The mean free path λ represents the average distance a gas molecule travels before colliding with another molecule. The molecules in the gas move randomly in all directions.

Let n denote the number density of molecules, defined as the number of molecules per unit volume:

$$n = \frac{N}{V} \tag{8.18}$$

where N is the total number of molecules and V is the volume.

Each molecule can be modeled as a hard sphere with a diameter d. Two molecules will collide if their center-to-center distance is less than d. The effective cross-sectional area for a collision between two molecules is given by:

$$\sigma = \pi d^2 \tag{8.19}$$

where σ is known as the "collision cross-section."

If v_1 and v_2 are the speeds of two different molecules, the relative speed v_{rel} can be calculated using the formula:

$$v_{\text{rel}} = \sqrt{v_1^2 + v_2^2} \tag{8.20}$$

The average relative speed is often approximated as $\sqrt{2} \times v_{\text{avg}}$, where v_{avg} represents the average speed of the molecules. The frequency of collisions ν is given as follows:

$$\nu = n \cdot \sigma \cdot v_{\text{rel}} \tag{8.21}$$

where v_{rel} is the relative speed, and σ is the cross-sectional area.

The mean free path λ is the average distance a molecule travels before it collides with another molecule and is given as follows:

$$\lambda = \frac{v_{\text{avg}}}{\nu} = \frac{v_{\text{avg}}}{n \cdot \sigma \cdot v_{\text{rel}}} \tag{8.22}$$

$$\lambda = \frac{v_{\text{avg}}}{n \cdot \sigma \cdot \sqrt{2} \times v_{\text{avg}}} = \frac{1}{\sqrt{2} \cdot n \cdot \sigma}$$

If d is the molecular diameter, then we have:

$$\lambda = \frac{1}{\sqrt{2} \cdot n \cdot \pi d^2} \tag{8.23}$$

This formula shows if either the number density increases or the diameter of the molecules gets larger, the mean free path becomes shorter. The number density n is related to the pressure p and temperature T by the ideal gas law:

$$p = nk_BT \tag{8.24}$$

Using the above two equations, the Knudsen number (Kn) is defined, which is a dimensionless number that helps us understand the flow characteristics of a gas and is given as follows:

$$\mathrm{Kn} = \frac{\lambda}{L} = \frac{k_B T}{\sqrt{2} \cdot p \cdot \pi d^2 \cdot L} \tag{8.25}$$

The Knudsen number helps us understand whether the flow can be considered a continuous fluid or if we need to consider the individual behavior of molecules. By understanding how these two numbers relate, we can choose the right models and assumptions to analyze fluid flows in different situations. It helps us determine the type of flow within a system depending upon its value as follows:

1. **Kn < 0.01 (continuum or viscous flow):** In this range, the mean free path of the gaseous molecules is much smaller than the characteristic length. The gas behaves as a continuous fluid, and the flow can be described using the Navier–Stokes equations. This type of flow is common in conditions where the pressure is high.
2. **0.01 < Kn < 0.1 (slip flow):** As the Knudsen number increases, the mean free path becomes less negligible. The flow remains primarily continuous, but there is noticeable slippage at the boundaries, necessitating adjustments to boundary conditions.
3. **0.1 < Kn < 10 (transition flow):** In the transition flow regime, the mean free path is comparable to the characteristic length. The flow cannot be fully described by either continuum or molecular models, requiring a more complex analysis.
4. **Kn > 10 (free molecular flow):** When the Knudsen number exceeds 10, the mean free path is much larger than the characteristic length. Gaseous molecules travel independently with minimal interaction, and the flow is governed by molecular kinetics rather than collective fluid behavior, which is common in high vacuum environments.

There is one more characteristic of the system called as the Reynolds number. It is mainly used to determine whether the flow of a fluid is smooth (laminar) or chaotic (turbulent). It does this by comparing the inertial forces to the viscous forces in the fluid. Reynolds number is a dimensionless quantity that indicates whether the flow of a fluid is laminar or turbulent. It is given by:

$$Re = \frac{\rho v L}{\mu} \tag{8.26}$$

where ρ is the fluid density, v is the fluid velocity, L is a characteristic length (such as the diameter of a pipe), and μ is the dynamic viscosity of the fluid.

Dynamic viscosity, denoted by μ, is a measure of a fluid's resistance to shear or flow. It quantifies the internal friction in a fluid, which arises when layers of the fluid move relative to each other. The dynamic viscosity is defined mathematically by Newton's law of viscosity:

$$\tau = \mu \frac{du}{dy} \tag{8.27}$$

where τ is the shear stress in the fluid (Pa), μ is the dynamic viscosity (Pa.s), and $\frac{du}{dy}$ is the velocity gradient perpendicular to the direction of the shear (1/s). In simpler terms, dynamic viscosity is a property that describes how much force is required to move one layer of fluid in relation to another. Higher viscosity implies greater resistance to flow. The dynamic viscosity of air at standard conditions (approximately 20°C or 68°F) is typically given as:

$$\mu \approx 1.81 \times 10^{-5}\ \text{Pa s} \tag{8.28}$$

Flow regimes:

- **Laminar flow:** $Re < 2000$
- **Transitional flow:** $2000 \leq Re \leq 4000$
- **Turbulent flow:** $Re > 4000$

Recalling, the mean free path λ, the average distance a molecule travels between collisions, is given by:

$$\lambda = \frac{1}{\sqrt{2} \cdot n \cdot \sigma} \tag{8.29}$$

where n is the number density of molecules and σ is the molecular collision cross-section. The average speed $\bar{v}$ of the gas molecules is:

$$\bar{v} = \sqrt{\frac{8k_B T}{\pi m}} \tag{8.30}$$

where k_B is Boltzmann's constant, T is the temperature, and m is the mass of a gas molecule.

The momentum flux, which is the momentum transfer per unit area per unit time, is given by:

$$\text{Momentum flux} = n \cdot m \cdot \lambda \cdot \frac{du}{dy} \cdot \bar{v} \tag{8.31}$$

Viscosity η is the proportionality constant relating momentum flux to the velocity gradient $\frac{du}{dy}$:

$$\eta = \frac{\text{Momentum flux}}{\frac{du}{dy}} = n \cdot m \cdot \lambda \cdot \bar{v} \tag{8.32}$$

However, since only one-third of the molecules, on average, move in the direction contributing to viscosity in three-dimensional space, the final expression for viscosity includes a factor of $\frac{1}{3}$:

$$\eta = \frac{1}{3} \cdot n \cdot m \cdot \lambda \cdot \bar{v} \tag{8.33}$$

In this derivation, n is the number density of molecules, m is the mass of a molecule, λ is the mean free path, $\bar{v}$ is the average molecular speed, k_B is Boltzmann's constant, T is the temperature, and η is the coefficient of viscosity. The units of viscosity η derived from this formula is kg/m/s, which correspond to the momentum transfer per unit area per unit time. This confirms that the formula is consistent with the concept of momentum transfer in gases, thereby validating the physical significance of η as a measure of internal friction due to molecular collisions.

Here are the main points related to the coefficient of viscosity:

1. The coefficient of viscosity η in gases reflects the internal friction due to collisions and momentum transfer between gas molecules.
2. Unlike liquids, viscosity in gases increases with temperature because higher molecular speeds lead to more frequent and energetic collisions.
3. Viscosity in gases is crucial for understanding flow behavior, impacting drag, heat transfer, and flow regimes such as the Reynolds number, especially in applications like aerodynamics.
4. In vacuum systems, as pressure decreases, the mean free path of gas molecules increases, resulting in fewer collisions, but the viscosity does not decrease significantly.
5. Gas viscosity primarily depends on momentum transfer during collisions rather than collision frequency, making it critical in maintaining low gas density and ensuring effective gas flow in high vacuum and UHV systems.
6. Understanding gas viscosity is essential for designing efficient vacuum pumps, predicting residual gas behavior, and ensuring the stability and performance of vacuum environments.

8.1.4 Conductance, Pumping Speed, Throughput, and Outgassing in Vacuum Systems

Conductance in a vacuum system refers to the ease with which gas can pass through a component or section of the system. It is analogous to electrical conductance, but instead of relating to electrical current, it pertains to gas flow. The conductance, denoted by C, is defined by the relationship:

$$C = \frac{Q}{\Delta P} \tag{8.34}$$

where:

- Q is the volumetric flow rate (typically measured in liters per second, L/s).
- ΔP is the pressure difference across the components (usually in Pascals, Pa).

Several factors influence conductance in a vacuum system:

- **Geometry:** The size, shape, and length of the vacuum components, such as pipes or apertures, play a critical role in determining conductance.
- **Flow regime:** Conductance is also dependent on the flow regime, which is characterized by the Knudsen number. In the free molecular flow regime, conductance is independent of pressure, whereas in the viscous flow regime, it varies with pressure.
- **Type of gas:** Different gases have varying molecular masses and velocities, which affect the conductance.

Optimizing conductance is essential in vacuum technology to achieve desired pressure levels effectively. Components with high conductance enable quicker evacuation and enhance the overall performance of the vacuum system.

In vacuum systems, conductance is a key parameter that determines the efficiency of gas flow through different components. When multiple vacuum elements are connected, their combined conductance is calculated differently based on whether they are arranged in series or parallel, similar to the principles used in electrical circuits.

When many vacuum components, such as pipes, valves, and apertures are connected in beamline, the overall conductance of the beamline is determined by the combined effect of each component's conductance. The total conductance C_{total} in a series arrangement is calculated using the following formula:

$$\frac{1}{C_{total}} = \frac{1}{C_1} + \frac{1}{C_2} + \frac{1}{C_3} + \cdots + \frac{1}{C_n} \tag{8.35}$$

where $C_1, C_2, \ldots, C_n$ represent the conductances of individual components in the series.

This relationship indicates that the total conductance is always lower than the conductance of the least conductive component in the series. In other words, the component with the smallest conductance will have the most significant impact on reducing the overall conductance of the system.

When vacuum components are arranged in parallel, the gas flow can take multiple paths through the system, which increases the overall conductance. The total conductance C_{total} for components connected in parallel is the sum of the individual conductances:

$$C_{total} = C_1 + C_2 + C_3 + \cdots + C_n \tag{8.36}$$

where:

- $C_1, C_2, \ldots, C_n$ are the conductances of the individual parallel components.

Here are some key points related to the conductance of the system:

1. In a parallel setup, the total conductance is higher than the conductance of any individual component, as multiple paths make it easier for gas to flow through the system.
2. Knowing how to calculate conductance in both series and parallel setups is essential for designing and optimizing vacuum systems.
3. Series configurations are used when precise control over gas flow is required, such as in restrictive orifice plates or long pipelines where the flow needs to be regulated.
4. Parallel configurations are used to increase the overall conductance, such as in manifold systems where multiple pumps or pathways are used to evacuate a chamber more quickly.
5. By arranging vacuum components in series or parallel, designers can meet specific performance, optimizing flow rates, pressure differentials, and evacuation times.

Some more useful quantities related to vacuum pumps are as follows:

Pumping speed measures how much gas a pump can remove over time, typically in L/s, m^3/h, or CFM. It is given by:

$$S = \frac{Q}{P} \tag{8.37}$$

where S is the pumping speed, Q is the gas throughput, and P is the pressure at the pump inlet.

Throughput refers to the amount of gas passing through a system per unit time, indicating how effectively the pump is removing gas.

Conductance C shows how easily gas flows through system components. The effective pumping speed S_{eff} at any point is:

$$\frac{1}{S_{eff}} = \frac{1}{S} + \frac{1}{C} \tag{8.38}$$

where S_{eff} is the effective speed, S is the actual pump speed, and C is the conductance.

S_{eff} is always less than or equal to S because C limits gas flow. Higher conductance leads to better system performance, so wide, short pipes are preferred.

Finally, it is important to know about **Outgassing in Vacuums**. It happens when materials in a vacuum chamber release gases or vapors as the pressure drops. This occurs because materials, such as metals, plastics, or glass, can trap gases like water vapor or air on their surfaces. When the pressure is reduced in a vacuum, these trapped gases escape into the chamber. Outgassing can cause several problems, such as contamination of sensitive equipment or samples, which can affect measurements and experiments. It can also raise the pressure inside the chamber, making it harder to maintain a vacuum and putting extra strain on the vacuum pumps. This can slow down the process and even damage equipment. To manage outgassing, it's important to use materials that release fewer gases, clean surfaces to remove trapped gases, and heat materials before putting them into the vacuum. Good vacuum pumps, like cryogenic or ion pumps, can help remove gases more efficiently. In industries where high or ultra-high vacuum is required, such as in electronics or scientific experiments, controlling outgassing is essential for achieving the best results.

8.2 Vacuum Accessories and Subcomponents

8.2.1 Fast Valves

1. A fast valve is essential in vacuum systems for quickly isolating or protecting components from sudden pressure changes or reverse flow.
2. These valves close rapidly to ensure the safety and stability of the vacuum environment.
3. Fast valves typically operate within less than 10 ms.
4. They are actuated by pneumatic or electromagnetic systems for quick response.
5. They function within high vacuum to UHV conditions.
6. Valve materials include stainless steel or aluminum, suitable for vacuum environments.
7. Seals are made from Viton, Buna-N, or other elastomers that are compatible with vacuum use.
8. Common connection types include ISO-KF, ISO-Universal, ConFlat (CF), or custom flanges.
9. They operate in a temperature range from −20 to +150°C, depending on the material.
10. Fast valves are available in various sizes, from DN16 to DN250 and beyond, depending on application needs.
11. They are crucial for protecting sensitive equipment and maintaining vacuum stability during critical experiments or processes.
12. Fast valves help prevent contamination and damage by providing immediate isolation when necessary.

8.2.2 Vacuum Bellows

1. Vacuum bellows are flexible, accordion-shaped components that manage movement, reduce vibrations, and correct misalignments in vacuum systems.
2. Their design allows bending, stretching, or compressing while maintaining a vacuum seal.
3. Typically made from stainless steel, they provide flexibility and strength, making them crucial in systems with movement or temperature changes.
4. Bellows are vital where rigid connections could fail, such as in cases of thermal expansion, mechanical motion, or vibration.
5. They connect to other vacuum system parts using various flanges, chosen based on system requirements.
6. ISO-KF flanges are used in low to medium vacuum systems for quick assembly and disassembly.
7. ISO-LF flanges handle larger connections in medium to high vacuum systems.
8. CF flanges are used in high vacuum and UHV systems, providing strong seals with metal gaskets.
9. ISO-BF flanges are used for large components in high vacuum and UHV systems, offering secure, bolted connections.
10. ASA flanges are common in industrial vacuum systems, handling large openings for high-pressure and vacuum applications.
11. Quick-flange (QF) offers even easier assembly and disassembly, ideal for laboratory use.
12. Welded flanges provide permanent connections for high vacuum or UHV systems by welding directly to the bellows.

8.2.3 Rubber Gaskets

1. Rubber gaskets create airtight or watertight seals, preventing leaks in various applications, including vacuum systems.
2. They fill gaps between surfaces, offering flexibility, durability, and tight seals, widely used across industries.
3. Besides sealing, they reduce vibrations, protect parts, and some provide thermal or electrical insulation.
4. Nitrile (Buna-N) gaskets resist oils, fuels, and chemicals, ideal for automotive and industrial use.
5. EPDM gaskets resist weathering, ozone, and UV radiation, making them suitable for outdoor applications.
6. Viton gaskets resist heat, chemicals, and oils and are used in chemical processing and aerospace.
7. Silicone gaskets offer flexibility across a wide temperature range and are used in food-grade and medical devices.
8. Neoprene gaskets resist weathering, oils, and chemicals, with flame-retardant properties, used in HVAC and marine environments.
9. Natural rubber gaskets offer high elasticity and abrasion resistance, and are used in shock absorbers and vibration isolators.

10. Polyurethane gaskets are durable with high abrasion resistance, suitable for dynamic applications like hydraulic seals.
11. Fluorosilicone gaskets combine flexibility with enhanced chemical resistance, used in aerospace and automotive industries.
12. Gasket selection depends on application needs like chemical compatibility, temperature range, and environmental conditions.

8.2.4 Metallic Gaskets

1. Metallic gaskets are essential for sealing in high-pressure, high-temperature, and corrosive environments where nonmetallic gaskets may fail.
2. Commonly made from stainless steel, copper, or aluminum, they ensure reliable seals under extreme conditions.
3. Widely used in the petrochemical, oil and gas, power generation, and aerospace industries for safety and efficiency.
4. Spiral wound gaskets, made from alternating layers of metal and filler (such as graphite or PTFE), offer flexibility and strength, ideal for high-pressure environments with temperature fluctuations.
5. Ring type joint gaskets are solid metal rings used in flanged connections, especially in oil and gas pipelines, to withstand high pressures.
6. Corrugated metal gaskets, with a corrugated metal core and soft sealing material, are suitable for thermal cycling applications like heat exchangers and boilers.
7. Metal-jacketed gaskets, consisting of a soft filler material wrapped in a metallic jacket, are ideal for sealing in high-temperature environments.
8. The choice of metallic gasket depends on the application's pressure, temperature, and chemical compatibility to ensure effective performance.

8.2.5 Vacuum Flanges and Adapters

1. Vacuum flanges connect various parts in vacuum systems, ensuring a secure, leak-tight connection to maintain vacuum integrity.
2. They are used in industries like semiconductor manufacturing, research labs, and aerospace, where high vacuum or UHV is critical.
3. The choice of flange depends on vacuum level, operating temperature, and material compatibility.
4. ISO-KF flanges, also called Klein flanges, are used in low to medium vacuum systems, allowing quick assembly and disassembly with a centering ring and O-ring seal.
5. ISO-LF flanges are similar to ISO-KF but designed for larger connections and higher vacuum levels.
6. CF flanges are used in high vacuum and UHV systems, known for low leak rates with a metal-to-metal seal using a copper gasket.
7. ASA flanges are used in high vacuum applications with large bore sizes, providing a strong, bolted connection.
8. Welded flanges offer a permanent connection, ideal for systems where disassembly is not needed.

9. Vacuum adapters connect different types of vacuum fittings, flanges, or tubing, enhancing system flexibility.
10. Reducing adapters connect components of different sizes, enabling diameter changes.
11. Flange-to-flange adapters connect flanges of different standards, such as ISO-KF to CF.
12. Elbow adapters change the direction of vacuum lines by 90° for a better layout.
13. The adapters allow branching of vacuum lines, useful in complex designs.
14. Quick-disconnect fittings enable easy attachment and detachment of hoses or tubing, ideal for laboratory use.
15. Choosing the right adapters and flanges ensures the vacuum system operates efficiently and maintains performance.

8.3 Vacuum Pumps

Vacuum pumps are known for their operating range to create a vacuum. First, there is a plot as shown in Figure 8.1 of various individual capabilities of vacuum pumps of different types. The various combinations of pumps that can be used to create UHV from atmospheric pressure as given below.

- **Rotary vane pump + turbomolecular pump**
- **Rotary vane pump + diffusion pump**
- **Diaphragm pump + Roots blower + turbomolecular pump**
- **Scroll pump + turbomolecular pump + ion pump**
- **Rotary vane pump + diffusion pump + cryopump**
- **Rotary vane pump + Roots blower + diffusion pump**

The typical pumping speeds of vacuum pumps vary by type and application. **Rotary vane pumps** operate at 1–100 m^3/h, suitable for rough vacuum applications, while **dry pumps** (e.g. scroll and screw types) range from 10 to 1000 m^3/h, offering oil-free operation. **Turbo pumps**, specified in liters per second (L/s), provide 50–500 L/s at high vacuum levels, with

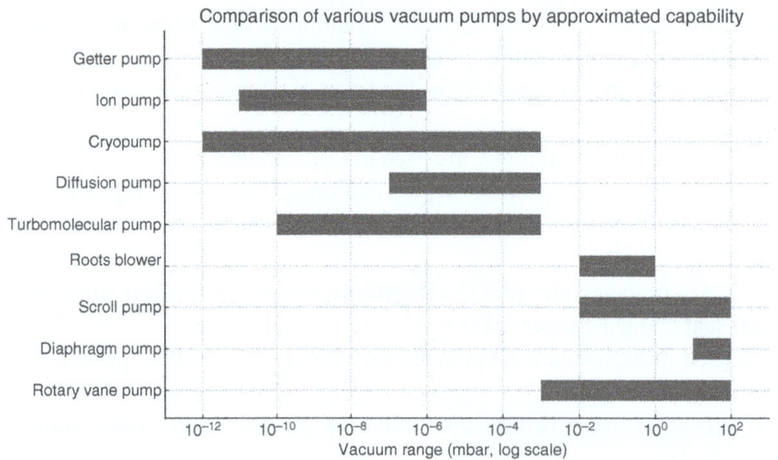

Figure 8.1 Comparison of various vacuum pumps in terms of their approximated capability.

performance dependent on gas molecular weight. **Cryogenic pumps** excel at condensable gas removal, with speeds from 100 to 10 000 L/s, while **ion pumps** offer 1–100 L/s, highly effective for reactive gases but less so for light gases like hydrogen. **Diffusion pumps**, ranging from 10 to 10 000 L/s, achieve high speeds but require baffles to minimize oil backstreaming. Pumping speed varies with the specific model, gas type, and the typical pumping speeds of vacuum pumps vary by type and application. **Rotary vane pumps** operate at 1–100 m^3/h, suitable for rough vacuum applications, while **dry pumps** (e.g. scroll and screw types) range from 10 to 1000 m^3/h, offering oil-free operation. **Turbo pumps**, specified in liters per second (L/s), provide 50–500 L/s at high vacuum levels, with performance dependent on gas molecular weight. **Cryogenic pumps** excel at condensable gas removal, with speeds from 100 to 10 000 L/s, while **ion pumps** offer 1–100 L/s, highly effective for reactive gases but less so for light gases like hydrogen. **Diffusion pumps**, ranging from 10 to 10 000 L/s, achieve high speeds but require baffles to minimize oil backstreaming. Pumping speed varies with the specific model, gas type, and operating pressure.

8.3.1 Oil-sealed Rotary Vane Pumps

It is used in general-purpose laboratory vacuum applications and creates vacuum by rotating vanes that trap and compress gas. The working principle is explained below stepwise.

They are a widely used type of mechanical vacuum pump, suitable for creating low to medium vacuum levels. These pumps are valued for their durability, long service life, and versatility, making them popular in various fields such as laboratory research, industrial production, and process engineering. Here is its working principle in steps:

1. The pump consists of a rotor with vanes that is eccentrically mounted inside a cylindrical chamber.
2. As the rotor turns, the vanes slide in and out of the rotor slots, maintaining contact with the chamber walls due to centrifugal force.
3. Gas enters the pump through the inlet and gets trapped in the spaces between the vanes and the chamber.
4. The trapped gas is compressed as the rotor continues to turn, reducing the volume of the space.
5. The compressed gas is pushed out through the exhaust valve, and the oil seal prevents backflow and ensures a tight seal during compression.
6. This process repeats continuously, creating a vacuum by removing gas from the system.

8.3.2 Turbomolecular Pump

It is used in high vacuum systems like electron microscopes and creates vacuum by accelerating gas molecules with rapidly rotating blades. The working principle is explained below stepwise:

1. The turbomolecular pump consists of a series of rapidly rotating blades, along with stationary blades, arranged in stages.
2. Gas molecules enter the pump and collide with the rotating blades, which impart momentum to the molecules.

3. The momentum causes the gas molecules to move from one stage to the next, progressively pushing them toward the exhaust.
4. Stationary blades are positioned between the rotating blades to guide the gas molecules in the desired direction, ensuring efficient pumping.
5. The gas is continuously transferred through the stages until it is expelled from the pump, creating a high vacuum in the system.

8.3.3 Ion Pump

It is used in UHV systems and creates vacuum by ionizing gas molecules and embedding them into a solid cathode. The working principle is explained below stepwise:

1. The ion pump creates a strong electric field inside the pump, which ionizes incoming gas molecules by stripping away their electrons.
2. The resulting positively charged ions are accelerated toward a negatively charged cathode due to the electric field.
3. When the ions collide with the cathode, they either get buried in the cathode material or react with it, becoming chemically bound.
4. Electrons released during ionization are guided by a magnetic field, causing them to spiral and increase the chances of further ionization.
5. This process continuously traps gas molecules, effectively lowering the pressure in the vacuum system.

8.3.4 Getter Pump

It is used in UHV systems and sealed-off devices and creates vacuum by chemically binding gases to a getter material. The working principle is explained below stepwise:

1. The getter pump contains a reactive material, often a metal, that is deposited on surfaces inside the pump.
2. When gas molecules enter the pump, they come into contact with the reactive material.
3. The reactive material chemically reacts with the gas molecules, binding them to the surface.
4. This reaction effectively removes the gas molecules from the vacuum system, lowering pressure.
5. Over time, the reactive material may become saturated, and it may need to be regenerated or replaced to maintain pumping efficiency.

8.3.5 Diaphragm Pump

It is used in applications requiring oil-free operation like medical and laboratory setups, and creates vacuum by mechanically flexing a diaphragm to move gases. The working principle is explained below stepwise:

1. The diaphragm pump has a flexible diaphragm that moves back and forth within a chamber.

2. When the diaphragm is pulled back, it creates a vacuum in the chamber, causing the inlet valve to open and draw fluid into the chamber.
3. As the diaphragm pushes forward, the inlet valve closes, and the fluid is compressed.
4. The compressed fluid forces the outlet valve to open, allowing the fluid to be discharged from the pump.
5. This cycle repeats, with the diaphragm's motion continuously drawing in and pushing out fluid, creating a steady flow.

8.3.6 Roots Blower

It is used in conjunction with other pumps for rough vacuum and creates vacuum by trapping gas between two rotating lobes and moving it from inlet to outlet. The working principle is explained below stepwise:

1. The Roots blower consists of two lobed rotors that rotate in opposite directions within a casing.
2. As the rotors turn, they trap a volume of gas in the space between the rotors and the casing.
3. The trapped gas is carried from the inlet side to the outlet side as the rotors continue to rotate.
4. When the trapped gas reaches the outlet, it is pushed out, creating a flow of gas.
5. This process repeats continuously, moving gas from the inlet to the outlet, and generating pressure or vacuum as needed.

8.3.7 Scroll Pump

It is used in oil-free vacuum applications like semiconductor manufacturing and creates vacuum by orbiting a spiral scroll to compress gas. The working principle is explained below stepwise:

1. The scroll pump consists of two spiral-shaped scrolls, one fixed and one orbiting.
2. As the orbiting scroll moves, it traps gas between the scrolls, creating a series of sealed pockets.
3. The trapped gas is gradually compressed as the orbiting scroll moves the pockets toward the center.
4. Once the gas reaches the center, it is discharged through an exhaust port.
5. This continuous movement of the scrolls creates a steady flow of gas from the inlet to the exhaust, resulting in vacuum generation.

8.3.8 Screw Vacuum Pumps

They are **positive-displacement pumps** that use rotating screws to trap and compress gas or material. These pumps do not need oil or liquid sealing, making them ideal for clean environments where contamination must be avoided. They work by using two synchronized screws that rotate in opposite directions inside a precisely designed housing. As the screws turn, they trap gas in the spaces between them and push it towards the exhaust, compressing it along the way. Screw pumps operate in the **rough to medium vacuum range**, typically from 1 to 10^{-3} mbar. Their main advantages include oil-free operation, making them clean and efficient; the ability to handle different types of gases, including corrosive and condensable ones; and low maintenance due to fewer moving parts. Because of these benefits, screw pumps are widely used in industries such as semiconductor and electronics manufacturing, pharmaceuticals, chemical processing, plasma coating, and vacuum packaging in the food industry.

8.3.9 Cryopump or Sorption Pump

It is used in high vacuum systems and creates vacuum by condensing gases onto a cold surface at cryogenic temperatures. The working principle is explained below stepwise:

1. The cryopump cools internal surfaces to very low temperatures, typically below 20 K, using a cryogenic fluid or refrigeration system.
2. Gas molecules from the vacuum chamber come into contact with these cold surfaces.
3. The gas molecules lose energy and condense or adsorb onto the cold surfaces, effectively trapping them.
4. As more gas molecules condense, they form a thin layer on the cold surfaces, reducing the pressure in the vacuum chamber.
5. The process continues until the surfaces become saturated, after which the cryopump needs to be regenerated by warming up to release the trapped gases.

8.3.10 Diffusion Pump

It is used in industrial and scientific high vacuum systems and creates vacuum by vaporizing oil or fluid and directing it downward to trap gas molecules. The working principle is explained below stepwise:

1. The pump's working fluid, typically oil, is heated in a boiler at the bottom of the pump.
2. The heated oil vaporizes and rises through the pump.
3. The vapor is directed through nozzles, forming high-speed jets that move downward.
4. Gas molecules from the vacuum chamber collide with the vapor jets, gaining momentum.
5. The gas molecules are driven toward the pump's outlet and are then captured by a backing pump.
6. The vapor reaches the cooled walls of the pump, where it condenses back into liquid form.
7. The condensed liquid returns to the boiler, allowing the cycle to repeat.

8.4 Vacuum Gauges

Vacuum gauges are important tools used to measure pressure in a vacuum system, helping to monitor and control the vacuum environment. They are widely used in areas like scientific research, industry, and semiconductor manufacturing, where accurate pressure readings are crucial. The effectiveness and safety of vacuum systems rely on the accuracy and sensitivity of these gauges. Regular calibration and upkeep are needed to keep vacuum gauges working properly. Knowing how different types of vacuum gauges work, and their limitations is key to choosing the right one for specific needs. Here are many different types of vacuum gauges.

1. **Pirani gauge**: Measures pressure using thermal conductivity, effective from 10^3 to 10^{-4} mbar.
2. **Capacitance manometer**: Detects pressure changes by measuring capacitance, working within 10^3 to 10^{-6} mbar.
3. **Thermocouple gauge**: Uses thermal conductivity for pressure measurement, covering 10^3 to 10^{-3} mbar.
4. **Ionization gauge (cold cathode)**: Measures very low pressures using ionization, effective from 10^{-2} to 10^{-11} mbar.
5. **Ionization gauge (hot cathode)**: Ionization-based gauge for ultra-low pressures, working within 10^{-3} to 10^{-12} mbar.
6. **McLeod gauge**: Measures low pressures by compressing gas, effective from 10 to 10^{-6} mbar.
7. **Penning gauge**: A cold cathode ionization gauge that measures pressures from 10^{-2} to 10^{-9} mbar.
8. **Bayard–Alpert gauge**: Hot cathode ionization gauge for extremely low pressures, effective from 10^{-3} to 10^{-12} mbar.
9. **Bourdon gauge**: Measures pressure through mechanical deformation, suitable for pressures from 10^3 to 10^{-1} mbar.

8.4.1 Pirani Gauge

A Pirani gauge is a thermal conductivity vacuum gauge used to measure pressures in the low to medium vacuum range. It operates by sensing the thermal conductivity of the gas in the vacuum chamber. The gauge has a heated filament, and as gas molecules collide with the filament, removing heat, which changes the filament's temperature. This temperature change alters the electrical resistance of the filament, which is related to the gas pressure. It is valued for its simplicity, reliability, and ability to measure a wide pressure range from approximately 10^3 to 10^{-4} mbar.

8.4.2 Capacitance Manometer

It is a precision vacuum gauge, which measures vacuum level by detecting changes in the capacitance of a diaphragm that responds to varying gas pressures. It consists of a flexible diaphragm positioned between two electrodes in the form of a capacitor. As gas pressure changes, the diaphragm deflects, altering the distance between the electrodes and changing the capacitance. The capacitance change is measured and converted into a corresponding vacuum level reading. They are highly accurate and can measure a wide range of pressures, from atmospheric levels down to UHV levels, typically between 10^3 and 10^{-6} mbar. They are used for precise and stable pressure measurements, such as in semiconductor manufacturing, research labs, and process control systems.

8.4.3 Thermocouple Gauge

It measures vacuum level by assessing the thermal conductivity of the gas in a vacuum system. It uses a thermocouple to detect temperature changes. It contains a filament heated by an electric current. The gas molecules collide with the filament and remove heat, which causes a temperature change. This change of temperature is detected by the thermocouple and is directly related to the system's pressure. The thermocouple gauge is particularly effective in the rough to medium vacuum range, typically from 10^3 to 10^{-3} mbar. They are known for simplicity, durability, and ease of usage for various industrial and laboratory vacuum applications.

8.4.4 Cold Cathode Ionization Gauge

A cold cathode ionization gauge measures a very vacuum level by ionizing gas molecules without a heated filament. It works by applying a high voltage between two electrodes, which creates a strong electric field to accelerate electrons. These high-energy electrons collide with gas molecules and produce ionization. The resulting positive ions are collected at the cathode, generating a current proportional to the gas pressure in the vacuum. They can measure pressures from 10^{-2} to 10^{-11} mbar, making them ideal for UHV applications. They are known for their durability, long lifespan, and ability to operate without the delicate components as required in hot cathode gauges.

8.4.5 McLeod Gauge

It is a specialized gauge used to measure vacuum level by compressing a known volume of gas and measuring the resulting increase in pressure. It isolates a gas sample from the vacuum system and compresses it within a sealed capillary tube. As the gas is compressed, its pressure increases proportionally to the actual pressure in the vacuum system. By measuring the height of the mercury column that compresses the gas, the pressure can be calculated using Boyle's law. It is highly accurate for measuring low pressures, typically ranging from 10 to 10^{-6} mbar. It is particularly useful for calibrating other vacuum gauges and applications where precise pressure measurements are required. It is less preferred today due to its manual operation and the availability of more convenient electronic gauges.

8.4.6 Penning Gauge

It is a cold cathode ionization gauge, which measures vacuum level by ionizing gas molecules using both electric and magnetic fields. The gauge operates by applying a high voltage between electrodes under a magnetic field, which traps electrons in a spiral motion, increasing their path length and the events of ionizing gas molecules. The ions are collected on the cathode, which generates a current proportional to the pressure in the vacuum system. They are effective over a pressure range from 10^{-2} to 10^{-9} mbar and are commonly used in applications requiring robust and reliable measurements of low pressures, such as in plasma physics experiments or in particle accelerators.

8.4.7 Bayard–Alpert Gauge

A Bayard–Alpert gauge is a highly sensitive vacuum gauge used for measuring extremely low pressures. It works on the ionization principle, where electrons are emitted from a heated filament and get accelerated toward a grid by an applied voltage. As the electrons go through the grid, they ionize the gaseous molecules in the vacuum. The resulting positive ions are attracted by a fine wire at the center of the grid, named as the ion collector. The current generated in the ion collector is proportional to the gaseous pressure in the vacuum chamber. This gauge minimizes the X-ray effect, to suppress inaccuracies at very low pressures, and hence can accurately measure pressures as low as 10^{-12} mbar. This gauge is widely used in UHV environments such as in surface science and semiconductor processing.

8.4.8 Bourdon Gauge

It is a mechanical device to measure pressure in gases and liquids across a wide range. It works on the principle of elastic deformation. It consists of a curved, hollow tube, known as the Bourdon tube, which straightens slightly when exposed to internal pressure. As the pressure increases inside the tube, its curvature decreases, and this movement is transmitted through a mechanical linkage to a pointer on a dial, providing a pressure reading. They are highly versatile, offering reliable measurements from atmospheric pressure down to around 10^{-1} mbar. Due to their robust design and ability to function without electricity, they are frequently used in industrial applications, including environments where electronic devices might not be suitable.

8.5 Helium Leak Detector: Mathematical Principles

A helium leak detector detects the amount of helium gas that escapes from a system, to indicate a vacuum leak. Helium gas is best for this purpose because it is a light gas that can easily enter through tiny leaks. Also, it is an inert gas, not reacting with materials or causing damage to the device being tested. It is a rare gas in the atmosphere, so detecting it in a vacuum system is straightforward, with minimal interference from ambient air. Lastly, it has low viscosity, which allows it to move quickly through leaks, making its detection faster and more accurate.

The leak rate Q is the amount of gas passing through a leak per unit of time, typically measured in units such as mbar·L/s or Pa·m^3/s. It can be calculated using the following equation:

$$Q = \frac{P_{\text{in}} \cdot V}{R \cdot T} \tag{8.39}$$

where Q is the leak rate (in mbar·L/s or Pa·m^3/s), P_{in} is the inlet pressure (in mbar or Pa), V is the system volume (in liters or cubic meters), and R is the universal gas constant, $R = 8.314\,\text{J/mol·K}$. T is the absolute temperature (in Kelvin).

A leak rate of 1 mbar·L/s means that every second, a volume of gas equivalent to 1 L is leaking through the system at a pressure difference of 1 mbar. This unit helps quantify the gas escaping from a vacuum system, which is crucial for understanding the severity of a leak and ensuring that vacuum systems are properly sealed and functioning efficiently. A leak rate of 10^{-9} mbar·L/s means that a very small amount of gas, equivalent to one-billionth of a liter per second, is leaking through the system at a pressure difference of 1 mbar. This is an extremely low leak rate, indicating that the system is very well sealed.

In vacuum systems, such a low leak rate is generally considered excellent, especially for high or UHV applications. A leak rate of 10^{-9} mbar·L/s order or below is recommended for such conditions. The helium partial pressure P_{He} in the leak detector is important in determining the measurement's sensitivity. It is calculated as:

$$P_{He} = \frac{Q_{He}}{S_{He}} \tag{8.40}$$

where P_{He} is the partial pressure of helium in the detector (in mbar or Pa), Q_{He} is the helium leak rate (in mbar·L/s or Pa·m^3/s), and S_{He} is the detector's pumping speed for helium (in L/s or m^3/s).

The sensitivity S_{det} of the helium leak detector is the minimum detectable leak rate, and it is calculated as:

$$S_{det} = \frac{P_{min} \cdot S_{He}}{P_{He}} \tag{8.41}$$

where S_{det} is the sensitivity of the leak detector (in mbar·L/s or Pa·m^3/s), P_{min} is the minimum detectable pressure change (in mbar or Pa), S_{He} is the pumping speed for helium (in L/s or m^3/s), and P_{He} is the helium partial pressure (in mbar or Pa).

In systems with multiple leaks or complex geometries, the overall leak rate Q_{total} is the sum of individual leak rates:

$$Q_{total} = \sum_{i=1}^{n} Q_i \tag{8.42}$$

where Q_{total} is the total leak rate (in mbar·L/s or Pa·m^3/s), Q_i is the leak rate of the i-th leak (in mbar·L/s or Pa·m^3/s), and n is the number of individual leaks.

Finally, the leak rate is a measure of the rate at which gas enters or escapes from a vacuum system through leaks or imperfections in the system.

8.5.1 Steps for Leak Detection

- Ensure the system is sealed and ready for testing. Evacuate the system using a vacuum pump to remove air and other gases.
- Introduce helium gas around the suspected leak points or throughout the system. Helium is used because it is a small, inert molecule that can easily penetrate small leaks.
- Connect the helium leak detector to the system's exhaust or an appropriate connection point. Monitor the vacuum environment for any traces of helium, indicating the presence of a leak.
- Move the helium probe around different parts of the system to locate the exact leak point. Higher helium concentration readings indicate proximity to the leak.
- The detector displays the leak rate, typically in units like mbar·L/s or Pa·m^3/s. This helps determine the severity of the leak.

- Record the leak rate and location for documentation. Analyze whether the leak is within acceptable limits or if it needs to be repaired.
- Repair the leak by sealing it or replacing faulty components. Retest the system with the helium leak detector to confirm the repair was successful.
- Perform a final leak detection test with full He spray to ensure the system is working correctly. Once the system passes with no detectable leaks, it is ready for operation.

To counter such issues, techniques like beam-based alignment and orbit correction algorithms are implemented, often supported by many beam optics simulation tools. So, it is a must to perform precise surveys, accelerator component alignments, and sensitivity analyses of beamline areas. Theodolites are mainly used for this task. They are precision devices that can remove the misalignment errors. One can align the beam optical component using markers provided by the manufacturer known as alignment marks or circles, etc. A theodolite is set up on a tripod and leveled using transverse spirit levels. It consists of a telescope, horizontal and vertical circles, trunnion axis, spirit levels, tripod, plumb bob or optical plummet, clamping and tangent screws, focusing knob, and an eyepiece, all working together for precise angular measurements. The telescope is mounted on a trunnion axis and can rotate horizontally and vertically. The user directs the telescope at a reference point or target or wall, fixes the line of sight/action by clamping the telescope, and then makes fine adjustments with tangent screws. Horizontal and vertical angles are measured by rotating the telescope horizontally and vertically and reading the angle on the horizontal and vertical circles, respectively. By iterative process, alignment is finalized for the beam optical component. By aligning the crosshairs of the theodolite with the fixed reference point, the height is transferred to the beamline components. Adjustments are made to the beamline until its height matches the reference point, making all the components as aligned as possible. Modern high precision theodolites have angular accuracy ranging from 0.5 to 2 arcseconds.

8.6 Numerical Problems

1. During a helium leak detection test in a vacuum system, helium is sprayed near a suspected leak point. The helium leak detector measures the leak rate of 2×10^{-8} Torr L/s. If the base pressure of the system is 1×10^{-9} Torr and the chamber volume is 200 L, determine the time required for the system's pressure to rise to 1×10^{-7} Torr due to the helium leak.
2. A helium leak detector is used to locate a leak in a 500 L vacuum chamber. Helium gas is introduced around the suspected leak area, and the detector registers a leak rate of 3×10^{-9} Torr L/s. If the pump is able to maintain a constant pumping speed of 100 L/s, calculate the steady-state pressure in the chamber due to the leak.
3. In a particle accelerator vacuum system, the pressure must be maintained at 1×10^{-8} Torr for optimal beam performance. If a small leak is detected with a leak rate of 5×10^{-6} Torr L/s, calculate how long it would take for the pressure to rise to 1×10^{-6} Torr in a chamber with a volume of 50 L.
4. A vacuum chamber in an accelerator with a volume of 100 Lis pumped by a system with a constant pumping speed of 100 L/s. If the outgassing rate is 1×10^{-10} Torr L/s, determine the final pressure that will be reached in the chamber.

5. During leak detection using helium gas, the leak detector measures a helium leak rate of 1×10^{-7} Torr L/s, while the background pressure in the system is 5×10^{-9} Torr shows how long it will take for the pressure to increase by two orders of magnitude if the helium continues to leak.
6. A particle accelerator requires a vacuum of 1×10^{-9} Torr to maintain its beam lifetime. If a small leak causes the pressure to rise to 5×10^{-8} Torr, find by what percentage the beam lifetime would decrease, assuming the lifetime is inversely proportional to the vacuum pressure.
7. An accelerator's vacuum chamber, with a surface area of $2\,m^2$, has an outgassing rate per unit area. Evaluate the total outgassing rate for the entire surface area and the required pumping speed to maintain a pressure of 1×10^{-8} Torr in a chamber volume of 200 L.
8. A vacuum system test reveals that a chamber with a volume of 80 L experiences a pressure rise from 1×10^{-9} to 3×10^{-6} Torr over a period of 10 minutes. Estimate the leak rate in Torr liters per second, and show whether this leak would significantly affect operations requiring a vacuum of 1×10^{-9} Torr.
9. An accelerator system starts at atmospheric pressure and needs to be pumped down to 1×10^{-9} Torr. The chamber volume is 150 L, and the pumping speed is 200 L/s. Estimate the time required to achieve this vacuum, assuming ideal conditions and no outgassing.
10. A vacuum chamber shows a slow pressure rise after pump-down, even though no external leaks are detected. If a virtual leak inside the chamber has a volume of 0.5 L and is initially at atmospheric pressure, calculate the time it would take for this virtual leak to equilibrate with the main chamber (at 1×10^{-8} Torr) with a pumping speed of 50 L/s.
11. In a large accelerator complex with multiple vacuum chambers, a leak in one chamber causes the pressure to rise from 1×10^{-9} to 1×10^{-7} Torr over 30 minutes in a chamber with a volume of 500 L. Calculate the total leak rate and analyze how this leak would impact other connected chambers.
12. A vacuum pipe with a conductance of 200 L/s connects two chambers in an accelerator. If the upstream chamber has a pressure of 1×10^{-5} Torr and the downstream chamber has a pressure of 5×10^{-7} Torr, determine the leak rate through the pipe and describe how this information can be used for troubleshooting and finding leaks.

A

Field-induced Breakdown in Accelerator Technologies

Breakdown in matter occurs when an insulating medium becomes conducting under the influence of an electric field. The electric field is the force per unit charge that accelerates electrons and ions. In accelerator systems, this process limits voltage holding in gas-filled tubes, radio frequency cavities, plasma sources, and vacuum components. The driving quantity across all regimes is the electric field, but the relevant form depends on the physical environment.

In gases and plasmas, breakdown is governed by the electric field present within the medium. This field energizes free electrons, which collide with neutral atoms to cause ionization and create an avalanche of charged particles. The breakdown voltage in this regime is described by Paschen's law, which relates the breakdown voltage $V_{\text{breakdown}}$ to the product of gas pressure p and electrode separation d [100, 101]:

$$V_{\text{breakdown}} = \frac{Bpd}{\ln(Apd) - \ln\left(\ln\left(1 + \frac{1}{\gamma}\right)\right)} \tag{A.1}$$

Here, A and B are constants specific to the gas, and γ is the secondary electron emission coefficient from the cathode. This relation applies to Tandem accelerator terminal spheres, SF6-filled acceleration tubes, and beamline gas strippers where macroscopic electric fields in the medium dominate discharge behavior.

In radio frequency accelerator structures, the critical quantity is not the average field in the volume but the local surface electric field. This is the electric field that exists at the boundaries of materials, often intensified due to microscopic surface features like sharp edges or particulate contamination. It is this enhanced surface field that determines the onset of phenomena such as field emission and multipactor. A widely used empirical limit in radio frequency design is the Kilpatrick criterion [94, 102], which estimates the maximum sustainable surface field in a vacuum at a given frequency:

$$f = 1.64\,E^2 \exp\left(\frac{-8.5}{E}\right) \tag{A.2}$$

Here, f is the frequency in megahertz, and E is the peak surface field in megavolts per meter. This constraint is essential in the design and operation of radio frequency quadrupoles, drift tube linear accelerators, superconducting cavities, and bunching structures where avoiding radio frequency breakdown is critical for continuous operation.

Charged Particle Beam Physics: An Introduction for Physicists and Engineers, First Edition.
Sarvesh Kumar and Manish K. Kashyap.

Companion Website: https://www.wiley.com/go/Kumar_1e

Plasma generation and maintenance depend on the energy gained by electrons from the applied electric field. When the power absorbed exceeds the ionization and transport losses, breakdown and plasma ignition occur. The threshold electric field for breakdown is given by [103]:

$$\mu_e E^2 \geq n_g \sigma_{\text{ion}} v_e \epsilon_{\text{ion}} \tag{A.3}$$

In this expression, μ_e is the electron mobility, n_g is the neutral gas density, σ_{ion} is the ionization cross section, v_e is the mean electron velocity, and ϵ_{ion} is the ionization energy. This condition governs breakdown thresholds in electron cyclotron resonance ion sources, plasma strippers, and radio frequency discharge chambers in accelerator front ends.

At high frequencies, breakdown can also result from the multipactor effect, a surface phenomenon driven by resonant secondary electron emission. In this case, electrons emitted from a surface return periodically due to the radio frequency field and generate more electrons upon impact, provided the timing and energy conditions are favorable. The resonance condition is given by [104, 105]:

$$T_{\text{flight}} = \frac{n}{2f}, \quad \delta > 1 \tag{A.4}$$

Here, T_{flight} is the transit time between surfaces, f is the radio frequency, and δ is the secondary emission coefficient. Multipactor breakdown is a risk in ceramic radio frequency windows, couplers, dielectric beamline components, and superconducting systems where surface electric fields govern secondary emission behavior.

In vacuum systems, where there is no medium to support traditional avalanche ionization, breakdown is triggered by field emission. This quantum mechanical process involves the tunneling of electrons through a surface potential barrier when the local electric field becomes strong enough. The emission current is given by the Fowler Nordheim relation [106, 107]:

$$J = AE^2 \exp\left(\frac{-B\phi^{3/2}}{\beta E}\right) \tag{A.5}$$

Here, ϕ is the material work function, β is the field enhancement factor due to surface roughness, and A, B are empirical constants. This mechanism dominates in electrostatic lenses, high-voltage extraction systems, and photoinjectors. Even at modest applied voltages, surface irregularities can locally produce strong electric fields that initiate breakdown. For this reason, surface treatment and cleanliness are essential in vacuum accelerator components.

In pulsed systems, breakdown behavior is affected by the pulse duration and voltage rise time. Short pulses reduce the time available for avalanche development and increase the effective breakdown threshold. This property is used in the operation of kicker magnets, spark gap switches, and fast beam extraction circuits.

In magnetized plasmas, the presence of a magnetic field alters electron motion and modifies the breakdown condition. When the collision frequency ν_c is much smaller than the cyclotron frequency ω_c, electrons become confined along magnetic field lines, and their mobility across the field is suppressed. This behavior modifies the breakdown threshold in electron cyclotron resonance sources and magnetic plasma confinement systems.

In all cases, whether in gases, plasmas, vacuum systems, or radio frequency structures, the electric field is the fundamental cause of breakdown. In many practical accelerator components, it is not the average field in the medium but the local surface electric field that sets the actual threshold. This surface field, often amplified by microscopic features, determines the onset of ionization, emission, and discharge. Understanding this field and its enhancement is essential for designing reliable high-field accelerator components.

Paschen's Law for the Breakdown Voltage of Gases

Paschen's law [100] describes the breakdown voltage $V_{\text{breakdown}}$ as a function of the product of the gas pressure p and the distance between two electrodes d. In a typical system, electrons are emitted from the cathode and move toward the anode through a gas medium under the influence of an electric field. The breakdown voltage depends on the ionization processes that occur in the gas, which are characterized by the Townsend ionization coefficients.

The first Townsend ionization coefficient, denoted as α, represents the number of ion pairs generated per unit distance traveled by an electron, and is given by:

$$\alpha = Ape^{-\frac{Bp}{E}}, E = \frac{V}{d} \tag{A.6}$$

where A and B are gas-specific constants, p is the pressure, and E is the electric field, which is related to the breakdown voltage V and the electrode separation d.

The second Townsend ionization coefficient γ represents the number of secondary electrons produced per ion impact on the cathode and is defined as:

$$\gamma = \frac{\text{Number of secondary electrons}}{\text{Number of incident ions}} \tag{A.7}$$

Now, let $n(x)$ represent the number of electrons after traveling a distance x, where n_0 is the initial number of electrons. The number of electrons after traveling a distance x is described by the equation:

$$n(x) = n_0 e^{\alpha x}, n(d) = n_0 e^{\alpha d} \tag{A.8}$$

For the gas discharge to be self-sustaining, the number of initial electrons must be equal to the number of secondary electrons produced at the cathode, which can be expressed as:

$$n_0 = \gamma (n(d) - n_0) \Rightarrow 1 = \gamma \left(e^{\alpha d} - 1\right) \tag{A.9}$$

Next, substituting α we obtain:

$$1 = \gamma\left(e^{Apde^{-\frac{Bpd}{V}}} - 1\right) \tag{A.10}$$

When the argument of the exponential is large, we approximate $e^x \approx x + 1$. Thus, above equation simplifies to:

$$e^{Apde^{-\frac{Bpd}{V}}} \approx Apde^{-\frac{Bpd}{V}} + 1 \Rightarrow \gamma\left(Apde^{-\frac{Bpd}{V}}\right) = 1 \tag{A.11}$$

Solving for $e^{-\frac{Bpd}{V}}$, we get:

$$Apde^{-\frac{Bpd}{V}} = \frac{1}{\gamma} \Rightarrow e^{-\frac{Bpd}{V}} = \frac{1}{Apd\gamma} \tag{A.12}$$

Taking the natural logarithm of both sides:

$$-\frac{Bpd}{V} = \ln\left(\frac{1}{Apd\gamma}\right) = -\ln(Apd\gamma) \Rightarrow \frac{Bpd}{V} = \ln(Apd) - \ln(\gamma) \tag{A.13}$$

Finally, solving for the breakdown voltage V, we find:

$$V = \frac{Bpd}{\ln(Apd) - \ln(\gamma)} \tag{A.14}$$

For a more precise description, we replace $\ln(\gamma)$ with $\ln\left(\ln\left(1+\frac{1}{\gamma}\right)\right)$, leading to the final form of Paschen's law:

$$V_{\text{breakdown}} = \frac{Bpd}{\ln(Apd) - \ln\left(\ln\left(1+\frac{1}{\gamma}\right)\right)} \tag{A.15}$$

This equation allows for the calculation of the breakdown voltage $V_{\text{breakdown}}$ for a given gas pressure p and electrode distance d, taking into account the constants A and B that are specific to the gas in question. The parameter γ, the secondary electron emission coefficient, also plays a significant role in determining the breakdown voltage.

Although Paschen's law is generally accurate for predicting breakdown voltages in gases, it may fail under certain extreme conditions [108], such as very high pressures or very small electrode separations. In these cases, the electric field may become sufficiently strong to induce field emission from the cathode, leading to measurable currents even before breakdown occurs. This field emission can generate ions that contribute to the space charge field, amplifying the electron current and causing breakdown. This explains why deviations from the Paschen curve can occur in conditions that are not captured by the classical law. For more details, refer to Ref. [108].

B

Panofsky–Wenzel Theorem in Accelerator Physics

The Panofsky–Wenzel theorem [109, 110] is crucial in accelerator physics, particularly for understanding the deflection of particle beams. A key conclusion from this theorem is that, in a transverse electric (TE) mode, the deflecting impulse caused by the electric field is exactly canceled by the impulse from the magnetic field.

The force acting on a charged particle, generalized for any charge q, due to electric and magnetic fields is given by the Lorentz force law:

$$\vec{F} = q(\vec{E} + \beta\hat{z} \times \vec{B}) \tag{B.1}$$

where q is the charge of the particle, $\vec{E}$ is the electric field, $\vec{B}$ is the magnetic field, and $\beta = \frac{v}{c}$ is the velocity of the particle as a fraction of the speed of light c. To calculate the divergence of this Lorentz force, we apply the divergence operator:

$$\nabla \cdot \vec{F} = q\left[\nabla \cdot \vec{E} + \beta\nabla \cdot (\hat{z} \times \vec{B})\right] \tag{B.2}$$

Using Gauss's law for electric fields, $\nabla \cdot \vec{E} = 4\pi\rho$, and the vector identity $\nabla \cdot (\hat{z} \times \vec{B}) = -\hat{z} \cdot (\nabla \times \vec{B})$, Equation B.2 becomes:

$$\nabla \cdot \vec{F} = q\left[4\pi\rho - \beta\hat{z} \cdot (\nabla \times \vec{B})\right] \tag{B.3}$$

Next, applying Ampère's law $\nabla \times \vec{B} = \frac{1}{c}\frac{\partial \vec{E}}{\partial t} + 4\pi\vec{J}$, where $\vec{J}$ is the current density, we substitute this into Equation B.3:

$$\nabla \cdot \vec{F} = q\left[4\pi\rho - \beta\left(\frac{1}{c}\frac{\partial \vec{E}}{\partial t} + 4\pi\rho\right)\right] \tag{B.4}$$

This simplifies to:

$$\nabla \cdot \vec{F} = q\left[4\pi\rho(1-\beta) - \frac{\beta}{c}\frac{\partial \vec{E}}{\partial t}\right] \tag{B.5}$$

For relativistic cases, introducing γ, where $\gamma = \frac{1}{\sqrt{1-\beta^2}}$, the divergence of the Lorentz force becomes:

$$\nabla \cdot \vec{F} = \frac{4\pi q\rho(1-\beta)}{\gamma^2} - \frac{q\beta}{c}\frac{\partial E_z}{\partial t} \tag{B.6}$$

Charged Particle Beam Physics: An Introduction for Physicists and Engineers, First Edition.
Sarvesh Kumar and Manish K. Kashyap.

Companion Website: https://www.wiley.com/go/Kumar_1e

To calculate the curl of the Lorentz force, we apply the curl operator to Equation B.1:

$$\nabla \times \vec{F} = q\nabla \times (\vec{E} + \beta\hat{z} \times \vec{B}) \tag{B.7}$$

Using Maxwell's third equation $\nabla \times \vec{E} = -\frac{1}{c}\frac{\partial \vec{B}}{\partial t}$ and applying the vector identity for the second term in Equation B.7, we obtain:

$$\nabla \times \vec{F} = q\left(-\frac{1}{c}\frac{\partial \vec{B}}{\partial t} + \beta\nabla \times (\hat{z} \times \vec{B})\right) \tag{B.8}$$

Expanding the second term using $\nabla \times (\hat{z} \times \vec{B}) = \hat{z}(\nabla \cdot \vec{B}) - \nabla(\hat{z} \cdot \vec{B})$, and since $\nabla \cdot \vec{B} = 0$, we have:

$$\nabla \times \vec{F} = -q\left(\frac{1}{c}\frac{\partial \vec{B}}{\partial t} + \beta\frac{\partial \vec{B}}{\partial z}\right) \tag{B.9}$$

The total impulse or deflection $\Delta\vec{p}(x, y, D)$ experienced by a particle at position (x, y, D) is obtained by integrating the Lorentz force over time:

$$\Delta\vec{p}(x, y, D) = \int_{-\infty}^{\infty} \vec{F}(x, y, D + \beta\, ct, t)\, dt \tag{B.10}$$

Substituting the expression for the Lorentz force into the integral, the curl of the impulse becomes:

$$\nabla \times \Delta\vec{p} = -q\int_{-\infty}^{\infty} \left(\frac{1}{c}\frac{\partial}{\partial t} + \beta\frac{\partial}{\partial z}\right)\vec{B}(x, y, z, t)\Big|_{z=D+\beta ct} dt \tag{B.11}$$

Evaluating the integral and assuming that the magnetic field $\vec{B}$ vanishes at infinity, we obtain:

$$\nabla \times \Delta\vec{p} = 0 \tag{B.12}$$

This result is the generalized Panofsky–Wenzel theorem, which states that the total transverse impulse imparted to a charged particle due to the electric and magnetic fields in TE mode is zero. In other words, any deflection caused by the electric field is canceled by the magnetic field, leading to no net transverse deflection.

To describe the longitudinal and transverse wakefields in an accelerator, we introduce the wake potential $W(x, y, D)$, such that the impulse can be written as:

$$\Delta\vec{p}(x, y, D) = -q\nabla W(x, y, D) \tag{B.13}$$

For the case where $\beta = 1$ (relativistic speed), the transverse wake potential satisfies the Laplace equation:

$$\nabla_{\perp}^{2} W = 0 \quad \text{or equivalently} \quad \frac{\partial^2 W}{\partial x^2} + \frac{\partial^2 W}{\partial y^2} = 0 \tag{B.14}$$

The Panofsky–Wenzel theorem provides a crucial insight into the connection between longitudinal and transverse wakefields in particle accelerators. As a particle beam passes through an accelerator structure, it generates wakefields that can affect both its energy (longitudinal wakefields) and its trajectory (transverse wakefields). According to this theorem, longitudinal wakefields are linked to transverse wakefields, meaning that any effect in the direction of motion is connected to sideways forces on the beam. This relationship is essential for designing accelerator structures to minimize unwanted beam deflection while maintaining precise control of the particle's energy.

C

Child–Langmuir Law and Richardson's Law

Consider a system consisting of two parallel electrodes separated by a distance d. Ions are produced at the cathode (located at $x = 0$) and are driven toward the anode (located at $x = d$) by an applied voltage V. The cathode is at potential $\phi(0) = 0$ and the anode at $\phi(d) = V$.

The Poisson's equation is:

$$\frac{d^2\phi(x)}{dx^2} = -\frac{\rho(x)}{\epsilon_0} \text{ and } J = \rho(x)v(x) \tag{C.1}$$

where $v(x) = \sqrt{\frac{2q\phi(x)}{m_i}}$ with q being the ion charge and m_i the ion mass.

Substituting $\rho(x)$ into Poisson's equation:

$$\frac{d^2\phi(x)}{dx^2} = -\frac{J}{\epsilon_0\sqrt{\frac{2q}{m_i}}\phi^{1/2}(x)} \tag{C.2}$$

Multiplying by $\frac{d\phi(x)}{dx}$ and integrating:

$$\frac{1}{2}\left(\frac{d\phi(x)}{dx}\right)^2 = -\frac{2J}{3\epsilon_0\sqrt{\frac{2q}{m_i}}}\phi^{3/2}(x) + C \tag{C.3}$$

Using boundary conditions $\phi(0) = 0$ and $\phi(d) = V$, we get:

$$\left(\frac{d\phi(x)}{dx}\right) = \sqrt{-\frac{4J}{3\epsilon_0\sqrt{\frac{2q}{m_i}}} \cdot \phi^{3/2}(x)} \tag{C.4}$$

Integrating over d:

$$d = \int_0^d dx = \int_0^V \frac{d\phi}{\sqrt{-\frac{4J}{3\epsilon_0\sqrt{\frac{2q}{m_i}}} \cdot \phi^{3/2}}} = \frac{1}{\sqrt{A}}\int_0^V \phi^{-3/4}d\phi \tag{C.5}$$

Charged Particle Beam Physics: An Introduction for Physicists and Engineers, First Edition.
Sarvesh Kumar and Manish K. Kashyap.

Companion Website: https://www.wiley.com/go/Kumar_1e

where $A = \frac{4J}{3\epsilon_0\sqrt{\frac{2q}{m_i}}}$.

The integral of $\phi^{-3/4}$ is:

$$\int \phi^{-3/4} d\phi = 4\phi^{1/4} \tag{C.6}$$

Finally the expression for the Child–Langmuir Law is:

$$J = \frac{4\epsilon_0}{9}\sqrt{\frac{2q}{m_i}}\frac{V^{3/2}}{d^2} \tag{C.7}$$

In accelerator physics, the Child–Langmuir Law is essential for determining the maximum current that can be extracted from ion or electron sources without beam spreading due to space charge effects.

Consider two planar metallic surfaces separated by d in vacuum. The current density J due to thermionic emission is given by:

$$J = q\int n(E)\, v_x(E)\, dE \tag{C.8}$$

where $n(E) = \rho(E)f(E)$, $\rho(E) = \frac{4\pi\sqrt{2m^3}\sqrt{E}}{h^3}$, and $f(E) = \exp\left(-\frac{E-\phi}{kT}\right)$. The velocity component normal to the surface is:

$$v_x(E) = \sqrt{\frac{2E}{m}} \tag{C.9}$$

Substituting the expressions for $n(E)$ and $v_x(E)$:

$$J = q\frac{8\pi mk^2T^2}{h^3}\exp\left(-\frac{\phi}{kT}\right) = A_0T^2\exp\left(-\frac{\phi}{kT}\right) \tag{C.10}$$

where $A_0^{\text{ion}} = \frac{8\pi q m_i k^2}{h^3}$ for ions and for electrons:

$$A_0^{\text{electron}} = \frac{8\pi e m_e k^2}{h^3} = 1.2\times10^6\ \text{A/m}^2\text{K}^2 \tag{C.11}$$

In accelerator physics, Richardson's Law is crucial for understanding thermionic emission, which governs electron or ion escape from heated surfaces in devices like electron guns and ion sources.

D

Larmor's Formula and Its Importance in Accelerator Physics

Let a charged particle with charge e be undergoing acceleration a. The particle radiates electromagnetic waves due to this acceleration. The system is observed in the far-field region, where the distance r from the accelerating charge to the observation point is much larger than the wavelength of the emitted radiation. In this region, the radiated electric field is primarily due to the transverse component of the acceleration, denoted by $a_{\perp}$. The electric field is $\mathbf{E}$ and the magnetic field is $\mathbf{B}$, both related through the Poynting vector $\mathbf{S}$, representing the energy flux.

To derive the Larmor formula for radiated power, consider the retarded potentials that describe the electromagnetic fields generated by the charge. The scalar potential ϕ and the vector potential $\mathbf{A}$ for a moving charge at the retarded time t_r are:

$$\phi(\mathbf{r}, t) = \frac{e}{4\pi\varepsilon_0} \frac{1}{|\mathbf{r} - \mathbf{r}'(t_r)|} \quad \text{and} \quad \mathbf{A}(\mathbf{r}, t) = \frac{\mu_0 e}{4\pi} \frac{\mathbf{v}(t_r)}{|\mathbf{r} - \mathbf{r}'(t_r)|} \tag{D.1}$$

These potentials represent the fields of the moving charge at the retarded time t_r, accounting for the finite speed of light. The electric field $\mathbf{E}$ and the magnetic field $\mathbf{B}$ are derived from the potentials. In the far-field region, the electric field is primarily determined by the time variation of the vector potential:

$$\mathbf{E} = -\nabla\phi - \frac{\partial \mathbf{A}}{\partial t} \approx -\frac{\partial \mathbf{A}}{\partial t} \tag{D.2}$$

The vector potential $\mathbf{A}$ is proportional to the velocity of the charge at the retarded time:

$$\mathbf{A}(\mathbf{r}, t) = \frac{\mu_0 e}{4\pi r}\mathbf{v}(t_r) \quad \text{so} \quad \frac{\partial \mathbf{A}}{\partial t} = \frac{\mu_0 e}{4\pi r}\mathbf{a}(t_r) \tag{D.3}$$

Substituting $\mu_0 = \frac{1}{\varepsilon_0 c^2}$, we get:

$$\mathbf{E} = \frac{e}{4\pi\varepsilon_0 c^2 r}\mathbf{a}(t_r) \quad \text{and in the far-field region:} \quad \mathbf{E} = \frac{e}{4\pi\varepsilon_0 c^2 r}\mathbf{a}_{\perp}(t_r) \tag{D.4}$$

The Poynting vector $\mathbf{S}$, representing the energy flux, is given by:

$$\mathbf{S} = \frac{\mathbf{E} \times \mathbf{B}}{\mu_0} \quad \text{and} \quad |\mathbf{S}| = \frac{|\mathbf{E}|^2}{\mu_0 c} \tag{D.5}$$

Charged Particle Beam Physics: An Introduction for Physicists and Engineers, First Edition.
Sarvesh Kumar and Manish K. Kashyap.

Companion Website: https://www.wiley.com/go/Kumar_1e

Since $\mathbf{B} = \frac{\mathbf{E}}{c}$ in the far-field region, substituting the expression for **E** into the Poynting vector gives:

$$|\mathbf{S}| = \frac{e^2 a_\perp^2}{16\pi^2 \varepsilon_0 c^3 r^2} \tag{D.6}$$

The total radiated power P_{rad} is found by integrating the Poynting vector over a spherical surface:

$$P_{\text{rad}} = \int |\mathbf{S}| dA = |\mathbf{S}| \times 4\pi r^2 = \frac{e^2 a_\perp^2}{6\pi \varepsilon_0 c^3} \tag{D.7}$$

This is the Larmor formula, giving the total power radiated by a nonrelativistic accelerating charge.

For a relativistic electron in a synchrotron, the power radiated and energy loss per turn can be expressed in terms of the fine-structure constant α, defined as:

$$\alpha = \frac{e^2}{4\pi \varepsilon_0 \hbar c} \quad \text{with power radiated} \quad P = \frac{2\alpha \hbar c}{3} \cdot \frac{\gamma^4}{R^2} \tag{D.8}$$

where γ is the Lorentz factor and R is the radius of the electron's orbit. The energy loss per turn is:

$$\Delta E = P \times \frac{2\pi R}{c} = \frac{4\pi \alpha \hbar c \gamma^4}{3R} \tag{D.9}$$

These expressions demonstrate the role of the fine-structure constant in determining the energy radiated and lost by relativistic electrons in synchrotron accelerators, reflecting quantum electrodynamic effects. Here are more conclusions:

1. The γ^4 dependence indicates that higher energy particles experience significantly greater energy losses per turn, limiting the beam's lifespan.
2. The equation is crucial in designing synchrotrons, as the power of the radio frequency cavities must compensate for the energy loss, ensuring the beam maintains its energy.
3. The rapid increase in energy loss with γ sets practical energy limits for circular accelerators, pushing the need for larger radii or alternative designs like linear accelerators.
4. Due to their low mass, electrons suffer greater synchrotron radiation losses than heavier particles (like protons), making this equation especially relevant in electron synchrotrons.
5. The equation helps control and optimize synchrotron radiation for light sources used in scientific research, such as X-ray and UV radiation facilities.

E

Stefan–Boltzmann Law and Its Applications in Particle Accelerators

Let a black body, which is a perfect emitter of radiation, be at a temperature T. The radiation it emits is spread across different frequencies. To calculate the total power radiated by the black body, begin with Planck's law, which describes the intensity of radiation emitted per unit area, per unit time, and per unit frequency at temperature T:

$$I(\nu, T) = \frac{2h\nu^3}{c^2} \cdot \frac{1}{e^{\frac{h\nu}{k_B T}} - 1} \tag{E.1}$$

The total power radiated by the black body, $P(T)$, is obtained by integrating this intensity over all frequencies ν:

$$P(T) = \int_0^\infty I(\nu, T) d\nu \text{ and } d\nu = \frac{k_B T}{h} dx \tag{E.2}$$

Let a dimensionless variable $x = \frac{h\nu}{k_B T}$ be introduced, and substituting this into the integral, we get:

$$P(T) = \frac{2(k_B T)^4}{h^3 c^2} \int_0^\infty \frac{x^3}{e^x - 1} dx \tag{E.3}$$

Since the integral $\int_0^\infty \frac{x^3}{e^x - 1} dx = \frac{\pi^4}{15}$, substituting this gives:

$$P(T) = \frac{2\pi^5 k_B^4}{15 h^3 c^2} T^4 \tag{E.4}$$

At this point, introduce the Stefan–Boltzmann constant σ, defined as:

$$\sigma = \frac{2\pi^5 k_B^4}{15 h^3 c^2} \text{ so } P = \sigma A T^4 \tag{E.5}$$

This is the Stefan–Boltzmann law, which shows that the power radiated by a black body is proportional to the fourth power of its temperature T. The Stefan–Boltzmann constant σ has a value of $5.670374419 \times 10^{-8}\ \mathrm{W\,m^{-2}\,K^{-4}}$. For real materials, introduce the emissivity ϵ, and the modified formula becomes $P = \epsilon\sigma A T^4$. This constant is crucial in calculating the total radiative power emitted by a surface, depending on its temperature.

Charged Particle Beam Physics: An Introduction for Physicists and Engineers, First Edition.
Sarvesh Kumar and Manish K. Kashyap.

Companion Website: https://www.wiley.com/go/Kumar_1e

In high-energy particle accelerators, beam dumps absorb intense energy from particle beams, and the Stefan–Boltzmann law helps in calculating the power radiated by the beam dump's surface to avoid damage due to heat. Similarly, in accelerator cavities, proper radiative cooling of walls heated by electromagnetic fields ensures stable operation, and the formula helps in designing the cavity's cooling systems.

Bibliography

1 Bartelle B. B., Barandov A., Jasanoff A. Molecular fMRI. Journal of Neuroscience, 36:4139–4148, 2016.
2 Van de Graaff R. J., Compton K. T., Van Atta L. C. The electrostatic production of high voltage for nuclear investigations. Physical Review, 43:149–157, 1933.
3 Cockcroft J. D., Walton E. T. S. Experiments with high velocity positive ions. (i) Further developments in the method of obtaining high velocity positive ions. Proceedings of the Royal Society A, 136:619–630, 1932.
4 Cockcroft J. D., Walton E. T. S. Experiments with high velocity positive ions. II. The disintegration of elements by high velocity protons. Proceedings of the Royal Society A, 137:229–242, 1932.
5 National Electrostatics Corporation. NEC Pelletron Accelerator. https://www.pelletron.com, 2025. Electrostatic accelerator systems, based on Pelletron voltage-multiplier technology.
6 Hinterberger F. Electrostatic Accelerators. CERN, 2006.
7 Abdel-Salam M. High-voltage Engineering: Theory and Practice, Revised and Expanded. CRC Press, 2018.
8 Lawrence E. O., Livingston M. S. The production of high speed lightions without the use of high voltages. Physical Review, 40:19, 1932.
9 Bethe H. E., Rose. M. E. The maximum energy usable in electrondiffraction. Physical Review, 52, 1937.
10 McMillan E. M. The synchrotron—a proposed high energy particle accelerator. Physical Review, 68:143–144, 1945.
11 Veksler V. I. A new method of acceleration of relativistic particles. Journal of Physics, 9:153–158, 1945.
12 Kerst D. W. The acceleration of electrons by magnetic induction. Physical Review, 60:47–53, 1941.
13 Kapitsa S. The microtron and areas of its application. Soviet Atomic Energy, 18(3):255–261, 1965.
14 Kapitza S. Modern developments of the microtron. In Proceedings of the Vth International Accelerator Conference on High Energy Accelerators, Frascati, 1965.
15 Jones L. W., Terwilliger K. M. A small model fixed field alternating gradient radial sector accelerator. Technical Report MURA-LWJ/KMT-5 (MURA-104), Midwestern Universities Research Association (MURA), Madison, Wisconsin, USA, Apr. 1956. Contains photos, scale drawings, and design calculations.
16 Trbojevic D., Courant E. D., Garren A. FFAG lattice without opposite bends. AIP Conference Proceedings, 530(1):333–338, 2000.
17 Mori Y. Ffag proton driver for muon source. Nuclear Instruments and Methods in Physics Research Section A: Accelerators, Spectrometers Detectors and Associated Equipment, 451(1):300–303, 2000.
18 Peach K., Cobb J., Sheehy S., Witte H., Yokoi T., Aslaninejad M., Easton M., Pasternak J., Barlow R., Owen H., et al. PAMELA: Overview and Status. JACoW, 2010.
19 Tajima T., Dawson J. M. Laser electron accelerator. Physical Review Letters, 43:267–270, 1979.
20 Chen P., Dawson J. M., Huff R. W., Katsouleas T. Acceleration of electrons by the interaction of a bunched electron beam with a plasma. Physical Review Letters, 54:693–696, 1985.
21 Brown I, et al. Physics and technology of ion sources. New York: John Wiley, 331, 1989.
22 Geller R. Electron Cyclotron Resonance Ion Sources and ECR Plasmas. IOP Publisher, 1996.
23 Geller R., Jacquot B. The multiply charged ion source minimafios. Physica Scripta, T3:19, 1983.

Charged Particle Beam Physics: An Introduction for Physicists and Engineers, First Edition.
Sarvesh Kumar and Manish K. Kashyap.

Companion Website: https://www.wiley.com/go/Kumar_1e

24 Geller R., Jacquot B., Pontonnier M. Status of the multiply charged heavy-ion source minimafios. Review of Scientific Instruments, 56(8):1505–1510, 1985.

25 Grunder H. A., Selph F. B. Heavy-ion accelerators. Annual Review of Nuclear Science, 27(1):353–392, 1977.

26 Hubbard E. L. Heavy-ion accelerators. Annual Review of Nuclear Science, 11(1):419–438, 1961.

27 Kumar S., Mandal A. Low energy ion beam dynamics of NANOGAN ECR ion source. Nuclear Instruments and Methods in Physics Research Section A: Accelerators, Spectrometers, Detectors and Associated Equipment, 814:73–81, 2016.

28 Kumar S., Sharma J., Sharma P., Sharma S., Mathur Y., Sharma D., Kashyap M. K. Experimental investigation of plasma instabilities by Fourier analysis in an electron cyclotron resonance ion source. Physical Review Accelerators and Beams, 21:093402, 2018.

29 Kumar S., Sharma J., Sharma S., Mathur Y., Nandi T., Sharma D., Kashyap M. K. Observations of parametric drift wave instabilities in an electron cyclotron resonance ion source. Plasma Physics and Controlled Fusion, 62:105013, 2020.

30 Geller R. Ecris sources for highly charged ions. Europhysics News, 22(1):8–11, 1991.

31 Geller R. Electron Cyclotron Resonance Ion Sources and ECR Plasmas. Routledge, 2018.

32 Geller R., Briand P., Jacquot B., Ludwig P., Melin G., Pontonnier M. ECR ion sources for accelerators. In: Proceedings of the 1988 Linear Accelerator Conference, 455, 1989.

33 Girard A., Hitz D., Melin G., Serebrennikov K. Electron cyclotron resonance plasmas and electron cyclotron resonance ion sources: physics and technology. Review of Scientific Instruments, 75(5):1381–1388, 2004.

34 Pardo R. Review of high intensity ion source development and operation. Review of Scientific Instruments, 90(12), 2019.

35 Caskey G. T., Douglas R. A., Richards H. T. Jr. Narrow 0+ state in 20ne and 0+6. Nuclear Instruments and Methods, 157:1, 1978.

36 Middleton R. Independent development of a sputter-type negative ion source (snics). In: Proceedings of the Symposium on Northeastern Accelerator Personnel, CONF-781051. Tennessee: Oak Ridge, 114, 1978.

37 Dudnikov V. Forty years of surface plasma source development. Review of Scientific Instruments, 83:02A708, 2012.

38 Hone M. A., Tranquille G. The duoplasmatron ion source for the new CERN linac preinjector. Technical Report PS/LR/Note 79-37. Geneva: CERN, 1979.

39 Morgan O. B., Kelley G. G., Davis R. C. Technology of intense dc ion beams. Review of Scientific Instruments, 38(4):467–480, 1967.

40 Lejeune C. Theoretical and experimental study of the duoplasmatron ion source. Part I: Model of the duoplasmatron discharge. Nuclear Instruments and Methods, 116(3):417–428, 1974.

41 Donets E. D. USSR Inventor's Certificate No. 248860, 1967. Issued 16 March 1967; Published in Bulletin of OIPOTZ, No. 24 (1969), p. 65.

42 Donets E. D., Ilushenko V. I., Alpert V. A. Ultrahigh vacuum electron beam ion source of highly stripped ions. In Proceedings of the First International Conference on Ion Sources, Saclay, France. 635, 1969.

43 Donets E. D., Ovsyannikov V. P. Ultrahigh charge state ion production in electron beam ion sources. Soviet Physics JETP, 53:466, 1981.

44 Donets E. D. Review of recent developments for electron-beam ion sources (EBIS) (invited). Review of Scientific Instruments, 67(3):873–877, 1996.

45 Zschornacka G., Schmidt M., Thorn A. Electron beam ion sources. arXiv preprint arXiv:1410.8014, 2014.

46 Levine M. A., Marrs R. E., Henderson J. R., Knapp D. A., Schneider M. B. The electron beam ion trap: a new instrument for atomic physics measurements. Physica Scripta, T22:157, 1988.

47 Vella M. Physics of the high current density electron beam ion source (EBIS). Nuclear Instruments and Methods in Physics Research, 187:313–321, 1981.

48 Chen F. K. Investigation of the striking characteristics of a penning ion source. Journal of Applied Physics, 56(11):3191, 1984.

49 Hoh F. C. Instability of penning-type discharges. Physics of Fluids, 6(8):1184, 1963.

50 Penning F. Die glimmentladung bei niedrigem druck zwischen koaxialen zylindern in einem axialen magnetfeld. Physica, 3(9):873–894, 1936.

51 Rohwer P., Baumann H., Schutze W., Bethge K. Studies of the center potential in a penning discharge. Nuclear Instruments and Methods in Physics Research, 211(2–3):543–546, 1983.

52 Tawara H., Itikawa T., Nishimura H., Yoshino M. Cross sections and related data for electron collisions with hydrogen molecules and molecular ions. Journal of Physical and Chemical Reference Data. 1990.
53 Sy A. V. Advanced penning-type ion source development and passive beam focusing techniques for an associated particle imaging neutron generator with enhanced spatial resolution. Berkeley: University of California, 2013.
54 Peacock N. J., Pease R. S. Sources of highly stripped ions. Journal of Physics D: Applied Physics, 2(12):1705, 1969.
55 Byckovsky Y. A., Eliseev V. F., Kozyrev Y. P., Silnov S. M. Laser generator of multi-charged ions for accelerators, 1969.
56 MacGill A. R. et al. Review of Scientific Instruments, 61(II):580, 1990.
57 Floettmann K. Some basic features of the beam emittance. Physical Review Accelerators and Beams, 6:034202, 2003.
58 Wille K. The Physics of Particle Accelerators: An Introduction. Oxford Academic, 2000.
59 Courant E., Snyder H. Theory of the alternating-gradient synchrotron. Annals of Physics, 3(1):1–48, 1958.
60 Steinhagen R. J. Tune and chromaticity diagnostics. 2009.
61 Verdier A. Chromaticity. 1992.
62 Lapostolle P. M. Possible emittance increase through filamentation due to space charge in continuous beams. IEEE Transactions on Nuclear Science, 18(3):1101–1104, 1971.
63 Struckmeier J., Klabunde J., Reiser M. On the stability and emittance growth of different particle phase-space distributions in a long magnetic quadrupole channel. arXiv preprint arXiv:2401.12595, 2024.
64 Kalvas T. Beam extraction and transport. arXiv:1401.3951, Conference: C12-05-29.2 (CERN-2013-007, pp. 537–564 [physics.acc-ph]), 537–564, 2014.
65 Reiser M. Theory and Design of Charged Particle Beams. John Wiley & Sons, 2008.
66 Billen J. H. Poisson superfish. LA-UR-96-1834. 1996.
67 Billen J. H., Young L. M. Poisson/superfish on pc compatibles. In: Proceedings of International Conference on Particle Accelerators. IEEE, 790–792, 1993.
68 Menzel M., Stokes H. K. Users guide for the poisson/superfish group of codes. Technical Report. Los Alamos, NM: Los Alamos National Lab (LANL), 1987.
69 Rogowski W. Die elektrische festigkeit am rande des plattenkondensators. Archiv für Elektrotechnik, 12:1–15, 1923. Original paper introducing the Rogowski profile for electrode edge shaping.
70 Dayton I. E., Shoemaker F. C., Mozley R. F. The measurement of two-dimensional fields. Part II: Study of a quadrupole magnet. Review of Scientific Instruments, 25(5), 1954.
71 Grivet P., Septier A. Les lentilles quadrupolaires magnetiques (i). Nuclear Instruments and Methods, 6:126–156, 1959.
72 Tjepkema O. G. Automated beam line tuning at the AGOR accelerator facility. PhD thesis. 2021.
73 Schmidt M., Peng H., Zschornack G., Sykora S. A compact electron beam ion source with integrated wien filter providing mass and charge state separated beams of highly charged ions. Review of Scientific Instruments, 80(6):063301, 2009.
74 Glaser W. Strenge berechnung magnetischer linsen der feldform h= h 0 1+(z/a) 2. Zeitschrift für Physikm, 117:285–315, 1941.
75 Sarma P., Bhandari R. Transfer matrix of a glaser magnet used for focusing charged particles in beam lines. Journal of Physics D: Applied Physics, 42(4):045508, 2009.
76 Goswami A., Babu P. S., Pandit V. S. Transfer matrix of a glaser magnet to study the dynamics of non-axisymmetric beam. Nuclear Instruments & Methods in Physics Research Section A-accelerators Spectrometers Detectors and Associated Equipment, 678:14–20, 2012.
77 Howells M., Kincaid B. M. The properties of undulator radiation. In New Directions in Research with Third-Generation Soft X-raySynchrotron Radiation Sources. Springer, 1994.
78 Reichel I. Insertion Devices. USPAS Lecture Slides, Michigan State University, 2012.
79 Ximen J. Aberration theory in electron and ion optics. In: Advances in Electronics and Electron Physics Supplement, 1986.
80 Berz M., Makino K. Cosy infinity. 10.2.2023.
81 Elkind M. M. Ion optics in long, high voltage accelerator tubes. Review of Scientific Instruments, 24(2):129–137, 1953.
82 Hellborg R. Electrostatic Accelerators. Springer, 2005.
83 Lapostolle P., Septier A. Linear Accelerators. Amsterdam: North-Holland, 1970.
84 Wangler T. Principles of RF Linear Accelerators. New York: Wiley, 1998.

85 Earnshaw S. On the nature of the molecular forces which regulate the constitution of the luminiferous ether. Transactions of the Cambridge Philosophical Society, 7:97, 1848.

86 Duff J. L. Longitudinal beam dynamics and stability. Small Accelerators Course, LAL-Orsay, CAS Zeegse, 24 May–2 June, 2005.

87 Sloan D. H., Lawrence E. O. The production of heavy high speed ions without the use of high voltages. Physical Review, 38:2021–2032, 1931.

88 Ben-Zvi I., Brennan J. The quarter wave resonator as a superconducting linac element. Nuclear Instruments and Methods in Physics Research, 212(1):73–79, 1983.

89 Padamsee H., Knobloch J., Hays T. RF Superconductivity for Accelerators, 2nd edition. Weinheim: Wiley-VCH, 2009.

90 Pukhnachov V. N., Kozyrev A. V. Theory of beam dynamics in a konus accelerator. Nuclear Instruments and Methods in Physics Research, 220(1):32–36, 1984.

91 Pukhnachov V. N., Kozyrev A. V. Nonlinear beam dynamics in konus accelerators. In Proceedings of the 1988 IEEE Particle Accelerator Conference. IEEE, 1789–1791, 1988.

92 Ratzinger U. The IH-structure and its capability to accelerate high current beams. In Proceedings of the 1991 IEEE Particle Accelerator Conference, 567–571, 1991.

93 Ratzinger U., Tiede. R. Status of the HIIF RF linac study based on H-mode cavities. Nuclear Instruments and Methods in Physics Research Section A, 415:229–235, 1998.

94 Wangler T. P. RF Linear Accelerators, 2nd edition. Wiley-VCH, 2008.

95 Loyer F. On line beam diagnostics at ganil. In Eleventh International Conference on Cyclotrons and their Applications, 449–452, 1986.

96 Forck P., Liakin D., Kowina P. Beam position monitors. arXiv:2009.10411, 2009.

97 Huang X. Beam-based Correction and Optimization for Accelerators. Taylor & Francis, 2020.

98 der Hochfrequenztechnik T. Hh meinke and f.-w. gundlach, 1986.

99 Takeuchi S., Shepard K. A sensitive beam-bunch phase detector. Nuclear Instruments and Methods in Physics Research Section A: Accelerators, Spectrometers, Detectors and Associated Equipment, 227(2):217–219, 1984.

100 Paschen F. Ueber die zum funkenübergang in luft: wasserstoff und kohlensäure bei verschiedenen drucken erforderliche potentialdifferenz. JA Barth, 1889.

101 Cobine J. D. Gaseous Conductors: Theory and Engineering Applications. Dover Publications, 1958.

102 Kilpatrick W. D. Criterion for vacuum sparking designed to include both RF and DC. Review of Scientific Instruments. 28(10):824–826, 1957.

103 Chen F. F. Introduction to Plasma Physics and Controlled Fusion, 3rd edition. Springer, 2016.

104 Myers A. C, Macek J. R. Multipactor discharge on dielectric surfaces. IEEE Transactions on Electron Devices, 38(10):2121–2125, 1991.

105 Ng K. Y. RF cavity design in particle accelerators. Reviews of Accelerator Science and Technology, 5:251–279, 2012.

106 Fowler R. H, Nordheim L. Electron emission in intense electric fields. Proceedings of the Royal Society A, 119:173–181, 1928.

107 Padamsee H, Knobloch J, Hays T. RF Superconductivity for Accelerators. Wiley-VCH, 2009.

108 Boyle W. S., Kisliuk P. Departure from Paschen's law of breakdown in gases. Physical Review, 97:255–259, 1955.

109 Wilson P. B. Introduction to wakefields and wake potentials. AIP Conference Proceedings, 184:525, 1989.

110 Panofsky W. K. H., Wenzel W. A. Some considerations concerning the transverse deflection of charged particles in radio-frequency fields. Review of Scientific Instruments, 27(11):967–967, 1956.

Index

Note: Page numbers in *italics* and **bold** refers to figures and tables respectively.

Charged Particle Beam Physics: An Introduction for Physicists and Engineers, First Edition.
Sarvesh Kumar and Manish K. Kashyap.

Companion Website: https://www.wiley.com/go/Kumar_1e

o

p

q

r

s

t

u

v

w